Advances in Intelligent Systems and Computing

Volume 541

Series editor

Janusz Kacprzyk, Polish Academy of Sciences, Warsaw, Poland
e-mail: kacprzyk@ibspan.waw.pl

About this Series

The series "Advances in Intelligent Systems and Computing" contains publications on theory, applications, and design methods of Intelligent Systems and Intelligent Computing. Virtually all disciplines such as engineering, natural sciences, computer and information science, ICT, economics, business, e-commerce, environment, healthcare, life science are covered. The list of topics spans all the areas of modern intelligent systems and computing.

The publications within "Advances in Intelligent Systems and Computing" are primarily textbooks and proceedings of important conferences, symposia and congresses. They cover significant recent developments in the field, both of a foundational and applicable character. An important characteristic feature of the series is the short publication time and world-wide distribution. This permits a rapid and broad dissemination of research results.

More information about this series at http://www.springer.com/series/11156

Fatos Xhafa · Srikanta Patnaik
Zhengtao Yu
Editors

Recent Developments in Intelligent Systems and Interactive Applications

Proceedings of the International Conference
on Intelligent and Interactive Systems
and Applications (IISA2016)

 Springer

Editors
Fatos Xhafa
Department of Computer Science
Universitat Politècnica de Catalunya
Barcelona
Spain

Srikanta Patnaik
SOA University
Bhubaneswar, Odisha
India

Zhengtao Yu
School of Information Engineering and
  Automation
Kunming University of Science and
  Technology
Kunming
China

ISSN 2194-5357   ISSN 2194-5365   (electronic)
Advances in Intelligent Systems and Computing
ISBN 978-3-319-49567-5   ISBN 978-3-319-49568-2   (eBook)
DOI 10.1007/978-3-319-49568-2

Library of Congress Control Number: 2016957289

Printed on acid-free paper

This Springer imprint is published by Springer Nature
The registered company is Springer International Publishing AG
The registered company address is: Gewerbestrasse 11, 6330 Cham, Switzerland

Preface

An *Interactive Intelligent System* (IIS) is an intelligent system that interacts with the users, or audience at large, which is designed to interact more with the user rather than with computer systems. The system utilizes the capabilities to perceive, interpret, learn, plan, decide, as well as use natural language and also reason, which has been already developed by field of artificial intelligence. Many of these techniques have been matured enough by now for specific functions such as "pattern recognition" or "Internet searching". It is difficult to understand interactive intelligent systems without examining the intelligent capabilities of machines as well as human interface.

There are very less number of interactive intelligent systems used presently. One of them is "PlateMate", a crowdsourcing nutrition analysis software, which allows users to take photos of their meals and receive estimates of food intake and composition from the photograph of foods. Accuracy of the prediction is depended on the information of the users for food logging via self-reporting, expert observation, and or algorithmic analysis. PlateMate crowdsources nutritional analysis from photographs using Amazon Mechanical Turk, automatically coordinating untrained workers to estimate a meal's calories, fat, carbohydrates, and protein.

Let us consider another example, a system that learns how to assist users in performing particular types of tasks, e.g., "SIRI". It is a computer program that works as an intelligent personal assistant and knowledge navigator of Apple IOS operating systems. It uses a natural language user interface to answer questions, make recommendations, and perform actions by delegating requests to a set of Web services. The software adapts to the user's individual language usage and individual searches with continuing use, and returns results that are individualized. In this case, while the system is learning, the users will in general also be learning: about the task itself, about the system and its learning, about how to act in such a way that the system learns more effectively.

The research on interactive intelligent systems has so far focused either on the realization of the systems' capabilities or on the cognitive processes and/or behavior of their users. The technical design which focuses only on the machine learning of the system or on that of the user interface has never understood the important opportunities for making the interaction among different learning agents of various applications. Design of interactive intelligent systems are fundamentally hard, because they require intelligent technology that is well suited for people's abilities, limitations, and preferences.

Interactive intelligent systems also require different kinds of interactions which can give the user a predictable and reliable experience despite the fact that the underlying technology is inherently proactive, unpredictable, and occasionally wrong. There are cases of such types of failure in the past, such as the accident occurred in Williston, Florida, on May 7, 2016, when the driver, Joshua Brown, 40, of Ohio put his Model S of Tesla into autopilot mode, which was controlling the driving of car on the highway. The autopilot mode of Tesla failed to distinguish a large white 18-wheel truck and trailer crossing the highway, against a bright spring sky. Thus, design of successful intelligent interactive systems requires intimate knowledge and ability to innovate in two disparate areas: human–computer interaction and artificial intelligence or machine learning.

There are various general issues and challenges in the area of *interactive intelligent systems*. The first issue is that in which way artificial agent and human intelligence can work together for better performance. In other words, how does intelligent processing yield the greatest benefits for interactive systems in comparison to other forms of computation? How do they allocate various processing between the human and the intelligent system, so as to enhance the performance for a mixed-initiative intelligent system.

The second issue is the study of possible negative side effects of the interactive intelligent systems, if it is not designed properly without giving proper attention to the cognitive processes of users. For instance, to study the reasons and situation when the users want to predict, understand, and control an intelligent system and how the designers will be able to provide the requirements of the user to the interactive intelligent systems.

Third issue pertains to the protection of users' privacy and also to restrict the intelligent systems not to know more than the required about the users, when they are interacting with the system.

Last but not least, the issue related to the method and techniques used for various types of interactive intelligent system, ranging from the method of understanding about the users' requirements to the techniques used of evaluating the success of a given combination of intelligent algorithms and interaction design.

Research on interactive intelligent systems is found in a considerable number of diverse research areas, which include autonomous systems, expert systems, mobile systems, recommender systems, knowledge-based and semantic web-based systems, human–computer interaction, virtual communication environments, environment-aware agents, agents for smart environments, intelligent robotics, methods in cognitive systems, mind, brain, and behavior, machine learning, natural language interaction, web intelligence, signal processing, speech technologies, audio and music processing, face and gesture analysis, computer vision and pattern recognition, data-driven social analytics, algorithm design, middleware agents, embedded agents, mobile agents, adaptive and personalized systems, business intelligence, e-learning, e-commerce and e-governance, and last but not least, decision support systems.

This volume contains 65 contributions from diverse areas of interactive intelligent systems, which has been categorized into five sections, namely: (i) Autonomous Systems; (ii) Internet & Cloud Computing; (iii) Pattern Recognition and Vision Systems; (iv) Mobile Computing and Intelligent Networking; (v) E-Enabled Systems.

(i) **Autonomous Systems**: This is one of the established areas of interactive intelligence system that typically consists of learning, reasoning, and decision-making which supports the system's primary function. There are 19 contributions composed of various algorithms, models and learning techniques.

(ii) **Internet and Cloud Computing**: It is one of the essential areas of IIS, which caters to enhance communication between the system and users, in a way which may not be closely related to the system's main function. This is commonly found in the areas of multimodal interaction, natural language processing, embodied conversational agents, computer graphics, and accessible computing. In this section there are 18 contributions related to microblogging, user satisfaction modeling to the design and construction of graphical cloud computing platform.

(iii) **Pattern Recognition and Vision Systems**: This is one of the primary functions of any interactive intelligent systems. There are 18 contributions in this section covering of the developments in this area of deep learning to binocular stereovision to 3D vision.

(iv) **Mobile Computing and Intelligent Networking**: This area is one of the leading areas of IIS, which covers ubiquitous or mobile computing and networking. This section contains five contributions.

(v) **E-Enabled Systems**: This is one of the essential areas of interactive intelligent system, as many interactive systems are now designed through Internet. It covers information navigation and retrieval, designing intelligent learning environments, and model-based user interface design. There are five contributions in this section.

Acknowledgments

The contributions covered in these proceedings are the result of the efforts of hundreds of researchers. We are thankful to the authors and paper contributors of this volume.

We are thankful to the Editor-in-Chief of the Springer Book series on **"Advances in Intelligent Systems and Computing"** Prof. Janusz Kacprzyk for his support to bring out the first volume of IISA-2016. It is noteworthy to mention here that this was a big boost to the newly started conference series "International Conference on Intelligent and Interactive Systems and Applications (IISA2016)", by creating awareness among the scholars who are working in this domain, attracting and managing numbers of interesting submissions, and bringing out a volume of 65 valued contributions.

We would like to appreciate the encouragement and support of Dr. Thomas Ditzinger, Executive Editor and his Springer publishing team.

We are thankful to our friend Prof. Madjid Tavana, Professor and Lindback Distinguished Chair of Business Analytics of La Salle University, USA, for is keynote address. We are also thankful to the experts and reviewers who have worked for this volume despite the veil of their anonymity.

We look forward to your valued contribution and support to the upcoming editions of the International Conference on Intelligent and Interactive Systems and Applications.

We believe that the readers would get immense ideas and knowledge from the first volume of the area of interactive intelligent systems and applications.

	Editors
Technical University of Catalonia, Spain	Fatos Xhafa
SOA University, India	Srikanta Patnaik
Kunming University of Science and Technology, China	Zhengtao Yu

Contents

Pattern Recognition and Vision Systems

E-Enabled Systems

Autonomous Systems

Testing Paper Optimization Based on Improved Particle Swarm Optimization

Xiang-Ran Du[✉], Shu-Jin Wu, and Yu-Lin He

Tianjin Maritime College, Tianjin, China
duxiangran1226@126.com

Abstract. The Computerized Examination System is an important part in computer aided education, which not only examines the learning outcome of every candidate, but also provides feedback for further improvement. The construction of the computerized examination system is time consuming and requires plenty of domain as well as pedagogy related information. This paper presents an examination sheet optimization based on the improved particle swarm optimization. The results obtained from the study show that the improved particle swarm optimization effectively enhance the effectiveness level of the computerized examination system without the help of the educational experts after a lot of training.

Keywords: Artificial intelligence · Evaluation function · Examination system · Particle swarm optimization · Self-learning

1 Introduction

The Computerized Examination System plays a significant role in Computer Assisted Instruction (CAI) which not only improves the efficiency in designing, testing paper, maintaining fairness of the examination and reduces usage of material and manpower resources but also discovers the problems of every student in the process of learning and identifies teaching deficiencies for the teachers. Some researchers have shown that the development of testing system not only needs an outstanding educational and technological team but also a powerful maintenance team is necessary to address contingent issues. In fact, the examination system suffers from an accurate evaluation function and enough developments on the system is has been pursued which remains inadequate to meet the educational requirement as usage time increases. To cope up with the problem of insufficient evaluation function and inadequate maintenance in computer examination system, an improved Particle Swarm Optimization (PSO) is proposed to optimize the evaluation function in the testing system. The results obtained from the study reveal that the novel approach effectively promotes the power of the evaluation function and self-optimizes the evaluation function as per the examination results without the help of the educational experts.

The paper is organized in six segments. Section 2 presents the background of the evaluation functions in computer examination system and introduces some relevant research outcomes of earlier studies. A testing paper optimization and the process of training the evaluation function in the examination system by the particle swarm

F. Xhafa et al. (eds.), *Recent Developments in Intelligent Systems and Interactive Applications*,
Advances in Intelligent Systems and Computing 541, DOI 10.1007/978-3-319-49568-2_1

optimization is showed in Sect. 3. In Sect. 4 the comparison experiment is carried out between the improved testing paper optimization and the unimproved one. Section 5 presents conclusion and the Reference section lists maintains the necessary references cited in this piece of research work.

2 Background and Relative Researches

In recent years, the study and examination modes have been radically changed due to the fact that the internet based information society has gained currency in an exponential rate. Some established academic institutions have introduced a wide spectrum of open online courses that attracts a great deal of students from across the globe. The students can seek knowledge and share the excellent education resources through the online courses at home instead of going to school. The track the progress of the study and information about every student with no face to face education is even more important. The online examination is a desirable method to continuously monitor the progress and recognize the learning bottlenecks for every student who pursues a course through the open online courses. There are two difficulties in the construction of the examination system. The first one is the development of high quality and intelligent evaluation functions which requires the expertise and experience of the educational experts. The second one is to reduce the level of difficulty in determining the exactness of the evaluation function of examination questions as it is highly influenced by the subjective consciousness of the experts. The present examination system with their static evaluation functions is unable to keep pace with the social and educational requirement and become obsolete in course of time. Some experiments have shown that the number of examining questions are proportionate to the difficulty of forming an optimal testing sheet, especially for the testing paper library is larger than five thousand.

Many researchers and experts have made significant contributions on the development of computerized Examination System.

A Maximin Model based on the practical constraints of test design is presented by Min J. Van Der Linden [1].

A CATES system based on complex learning environments is proposed by Chou [2], the Interactive Examination system is a collective and collaborative project.

Hong Duan and Wei Zhao used Analytic Hierarchy Process and Hybrid Meta-heuristic Algorithm to compose the test-sheet [3].

Gwo-Jen Hwang and his colleague put forwarded an optimizing auto-reply accuracy approach based on an enhanced genetic in an e-learning system [4].

Gui-Xia Yuan successfully designed a multi-objects and multi-constraints mathematical model for the purpose of composing examination sheets and an improved genetic algorithm is used to solve the model [5].

Ren-jie Wu put forwarded an Ant Colony Algorithm to optimize the test paper generation [6].

Peng-Yeng Yin and Kuang-Cheng Chang realized the composing examination sheets algorithm based on the Particle Swarm Optimization [7].

3   Testing Paper Optimization

Particle Swarm Optimization (PSO) is one of the famous optimization algorithms first proposed by professor Kenndy and Eberhart in 1995 [8]. Like Genetic Algorithm that has applied into many regions and made a plenty of successful cases [10], the PSO algorithm can dispose of some phenomenon or problems in real society by near-optimal solution where no optimal solution can solve [9]. The research of the PSO has become an important field of study and attracted many researchers in recent years, Although the development of the particle swarm optimization is later than the genetic algorithm. The advantages of the Particle Swarm Optimization are high efficiency, easy realization and comprehension.

3.1   Particle Swarm Optimization (PSO)

A TPOPSO (Testing Paper Optimization based on PSO) algorithm is proposed here to deal with the poor quality testing paper caused by inaccurate attributes that have direct effect on the testing sheet or lack of the help from educational experts. The composing testing paper algorithm with high efficiency and quality cannot generate a satisfactory testing paper without the accurate attributes.

3.2   Parameter Initialization

Every particle there exists a list of parameters needed to be optimized and expressed by three-dimensional vectors $p = [p_{111}, p_{112}, p_{113}... p_{NSP}]$. Parameter N is the number of the examination type in a testing paper, the S is the number of the examination question for every examination type and the P is the attribute for every examining question. The p_{ijk} is the k th attribute of the j th examining question in the i th examination type. These parameters are significant to compose a high quality testing paper that can exactly estimate the real ability of the examinee.

The parameters initialization is accomplished by random assignment from 0 to 1. The velocity information decides the updated speed of the parameter, which is high speed in the initial state and become slower with the optimization.

3.3   Evaluation Function of the TPOPSO

The evaluation function in the TPOPSO is to estimate the every particle expressed testing papers that do not include every examination question in the testing question library. The k th testing paper in the testing paper library is expressed by decision vector x_k, $x_k = [x_{11}, x_{12}, x_{13}, ..., xNt, x_{Nts}]$. If the j th examination question of the i th examination type is in the examination paper, $x_{ij} = 1$, else $x_{ij} = 0$. The product of the decision vector and these parameters can evaluate a testing paper.

The evaluation function estimates a testing paper from four aspects including the Difficulty Deviation, Discrimination Deviation, Reliability Deviation and Time Error respectively represented by F_1, F_2, F_3 and F_4 in the following formula:

$$\text{MIN } F(x) = w_1 * F_1 + w_2 * F_2 + w_3 * F_3 + w_4 * F_4 \tag{1}$$

The parameter w_i ($i = 1, 2, 3, 4$) are weight of the four parameters decided which weight is more important than others. The ranges of four weights are from 0 to 1 and these weights satisfy constraint formula as presented below:

$$w_1 + w_2 + w_3 + w_4 = 1 \tag{2}$$

The Difficult Deviation of a testing paper is the absolution of the difference be-tween the evaluative difficulty and the testing difficulty given by the analysis of the result of a testing examination. The formula of the difficult deviation is shown at (3).

$$F_1 = \left| \left(\sum_{i=1}^{N_t} \sum_{j=1}^{N_{ts}} \left(d_{ij} \cdot s_{ij} \cdot x_{ij} \right) \right) \middle/ TS - D \right| \tag{3}$$

The actual difficulty, discrimination, reliability and the time are acquired by the plenty of the analysis on the examination results. These examinations are accomplished by the various candidates tautologically testing diverse questions at the different places and time.

3.4   The Global and Local Optimum

The global optimum and the local optimum are two elements in the Particle Swarm Optimization decided the speed and the quality of the optimization in the particles. The global optimum is the particle with the highest evaluation and is only one in the particle swarm. Every particle has a local optimum that is best evaluation in the process of optimization.

In the TPOPSO, the global optimum is the particle with the minimum absolute difference in the difficulty, discrimination, reliability and time and the local optimum for every particle is regarded as the minimum absolute difference in the process of updating itself. The local optimum only emphasizes on the individual instead of pondering upon the local scope in the standard Particle Swarm Optimization, which makes the particles easily fall into the best value in the local areas. In order to overcome the shortcoming of the PSO, an Improved Particle Swarm Optimization (IPSO) based on the K-nearest neighbor rule is applied to improve the local optimum of the Particle Swarm Optimization. The method of computing the local optimum in the improved PSO is to average the local optimum of the K-nearest particles as narrated in the following formula:

$$LocalBest_j = \frac{\sum_{i=1}^{k} LocalBest_i}{k} \tag{4}$$

4 Experiment and Discussion

The constitution of the examination question database and a testing paper is illustrated and an experiment is conducted to compare the performances between the Improved Particle Swarm Optimization (IPSO) and the unimproved one in the TPOPSO.

4.1 Examination Question Database

The experiment accomplishes on the fundamental application of computer technology by focusing on the computer basic knowledge and the ability of using the office-software. The examination question database includes only-choice questions, multiple-choice questions and judgment questions and fills questions and operating que-tions. The Table 1 shows the number of question type in examination question library.

Table 1. Testing questions distribution

Question types Distribution	Only-choice questions	Multiple-choice questions	Judgment questions	Fill questions	Operating questions
Number	186	172	178	182	166
Score	2	4	2	2	10

4.2 The Performance of the TPOPSO

To examine the improving effect, a comparison experiment is performed between the TPOPSO and the improved TPOPSO based on a new Particle Swarm Optimization. The parameters in this experiment are the $C1 = C2 = 1.5$ and rand = 0.3. The comparative result is showed in the Fig. 1(a) and (b) are the optimizing performances before 1000 generation and from 1000 generation to 2000 generation respectively.

From the Fig. 1(a), it is observed that the performance of the improved TPOPSO is not better than the unimproved one before 400 generation that the best fitness re-bounds around the 42, the best fitness of the unimproved has clearly declined to about 38 on the 400 generation at the same generation. When the optimization arrives to the 420 generation, the performance of the improved algorithm begins to difference and the global optimum in particle swarm turns to decline. From the 420 generation to the 720 generation, the best fitness of the unimproved TPOPSO is more excellent than the improved one, however the optimizing speed of the latter is faster than the former. When the optimization reaches to the 720 generation that is cut-off point in the train-ing process, the best fitness of them is almost same at 14.5. After the 720 generation, the global optimum of the improved TPOPSO is superior to the unimproved TPOPSO, the advantage of the improved TPOPSO is more and more obvious as the optimization continuous. The global optimum of the improved TPOPSO and unimproved one are 10.5 and 4.99 respectively when the training experiment has optimized 1000 generation.

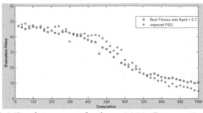
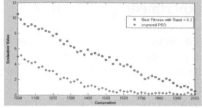

(a) Performance before 1000 Generation   (b) Performance after 1000 Generation

Fig. 1. Comparison of the performances

The Fig. 1(b) shows the optimizing situation from the 1000 generation to the 2000 generation. The performance of the improved TPOPSO distinctly decreased from about 4.5 to 0.5 with little zigzag phenomenon before the 1700 generation. After that the best fitness is finally leveling out at around 0.2 until the 2000 generation. Although the training performance of the unimproved TPOPSO is more and more excellent from the 1000 generation to the 2000 generation and the global optimum clearly reduced from 10 to 0.5, the difference on the optimizing speed and stability for the unimproved TPOPSO compared with the improved TPOPSO is apparent.

5   Conclusion

An intelligent examination system having the ability on optimizing the evaluation function of examining questions without help of the educational experts is proposed and empirically examined. The experiment shows that the evaluating accuracy of the testing paper can be effectively improved by the improved TPOPSO.

References

1. van Der Linden, W.J., Boekkooi-Timminga, E.: A maximin model for test design with practical constraints. Psychometrika **54**(2), 237–247 (1989)
2. Chou, C.: Constructing a computer-assisted testing and evaluation system on the World Wide Web the CATES experience. IEEE Trans. Educ. **43**, 266–272 (2000)
3. Hong Duan, T., Zhao, W., Wang, G., Feng, X.: Test-Sheet Composition Using Analytic Hierarchy Process and Hybrid Metaheuristic Algorithm TS/BBO, vol. 7, pp. 1–22. Hindawi Publishing Corporation Mathematical Problems in Engineering (2012)
4. Hwang, G.-J., Lin, B.M.T., LIn, T.-L.: An effective approach for test-sheet composition with large-scale item banks. Comput. Educ. **46**, 122–139 (2006)
5. Yuan, G.-X.: Modeling and research on computer composing test paper intelligently system. Comput. Simul. **11**, 370–373 (2011)
6. Ren-Jie, W.: Study on intelligently composing test paper based on ant colony optimization. Comput. Simul. **8**, 380–384 (2011)

7. Yin, P.-Y., Chang, K.-C., Hwang, G.-J., Hwang, G.-H., Chan, Y.: A particle swarm optimization approach to composing serial test sheets for multiple assessment criteria. Educ. Technol. Soc. **9**, 3–15 (2006)
8. Spink, A.: Term relevance feedback and mediated database searching: implications for information retrieval practice and systems design. Inf. Process. Manage. **31**, 161–171 (1995)
9. Xiangran, D., Zhang, M., Wang, X.: Self-optimizing evaluation function for Chinese-chess. Hybrid Inf. Technol. **7**(4), 163–172 (2014)
10. De Castro, L.N., Von Zuben, F.J.: Learning and optimization using the clone selection principle. IEEE Trans. Evol. Comput. **5**, 239–251 (2002)

The Null Space Pursuit Algorithm Based on an Arbitrary Order Differential Operator

Weiwei Xiao and Taotao Xing[✉]

North China University of Technology, Beijing, China
1256345607@qq.com

Abstract. The Null Space Pursuit (NSP) algorithm based on the differential operator is an important method of signal denoising and signal separation. In this paper, we propose an arbitrary order differential operator and use it in a complex signal which is a sum of simple signal. By solving an optimization problem, we also estimate the parameters of the differential operator. Finally, we confirmed the practicability of the algorithm through experimental simulation.

Keywords: Null space pursuit · Differential operator · Signal separation

1 Introduction

In the recent years, the methods of signal denoising and the separation of signal have drawn greater attention of scholars. The process of signal separation involves breaking down a complex signal into a sum of simple signals. The methods used to separate signals vary because of the construct of the simple signal which is decomposed from complex signal are different. For instance in the empirical mode decomposition (EMD) method [1, 2], an oscillatory signal is resolved into a sum of intrinsic mode functions (IMFs) and in the Matching Pursuit (MP) method [3], a signal is resolved into a total of time-frequency atoms. The Null Space Pursuit (NSP) algorithm on the basis of a differential operator is of particular interest to us in all methods of signal separation.

Silong Peng and Wen-Liang Hwang (2008) conceptualized the NSP algorithm based on a differential operator [4], two years later during 2010 in the next course of attempt they further improved the NSP algorithm. This method makes use of an adaptive operator to separate a complex signal into a sum of simple signals, and these simple signals belong to the null space in the above. The important steps of the method include estimating the adaptive operator T_s from the input signal S and decomposing S into R and U, where R is the residual signal and U is extracted signal, which $T_s(U) = 0$. In [5], they developed the following second-order differential operator:

$$T_s = \frac{d^2}{dt^2} + \omega^2(t).$$

which can eliminate the FM signal $\cos(\phi(t))$, where $\phi(t)$ is a local linear function.

The NSP algorithm on the basis of a differential operator has attracted wide attention due to its adaptability. In 2011, Xiyuan Hu [6] put forwarded the null space

© Springer International Publishing AG 2017
F. Xhafa et al. (eds.), *Recent Developments in Intelligent Systems and Interactive Applications*,
Advances in Intelligent Systems and Computing 541, DOI 10.1007/978-3-319-49568-2_2

pursuit algorithm and then further expanded the range of signal that could be decomposed. He empirically established the following second-order differential operator:

$$T_s = \frac{d^2}{dt^2} + P(t)\frac{d}{dt} + Q(t).$$

It can eliminate an AM-FM signal.

In this paper, we improve the algorithm [7] to use an arbitrary order differential operator. It can annihilate the signal:

$$A_1(t)\cos(\phi_1(t)) + A_2(t)\cos(\phi_2(t)) + \cdots + A_m(t)\cos(\phi_m(t)).$$

What's more,
This algorithm expands the range of signal that could be decomposed.

2   The Null Space Pursuit Algorithm Based on an Arbitrary Order Differential Operator

In this paper, we propose the following arbitrary order differential operator:

$$T_s = \frac{d^m}{dt^m} + a_{m-1}(t)\frac{d^{m-1}}{dt^{m-1}} + \cdots + a_1(t)\frac{d}{dt} + a_0(t). \tag{1}$$

This algorithm can estimate the orders of the above operator and the values of parameters $a_0, a_1, \cdots a_{m-1}$.

In a discrete case, the form of the operator is expressed as:

$$T_s = D_m + P_{a_{m-1}(t)}D_{m-1} + \cdots + P_{a_1(t)}D + P_{a_0(t)} \tag{2}$$

Where P_{a_i} is a diagonal matrix whose diagonal elements are a_i, where $i = 0, 1, \cdots m - 1$, and $D_m(m = 1, 2, 3, \cdots)$ is the matrix of the n-order difference.

Then, the values of parameters $a_0, a_1, \cdots a_{m-1}$ and the signal R is estimated by the following optimization method that minimizes the problem:

$$\min_{a_0,\cdots,a_{m-1},R,\lambda_1,\gamma,\lambda_2} \left\{ \|T_s(S - R)\|^2 + \lambda_1\left(\|R\|^2 + \gamma\|S - R\|^2\right) \right.$$
$$\left. + \lambda_2\left(\|D_{m-1}a_{m-1}\|^2 + \cdots + \|D_1a_1\|^2 + \|a_0\|^2\right) \right\} \tag{3}$$

where S is the input signal, R is the residual signal, γ is the leakage parameter and λ_1, λ_2 are Lagrange parameters. Let

$$F(a_0, \cdots a_{m-1}, R) = \|T_s(S-R)\|^2 + \lambda_1 \left(\|R\|^2 + \gamma\|S-R\|^2 \right)$$
$$+ \lambda_2 \left(\|D_{m-1}a_{m-1}\|^2 + \cdots + \|D_1 a_1\|^2 + \|a_0\|^2 \right) \} \tag{4}$$

For convenience, we let ϕ be the column vector as follows:

$$\phi = \left[a_{m-1}^T, a_{m-2}^T, \cdots a_1^T a_0^T \right]^T.$$

Then (4) becomes

$$F(\Phi, R) = \|(D_m + B_\Phi M_1)(S-R)\|^2 + \lambda_1 \left(\|R\|^2 + \gamma\|S-R\|^2 \right) + \lambda_2 \left(\|M_2\Phi\|^2 \right) \tag{5}$$

where $B_\Phi = [P_{a_{m-1}}, P_{a_{m-2}}, \cdots P_{a_1}, P_{a_0}]$ $M_1 = \left[D_{m-1}^T, D_{m-2}^T, \cdots D_2^T, D_1^T, E^T \right]^T$ and

$$M_2 = \begin{pmatrix} D_{m-1} & & & & \\ & D_{m-2} & & & \\ & & \ddots & & \\ & & & D_1 & \\ & & & & E \end{pmatrix}, \text{ in which } E \text{ is the identity matrix.}$$

For convenience, we rewrite the first item of (5) as the following:

$$(D_m + B_\Phi M_1)(S-R) = D_m(S-R) + B_\Phi M_1(S-R)$$
$$= D_m(S-R) + [P_{a_{m-1}}, P_{a_{m-2}} \cdots, P_{a_1}, P_{a_0}][D_{m-1}(S-R), \cdots, D_1(S-R), E(S-R)]^T$$
$$= D_m(S-R) + P_{D_{m-1}(S-R)}a_{m-1} + \cdots + P_{D_1(S-R)}a_1 + P_{(S-R)}a_0$$
$$= D_m(S-R) + \left[P_{D_{m-1}(S-R)}, \cdots, P_{D_1(S-R)}, P_{(S-R)} \right] [a_{m-1}, \cdots, a_1, a_0]^T$$
$$= D_m(S-R) + A\phi$$

where $A = [P_{D_{m-1}(S-R)}, \cdots, P_{D_1(S-R)}, P_{(S-R)}]$. Then (5) becomes

$$F(\phi, R) = \|D_m(S-R) + A\phi\|^2 + \lambda_1 \left(\|R\|^2 + \gamma\|S-R\|^2 \right) + \lambda_2 \left(\|M_2\phi\|^2 \right). \tag{6}$$

We assume

$$\frac{\partial F}{\partial \phi} = 2A^T(D_m(S-R) + A\phi) + 2\lambda_2 M_2^T M_2 \phi = 0. \tag{7}$$

Then

$$\hat{\phi} = -\left(A^T A + \lambda_2 M_2^T M_2 \right)^{-1} A^T D_m(S-R). \tag{8}$$

Similarly, we let $\frac{\partial F}{\partial R}\Big|_{\phi=\hat{\phi}} = 0$ to estimate $\hat{R}$ and obtain

$$\hat{R} = \left(T_s^T T_s + \lambda_1(1+\gamma)E\right)^{-1}\left(T_s^T T_s S + \lambda_1 \gamma S\right)$$
$$= M\left(\hat{\lambda}_1, \hat{\gamma}\right)\left(T_s^T T_s S + \lambda_1 \gamma S\right) \tag{9}$$

Where $T_s = D_m + P_{a_{m-1}}D_{m-1} + \cdots + P_{a_1}D + P_{a_0}$, $M\left(\hat{\lambda}_1, \hat{\gamma}\right) = \left(T_s^T T_s + \lambda_1(1+\gamma)E\right)^{-1}$.
For the NSP algorithm, parameters λ_1 and γ can be counted as follows:

$$\lambda_1 = \frac{1}{1+\hat{\gamma}}\frac{S^T M(\lambda_1,\hat{\gamma})^T S}{S^T M(\lambda_1,\hat{\gamma})^T M(\lambda_1,\hat{\gamma})S} \tag{10}$$

$$\gamma = \frac{(S-\hat{R})^T S}{\|S-\hat{R}\|^2} - 1 \tag{11}$$

However, the optimal value of λ_2 cannot be estimated by the above procedure. So, we can try more than one value of λ_2, according to the result to select the optimal solution. In practice, we find the best solution of (4) is not sensitive to the value of λ_2. Thus, the value of λ_2 can be fixed.

3   Experiment Result and Analysis

We assume that the value of λ_2 is given.

(1) Input: the signal $S(t)$, the initial values of λ_1^0 and γ^0, give the stopping threshold ε; Let $k = 1$;
(2) Let $j = 0, \hat{R}_j = 0, \lambda_1^j = \lambda_1^0, \gamma^j = \gamma^0$;
(3) Compute ϕ_j according to (8) and denoted by signal $\hat{R}_j$; Then, obtain the values of $\hat{a}_0, \hat{a}_1, \cdots, \hat{a}_{m-1}$;
(4) Compute λ_1^{j+1} according to (10) and denoted by $\phi_j, \hat{R}_j, \lambda_1^{j+1}, \lambda^j$;
(5) Compute $\hat{R}_{j+1}$ according to (9) and denoted by $\phi_j, \gamma^j, \lambda_1^{j+1}$;
(6) Compute γ^{j+1} according to (11) and denoted by $\hat{R}_{j+1}$, then set $j = j+1$;
(7) If $\|\hat{R}_{j+1} - \hat{R}_j\| > \varepsilon$, go to step (3); otherwise, go to the next step;
(8) Output the extracted signal $\hat{U} = (1+\gamma^{j+1})(S-\hat{R}_j)$ and the residual signal $\hat{R} = S - \hat{U}$;
(9) If $\|\hat{R}\|^2 > \varepsilon\|S\|^2$, set $k = k+1$ and go to step (2); otherwise, stop this program.

We illustrate several examples to demonstrate the results achieved by our algorithm. In the first example, the input signal is the signal $t\cos(t) + 3t\cos(3t) + 5t\cos(5t)$ in additive Gaussian random noise, as shown in Fig. 1. By running a Matlab program, we obtain the PSNR of input signal is 9.42217 dB, and the PSNR of extracted signal is 16.1682 dB.our algorithm have a good effect to signal denoising.

In this example, $\gamma^0 = 1, \lambda_1 = 0.0001, \lambda_2 = 1000, \varepsilon = 0.285$, and the orders of differential operator is six.

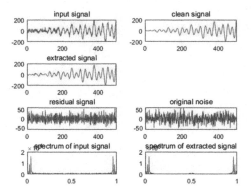

Fig. 1. Signal denoising

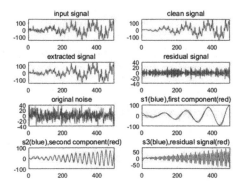

Fig. 2. Signal denoising and signal separation

In the second example, our algorithm can denoise from a noisy chirp signal and make the extracted signal into its coherent subcomponents. First, we input the signal $s1 + s2 + s3$ in additive Gaussian random noise, which $s1 = 4t \cos t$, $s2 = 3t \cos 5t$, $s3 = 2t \cos 15t + t \cos 20t$. We can see that the extracted is closed to the clean signal in Fig. 2. And it can estimate a eighth-order differential operator. In this example, we can obtain the PSNR of input signal is 11.5881 dB, and the PSNR of extracted signal is 13.6667 dB by running a Matlab program. In this process, let $\gamma^0 = 1, \lambda_1 = 0.1, \lambda_2 = 0.1, \varepsilon = 0.16$. Then, we separate the signal $s1 + s2 + s3$ into three subcomponents. In Fig. 2, the first component is closed to the signal $s1$, the second component is closed to the signal $s2$ and the residual signal is closed to the signal $s3$. The values of parameters for extraction of the first and second subcomponents are set at $\gamma^0 = 1, \lambda_1 = 0.0001, \lambda_2 = 0.1, \varepsilon = 0.01$ and $\gamma^0 = 1, \lambda_1 = 0.001, \lambda_2 = 0.01, \varepsilon = 0.1$.

4 Summary

In the paper, we put forward an NSP algorithm based on an arbitrary order differential operator. It improves the order of differential operator and expands the scope of signal that could be decomposed. In our future work, we will extend the method to images.

Acknowledgement. This work was supported by the Beijing Natural Science Foundation (1152001) and National Natural Science Foundation of China (11126140,11201007).

References

1. Rilling, G., Flandrin, P.: One or two frequencies the empirical mode decomposition answers. IEEE Trans. Signal Process. **56**, 85–95 (2008)
2. Wu, Z., Huang, N.E.:A study of the characteristics of white noise using the empirical mode decomposition method. In: Proceedings of the royal society of London (series A)460, 1597–1611(2004)
3. Mallat, S., Zhang, Z.: Matching pursuits with time-frequency dictionaries. IEEE Trans. Signal Process. **41**, 3397–3415 (1993)
4. Peng, S.L., Hwang, W.L.: Adaptive signal decomposition based on local narrow band signals. IEEE Trans. Signal Process. **56**, 2669–2676 (2008)
5. Peng, S.L., Hwang, W.L.: Null space pursuit: an operator-based approach to adaptive signal separation. IEEE Trans. Signal Process. **58**, 2475–2483 (2010)
6. Hu, X.Y.: Adaptive signal and image separation and its application, Dissertation for the degree of doctor of philosophy. Institute of Automation Chinese Academy of Sciences. Beijing (2011)
7. Xiao, W.W., Guo, Y.X.: The null space pursuit algorithm based on an arbitrary even-order differential operator. In: 2014 International Conference on Industrial Engineering and Management Science. Tianjin, China (2014)

Parameters Estimation of Regression Model Based on the Improved AFSA

Yujia Jin[1,2], Zhuoxi Yu[1,2(✉)], and Milan Parmar[1]

[1] School of Management Science and Information Engineering,
Jilin University of Finance and Economics, Changchun, China
yzx8170561@163.com
[2] Jilin Province Key Laboratory of Internet Finance, Changchun, China

Abstract. This paper aims at improving the AFSA algorithm. The improved AFSA algorithm is applied on estimation of parameters for the multiple linear regression models. Comparing the AFSA and the Least Squares, the results of simulation experiments verify that the estimating performance of the improved algorithm is better than the AFSA and the Least Squares. Thus, a noble approach for estimation of parameters is proposed in this research work.

Keywords: AFSA · Multiple linear regression model · Parameter estimation

1 Introduction

Multiple Linear Regression analysis [1] is an important data analysis method, widely used in industry, agriculture, medicine, social surveys, biological information processing and in a number socio-economic spheres. Estimation of Parameters is of paramount importance while solving the problem of multiple linear regression analysis. In the recent times, the most common method is the Least Squares. But this method suffers from computational complexity and its program structure is not universal. With the development of Genetic Algorithm (GA), Fish Algorithm (FA) and other Intelligent Algorithm Technology, applying Intelligent Optimization Algorithms to parameters estimation in regression model has gained substantial momentum in the recent years. Liu Jin-ping [2] uses the improved particle swarm algorithm for parameter estimation in multiple linear regression model. Huo Qian et al. [3] use genetic algorithm to solve parameters of nonlinear multivariable regression model and verify the effectiveness and practicality of the algorithm based on specific examples. Zhang Jiao-Ling et al. [4] use artificial Bee Colony Algorithm to estimate parameters of multiple linear regression. Sun Hui et al. [5] apply the PSO algorithm to improve the multiple linear regressions. However, the encoding and decoding process of Genetic Algorithms is very complicated and it also affects the optimization efficiency. Hence PSO may make the parameter optimization process into a local optimum due to the initialized parameter settings and cannot get the real optimal solution.

Due to the deficiency in obtaining exact solutions of AFSA, the authors in this research study attempt to improve the algorithm parameters and the swarming behavior of the artificial fish and apply the improved AFSA to the parameters estimation in

© Springer International Publishing AG 2017
F. Xhafa et al. (eds.), *Recent Developments in Intelligent Systems and Interactive Applications*,
Advances in Intelligent Systems and Computing 541, DOI 10.1007/978-3-319-49568-2_3

multiple linear regression models. Through the contrast experiments, the improved AFSA is effective, practical and simple.

2 Multiple Linear Regression Model

In multiple linear regression analysis, when predicting random variables Y, there exists a number of factors affecting its future value. When there is a basic linear relationship between Y and $x_1, x_2, \ldots, x_p (p > 1)$, the direction and magnitude of the relation can be predicted by applying linear regression equation. In the case of n arguments, multiple linear regression equation is:

$$Y = \beta_0 + \beta_1 x_1 + \beta_2 x_2 + \ldots + \beta_p x_p + \varepsilon \tag{1}$$

Among them, $\beta_0, \beta_1, \beta_2, \ldots, \beta_p$ is $p + 1$ parameters are to be estimated.

3 AFSA

AFSA [6] (Artificial Fish-swarm Algorithm) is a new efficient optimization method based on swarm intelligence developed by Li Xiao-Lei, who made an astute observation of habits and activity characteristics of fish and proposed this new type of swarm autonomous agent optimization method.

The state of artificial fish individuals can be expressed as a vector $X = (x_1, x_2, \ldots, x_n)$, among them $x_i(i = 1, 2, 3, \ldots, n)$ are optimization variables;

Food Concentration of artificial fish's current location is expressed as $Y = f(X_i)$, where Y is objective function;

The Distance between artificial Fish individuals is expressed as $d_{i,j} = \|X_i - X_j\|$; Visual is the feeling range of artificial Fish;

Step is the moving step of artificial fish;

δ is the crowding factor.

AFSA is to initialize a group of artificial fish (random solution), and search the optimal solution iterative methods in which artificial fish update themselves through feeding, clusters and rear-end and other acts in order to achieve optimization.

4 The Improved AFSA

We improve the algorithm parameters and the swarming behavior of the AFSA algorithm for its deficiency of optimization efficiency and accuracy.

4.1 The Improvement of Algorithm Parameters

Introducing the Kernel to adjust the visual and the step, you can make an adaptive adjustment for the step and the visual without making too much adjustment for the entire algorithm [7]. The specific method is to introduce the kernel to adjust the time parameters of the function and bring into the following function to adjust the visual and the step.

$$K(x) = \frac{15}{16}(1 - x^2)^2 \tag{2}$$

$$x = \frac{t}{t_{\max}} \tag{3}$$

$$Visual_{k+1} = Visual_{\max} + K(x) \times Visual_k \tag{4}$$

$$Step_{k+1} = Step_{\max} + K(x) \times Step_k \tag{5}$$

$Visual_{\max}$ is the maximum visual, $Step_{\max}$ is the maximum step. Setting the initial Visual and the Step are at maximum which is beneficial to the fast convergence and to avoid local extremum. With the running of algorithm, $K(x)$ is gradually decreasing and eventually to zero. At this point, the Visual and the Step come to the minimum which is beneficial to exact convergence in the late algorithm. The Step and Visual here only affect the foraging behavior, they do not affect the clustering behavior and the rear-end behavior.

4.2 The Improved Swarming Behavior

We improve the swarming behavior of the artificial fish [8]. Let $X_i(t)$ be the current state of artificial fish, the i-th artificial fish represents a feasible vector solution X_i, $X_i = (x_1^i, x_2^i, \ldots, x_n^i)$(n is the dimension), nf is the number of partners within the field of view, $X_c(t)$ is the central location of the fish-swarm, Rand represents a random number between 0 and 1. If $f(X_c(t)) \cdot nf < \delta \cdot f(X_i(t))$, it means that the center of its partners has more food and the crowding factor is not high in the field, and then the artificial fish make one step to the global optimal location X_{best}; if $f(X_c(t)) \cdot nf > \delta \cdot f(X_i(t))$, then the fish turn to the foraging behavior, the expression is as follows:

$$X_i(t+1) = X_i(t) + \frac{(X_c(t) - X_i(t)) + (X_{best} - X_i(t))}{\|(X_c(t) - X_i(t)) + (X_{best} - X_i(t))\|} \cdot step \cdot Rand() \tag{6}$$

4.3 The Definition of Artificial Fish Individuals and Fitness Function

We use the improved AFSA to estimate parameters of the multiple linear regression model [9]. Considering a set of parameters of the multiple linear regression model as an artificial fish, each artificial fish represents a candidate solution of the model estimation, the first i artificial fish expresses as:

$$X_i = (x_1^i, x_2^i, \ldots, x_n^i) \tag{7}$$

Using the fitness function to evaluate each artificial fish, the fitness function is defined as follows:

$$\phi(\theta) = \sum_{i=1}^{n}[y_i - (\beta_0 + \beta_1 x_{i1} + \beta_2 x_{i2} + \ldots + \beta_{in} x_{in})]^2 \tag{8}$$

Calculating the minimum value of the fitness function, you can get a set of estimating parameters namely the optimal solution.

4.4   The Process of the Improved AFSA

The main steps of the improved AFSA for parameters estimation are as follows [10]:

1. To initialize parameters, including initial position of artificial fish, the step of artificial fish, the visual of artificial fish, the number of artificial fish-swarm, the maximum number of iterations and etc.
2. To calculate the fitness function of each artificial fish, and select the artificial fish that have the optimal objective function and then recording on the bulletin board
3. The following operation is performed on each artificial fish:
   1. After the foraging behavior, the rear-end behavior and the clustering behavior, select the most improved behavior as the execution behavior of the artificial fish
   2. Comparing the fitness function value of the artificial fish which obtained from (1) with the record in the bulletin board, so that the minimum value is always in the bulletin board
   3. To judge the terminal conditions of the algorithm, if they are satisfied, output the global optimal solution and terminate the program; otherwise, jump to (2) and cycle again until it satisfies the terminal conditions.

5   Results and Conclusion

In order to verify the application of the improved AFSA in parameters estimation in multivariate linear regression model, the dataset presented in Table 1 involves the variable y which is influenced by the variables x_1, x_2 and x_3 leads to establishing the following multiple linear regression model:

$$y = \beta_0 + \beta_1 x_1 + \beta_2 x_2 + \beta_3 x_3 \tag{9}$$

In the Matlab, setting parameters of the AFSA: population size $n = 20$, iterations $= 100$, continuous operation 20 times, $Visual = 1.5$, the maximum moving $step = 0.25$, crowding factor $\delta = 0.2$.

The comparing results of the improved AFSA, the AFSA and the Least Squares including the parameters estimation, residual sum of squares, optimal solution of the linear regression model and run time of the algorithm are shown in Table 2.

Table 1. Parameters estimation data

y	x_1	x_2	x_3
124.9	131.8	127.1	111.3
125.1	122.9	122.5	130.6
121.3	127.6	115.4	115.4
124.8	119.9	123	120.3
...	...	...	...
124.3	121.7	123	119.4
123.3	117.2	119	125.5
125.4	121.8	124.1	129.3
123.6	124.8	129.1	112.1

Table 2. Comparison of results

Algorithm	β_0	β_1	β_2	β_3	Residual sum of squares	Optimal estimation	Run time
The improved AFSA	−0.521	0.343	0.423	0.241	0.996	125.664	1.132
AFSA	−0.497	0.341	0.437	0.225	0.993	124.991	1.236
Least squares	−0.436	0.332	0.423	0.221	0.991	124.836	1.712

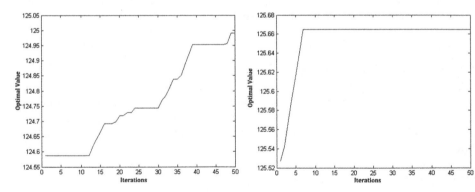

Fig. 1. The typical experimental curve with the AFSA

Fig. 2. The typical experimental curve with the improved AFSA

From Table 2, the application of the improved AFSA in parameters estimation is better than the results of the AFSA and the Least Squares, and it can also improve the run time.

At the same time, the typical experimental curve using the AFSA and the improved AFSA are respectively shown in Figs. 1 and 2.

From the comparison of Figs. 1 and 2, the optimal estimation using the AFSA tends to fall a local optimum in 124.588. However, the improved AFSA has the ability to break through the local extremum, and search the global optimum.

Acknowledgment. The research was supported by National Natural Science Foundation of China. (Grant Nos. 11571138).

References

1. Yu, X.-L., Ren, X.-S.: Multivariate Statistical Analysis, 2nd edn. Statistics Press, London (2006). pp. 20–25
2. Liu, J.-P., Yu, J.-X.: Parameter estimation of multiple linear regression models based on the improved particle swarm optimization algorithm. Comput. Eng. Sci. **32**(4), 101–105 (2011)
3. Qian, H., Li, S.-Q., Wang, W.-Y.: The study on determination of multivariate nolinear regression coefficient about genetic algorithm. J. Agric. Univ. **25**(2), 07–14 (2002). Hebei
4. Zhang, J.-L.: Application of artificial bee colony algorithm to multiple linear regression. J. Guangdong Polytech. Normal Univ. **31**(3), 31–33 (2011)
5. Sun, H., Zhang, Z.-M., Ge, H.-J.: Application of PSO to improve multiple linear regression. Comput. Eng. Appl. **43**(3), 43–44 (2007)
6. Li, X.-L., Shao, Z.-J., Qian, J.: An optimizing method based on autonomous animals: fish-swarm algorithm. Syst. Eng. Theor. Pract. **22**(11), 76–82 (2002)
7. Jiang, M.Y., Mastorakis, N.E., et al.: Multi-threshold image segmentation with improved artificial fish swarm algorithm. In: Proceedings of European Computing Conference (ECC 2007), pp. 117–120 (2007)
8. Chen, F.: Analysis and reaearch on improved Artificial Fish Swarm Algorithm. Xidian University, Xi'an (2012)
9. Xiao, J.M., Zheng, X.M., Wang, X.H., et al.: A modified artificial fish-swarm algorithm. In: Proceedings of the 6th World Congress on Intelligent Control and Automation, pp. 3456–3460 (2006)
10. Li, Y.: Solution to problems concerning AFSA based multiple linear regression analysis. J. Bohai Univ. **32**(2), 168–171 (2011). (Natural Science Edition)

Research on Hysteresis Modeling for Piezoelectric Displacement Actuators

Wang Rongxiu[1]([✉]), Jiang Chaoyuan[1], Cheng Yao[2], Hu Weijun[1], and Wang Bo[2]

[1] Chongqing Technology and Business University, Chongqing 400067, China
wrx_07@163.com
[2] Chongqing University of Technology, Chongqing 400050, China

Abstract. This paper introduces a Hysteresis Model for Piezoelectric micro-displacement actuators by the investigation of its geometrical features of Piezoelectric Hysteresis behavior. This model is based on the similarities and symmetries showed between the hysteresis loop curves with centrosymmetrical centers. The variations and commutations of loops are marked with foldback points which identify the centrosymmetrical centers, and simplify the acquirement of the wanted loop functions by the replacement of variables. The comparisons to experimental results verify the model's validity, accuracy and demonstrate the fineness in depiction and easy computation to Piezoelectric Displacement Actuators with hysteresis behaviors.

Keywords: Piezoelectric ceramics · Hysteresis model · Centrosymmetry · Micro-displacement actuator

1 Introduction

Piezoelectric Ceramics owing to the excellent characteristics, find their applications in micro-structures, microelectronics, precision machining, optical devices, robots, or a host of other fields [1, 2]. Piezoactive Micro Displacement Positioning systems in an Atomic Force Microscopy have the advantages of high resolution in the power of nanometer, sufficient output force, instant response, noise free, no mechanical return gap and friction free. In spite of those merits, they also present some drawbacks, mainly as hysteresis, creep deformation and nonlinearity, in which the hysteretic behavior is the major nonlinearity that limits their applications and performances in Piezoactive Micro and Nano Systems [3].

Models of non-linear hysteretic behavior of Piezoelectric materials are abundant in literatures, and can be divided into two groups based on their approaches. Microscopic models primarily stem from energy relations applied at the atomic or molecular level, try to establish hysteretic constitutive relations to describe Piezoelectric materials by understanding the causes of hysteresis with the help of fundamental physical principles, as Bassiouny [4, 5] and Kamlah [6]. Macroscopic models often use empirical relations to describe the behaviors of piezoelectric materials, and do not consider the underlying physics. They rely mainly on the input and output characteristics to get the mathematical fitting curves, as Preisach model and PI model [7, 8].

F. Xhafa et al. (eds.), *Recent Developments in Intelligent Systems and Interactive Applications*,
Advances in Intelligent Systems and Computing 541, DOI 10.1007/978-3-319-49568-2_4

The microscopic models are essentially the fundamental ones, however the mechanism is very complicated and microscopic models require a great number of parameters, often not available. The final models are inevitably approximate to reality. As an alternative choice, macroscopic models are preferred for their advantages of simple and direct descriptions, easy to establish the final practical models. The existent macroscopic models have drawbacks such as heavy computation and complex algorithm. Preisach model, for instance, applied Preisach operator to obtain the hysteretic model by quadratic integral, resulting in large amount of computation. PI model improved Preisach Model using many PI operators to replace Preisach operator, which similarly led to complexity because of a lot of integrations, and the reverse model is difficult to develop [9, 10].

This paper presents a simple macroscopic hysteretic model based on the input-output features of Piezoelectric Ceramic Micro Displacement Actuators, utilizing the similarity between hysteresis loop curves and their geometric symmetrical centers. The computational results from this model are compared with experimental data which verifies the validity and show that it provides precise description for hysteretic behaviors of the real ceramic micro displacement actuators.

2   The Symmetry and Features of Hysteretic Loop Curves for a PZT Patch

2.1   Hysteretic Loop and Central Symmetry

Figure 1 shows that the real hysteresis loop curves under various voltages for a poled PZT-5 displacement actuator. The curve AbaBecA is denoted as the zeroth class loop curve (or the outmost loop); Curves as BecfB, AbadA are called first class loop curves; the curves between points a, d belong to the second class loop (loops with higher class than zero are also labeled as inner loops). Figure 1 also demonstrates that the hysteresis increases as the drive voltage gets higher with larger enclosed area, however the centrosymmetry or similarity remains unchanged.

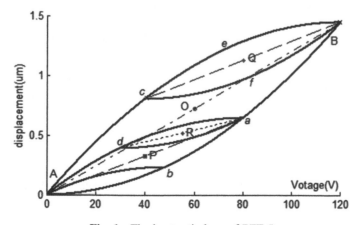

Fig. 1. The hysteretic loop of PZT-5

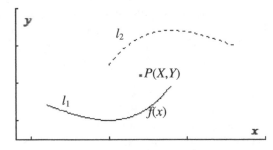

Fig. 2. Two centrosymmetric curves to center P

The similarities presented by loops make it reasonable to assume that there is a centrosymmetric center for the concave up and concave down segments of one loop. Take the zeroth class loop curve AbaBecA as an example, the forward segment AbaB is monotonically increasing and concave down while the backward segment BecA is monotonically increasing and concave up. BecA is centrosymmetric to AbaB with the point O that is the area center enclosed by AbaBecA, as shown in Fig. 1. The curve BecA can be depicted by the segment AbaB by 180 degree rotation in respect of the center O, vice versa.

In Fig. 2, two line l_1 and l_2 are centrosymmetric to the point P, and l_2 can be written as $y = 2Y - f(2X - x)$ if the function for l_1 is $y = f(x)$, that means the relation of the variables for l_2 satisfies the equation $2Y - y = f(2X - x)$, indicating the substitution of $2X$-x, $2Y$-y for x, y in the expression $y = f(x)$.

2.2   Foldback Point and Centrosymmetric Center

The centrosymmetries and similarities of various loop segments permit a simple way to identify a loop if its forward segment or backward one and the centrosymmetric center are known. A centrosymmetric center is the same point as a geometrical center of a loop locating at the midpoint of the line connecting the endpoints of segments, as the midpoint O of line AB. Geometrically, a foldback point is here defined as the turning point as the curve concavity changes from up to down or from down to up. In other words, a foldback point is the crossover point of a concave up curve and a concave down one, as the points A,B,a,b,c and the like in Fig. 1. When the drive voltage is turned from increasing to decreasing or from decreasing to increasing a foldback point appears at the moment. The newly appeared foldback point makes a pair with the previous one and determines the centrosymmetric center for the current loop. Therefore, foldback points appear and disappear continuously as the process goes on with various loops. The centrosymmetry of present loop is only decided by the latest two foldback points.

The point A is special in the zeroth class loop AbaBecA since A is the starting point and the final ending point for all loops, it can be viewed as a fixed foldback point. It is always convenient to set A as the origin of coordinate system.

2.3 The Variation of Loops, Their Upgrading/Degrading and Emerging/Vanishing

A complete loop is an enclosed curve consisting of two foldback points, a centroymmetric center and two monotonic segments with concavity or convexity. The upgrading of a loop happens if it changes itself into a higher class loop, from a zeroth class loop to a first class loop or from a one class to a two class. Similarly the degrading of a loop takes place while it changes itself into a lower class loop.

The ending point of a loop is a previous foldback point. After the fulfillment of a loop its center and two foldback points vanish and the subsequent loop is determined by the remaining foldback points and the fixed point A.

The zeroth class loop encloses the largest area and all other loops lie within it. Its curve provides entrances or exits for other loops. If the zero class loop behavior is known all other loops are determined according to the centrosymmetry described above.

3 Mathematic Model

For convenience, x denotes drive voltage; y represents the displacement output from a Piezoelectric actuator; c_x, c_y, denote the voltage and displacement in point c. Function $y = f(x)$ describes the behavior of segment AbaB in the zeroth class loop, the function for the ongoing loop curve can be acquired by the following replacement in $y = f(x)$ before a new foldback point is produced.

$$x \rightarrow X_{i-1} + X_i - x, \ y \rightarrow Y_{i-1} + Y_i - y \tag{1}$$

Where X_i, Y_i are the coordinates of the new foldback point and X_{i-1}, Y_{i-1} represent the pervious foldback point.

The forward segment da, for instance, is formed by three foldback points A, a and d appeared successively so the procedure of the replacement of variable is as bellows.

$$y \rightarrow A_y + a_y - y \rightarrow A_y + a_y - (a_y + d_y - y)$$
$$x \rightarrow A_x + a_x - x \rightarrow A_x + a_x - (a_x + d_x - x)$$

Their function replacements become accordingly as:

$$y = f(x) \rightarrow y = A_y + a_y - f(A_x + a_x - x) \rightarrow y = d_y - A_y + f(A_x - d_x + x)$$

The upgrading of a loop is accompanied by addition of new foldback points, yet the pattern of variable replacement remains the same and can be generalized as a basic rule suitable to be applied to the case of loop degrading. When a foldback point disappeared the replacement takes the contrary order. To make it clearer, see Fig. 1. The equation for the segment cfB in loop BecfB is $y = c_y - A_y + f(A_x - c_x + x)$. After reaching B, the foldback point c vanishes and the loop BecfB is finished so the replacement leads to

$x \rightarrow B_x + c_x - x$, $y \rightarrow B_y + c_y - y$, and the equation for segment Bec results in $y = A_y + B_y - f(A_x + B_x - x)$.

The function to judge the emerging of a foldback point is defined as

$$I(x_i) = \text{sgn}(\Delta x_{i-1} \Delta x_i) \tag{2}$$

Where $\Delta x_i = x_i - x_{i-1}$ and x_i is the ith input variable; "sgn" is the signum function. In the case $I < 0$, point (x_i, y_i) becomes a foldback point and is denoted by (X_α, Y_α) to indicate the α th foldback point. The equation to decide the disappearance of the α th foldback point can be written as

$$D(X_\alpha) = X_{\alpha-1} \tag{3}$$

4   Experimental Results

Take a poled PZT-5 stack as the experimental research object to test the algorithm. The piezoelectric ceramic stack consists of 36 wafers with thickness 0.33 mm, piezoelectricity constant 450×10^{-12} C/N, maximum voltage 135 V, and maximum displacement stroke 1.5 um. The forward segment of the zeroth class loop is measured with experiment, and then find out the function of $y = f(x)$ approximately by application of polynomial   curve   fitting   with   residue $|R(x)| < 0.002$,   it   is $y = 0.5 \times 10^{-4}x + 1.02 \times 10^{-4}x^2 - 2.25 \times 10^{-7}x^3 + R(x)$.

The backward segment of the zeroth loop is compared as in Fig. 3, where the solid represents model results and the dashed comes from experiment. It is clear to see that the difference is very small, the error < 0.001 um.

Figure 4 shows the comparison of the higher loop, demonstrates the desired performance of the proposed model. The itinerary is $abcdefghea$, the dashed line is part of the zeroth class loop in Fig. 3, the crossed curve represents experimental results and the solid curve is drawn with model calculation. In Fig. 4, the segment abc comprises of one part of the zeroth class loop. Point a is the fixed foldback point. abc turns into the first class loop segment $cdea$ at c and changes to efg by e, and next transforms to ghe

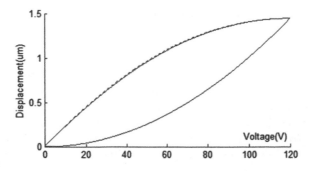

Fig. 3. The comparsion of the backward segment of the zeroth class loop

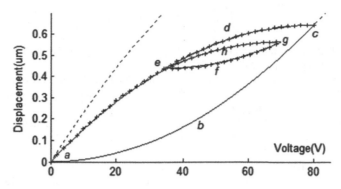

Fig. 4. Comparision of some higher loops

Table 1. Evolvement of the function of curve *abcdefghea* in Fig. 4

Segment	Point/point group	Function	Center
abc	*a*	$y = f(x)$	none
cdea	(a,c)	$a_y + c_y - y = f(a_x + c_x - x)$	one
efg	$(a,c);(c,e)$	$a_y + c_y - (c_y + e_y - y) = f[a_x + c_x - (c_x + e_x - x)]$	two
ghe	$(a,c);(c,e);(e,g)$	$a_y + g_y - y = f(a_x + g_x - x)$	three
ea	(a,c)	$a_y + c_y - y = f(a_x + c_x - x)$	one

through *g*, and finally returns to *cdea* at *e*. The evolvement of the function for all the segments is shown in Table 1 according to the mathematical model discussed in previous sections.

5   Conclusions

From the geometric characteristics point of view, the Piezoelectric hysteresis loops are discussed and analyzed, and a mathematic model is established to describe the hysteresis behaviors. The key algorithm for this model is based on the similarities and centrosymmetries exhibited between various curves of all loops. The centrosymmetry permits a simple way to identify an unknown loop segment function by the calculation of its centrosymmetric center. The upgrading or degrading of a loop curve is therefore associated with the input and output variable replacement in the foregone function by means of produced foldback points. The comparisons with experiments verified that this model can be used to describe the hysteresis behaviors of Piezoelectric ceramics with simplicity, precision and ease of computation. It can also be applied to approximate higher class loops formed after many foldback points appeared.

Acknowledgements. This work was supported by the Educational Foundation of Chongqing (Grant No. KJ1500617,KJ1500935)

References

1. Hocken, R.J.: Nanotechnology and its impact on manufacturing. In: Sympositum on Flexible Automation, Tokyo, Japan, USA, pp. 7–15 (1993)
2. Wang, G.: Introduction to the leading edge of Nanofabrication. Science Press, Peking (2009)
3. Bashash, S.: Modeling and control of piezoactive micro and nano systems, Clemson University, Clemson, South Carolina (2008)
4. Bassiouny, E., Ghaleb, A.F., Maugin, G.A.: Thermo dynamical formulation for coupled electromechanical hysteresis effects-I. Basic equations. Int. J. Eng. Sci. **26**(12), 1279–1295 (1988)
5. Bassiouny, E., Ghaleb, A.F., Maugin, G.A.: Thermo dynamical formulation for coupled electromechanical hysteresis effects-II. Poling of ceramics. Int. J. Eng. Sci. **26**(12), 1297–1306 (1988)
6. Kamlah, M., Bohle, U.: On a non-linear finite element method for piezoelectric structures made of hysteretic ferroelectric ceramics. Comput. Mater. Sci. **19**(1), 81–86 (2000)
7. Wang, Y., Zhao, X.: Inverse control algorithm to compensate the hysteresis and creep effect of piezoceramic. Opt. Precis. Eng. **2006**(6), 1032–1040 (2006)
8. Ge, P., Jouaneh, M.: Generalized preisach model for hysteresis nonlinearity. Precis. Eng. **20**, 99–111 (1997)
9. Wang, X., Guo, S., Ru, C., Ye, X.: A simple modeling method for the piezoelectric hysteresis behavior. J. Harbin Eng. Univ. **31**(1), 271–276 (2010)
10. Haichen, Q., Zhouping, Y.: Research on hysteresis contstitutive relation on piezoceramic crystals. China Mech. Eng. **25**(15), 2059–2063 (2014)
11. Cui, Y., Sun, B., Dong, W.: Causes for hysteresis and nonlinearity of piezoelectric ceramic actuators. Opt. Precis. Eng. **11**(3), 270–275 (2003)

An Iterative Method for the Least Squares Anti-bisymmetric Solution of the Matrix Equation

Li Lin$^{(\boxtimes)}$, Yuan Xiujiu, Liu Jing, and Su Dongqing

Science College, Air Force Engineering University, Xi'an 710051, China
156171643@qq.com

Abstract. In this paper the authors on the basis of Conjugate Gradient Method of solving linear algebraic equations, using special transformation and approximate disposal, proposed an Iterative Method, which can solve the least squares anti-bisymmetric solution of the matrix equation $AXB + CXD = F$. Using this iterative method, for an arbitrary initial anti-bisymmetric matrix, we can obtain a solution within finite iterative steps in the absence of round off errors. Further this solution with least norm can be obtained by choosing a special initial matrix. In addition, the expression of its optimal approximation solution to a given matrix can be obtained.

Keywords: Matrix equation · Iterative method · Anti-bisymmetric matrix · Least squares solution Optimal approximation

1 Introduction

$R^{m \times n}$ and $SR^{n \times n}$ are denoted by the set of $m \times n$ real matrices and the set of $n \times n$ real symmetric matrices, respectively. The superscripts T and + is the transpose and Moore-Penrose generalized inverse of matrices, respectively. $A \otimes B$ is the Kronecker product of matrices A and B. In space $R^{m \times n}$, we define inner product as: $(A, B) = \mathrm{tr}(A^T B)$ for all $A, B \in R^{m \times n}$. Then the norm of a matrix A generated by this inner product is, obviously, Frobenius norm is denoted by $\|A\|$. Let $R(A)$ stand for its column space. Let vec(.) represent the vec operator, i.e. $\mathbf{vec}(A) = (a_1^T, a_2^T, \cdots a_n^T)^T$ for $A = (a_1, a_2, \cdots, a_n) \in R^{m \times n}$, $a_i \in R^m$ $(i = 1, \cdots, n)$. I_n denotes $n \times n$ the unit matrix. e_i denotes the ith column of the identity matrix I_n and $S = (e_n, e_{n-1}, \cdots, e_1)$.

Definition 1. An $n \times n$ matrix X is called an anti-bisymmetric matrix if $X = -X^T$ and $X = SXS$. We denote by BASR$^{n \times n}$ the set of all $n \times n$ anti-bisymmetric matrices.

We consider the following two problems.

Problem I. Given $A, C \in \mathbf{R}^{m \times n}, B, D \in \mathbf{R}^{n \times p}, F \in \mathbf{R}^{m \times p}$, find $X \in \mathbf{BASR}^{n \times n}$ such that

$$\|AXB + CXD - F\| = \min$$

F. Xhafa et al. (eds.), *Recent Developments in Intelligent Systems and Interactive Applications*,
Advances in Intelligent Systems and Computing 541, DOI 10.1007/978-3-319-49568-2_5

Problem II. Given $\overline{X} \in \mathbf{R}^{n \times n}$, find $\widehat{X} \in S_E$ such that

$$\left\|\widehat{X} - \bar{X}\right\| = \min_{X \in S_E} \left\|X - \bar{X}\right\|$$

Where S_E the solution is set of Problem I and $\|\bullet\|$ is Frobenius norm.

The problem of solving the special least squares solutions of the famous matrix equation $AXB + CXD = F$ has been widely applied in many fields, such as structure design, system engineering, network programming, reconnaissance, remote sensing, autocontrol theory, molecule spectroscopy, civil engineering, and so on. Research on solving the problem has been actively ongoing for the past 40 or more years. For instance, some important results have been obtained in [2, 5–7], but some problems have not settled up to now. The anti-bisymmetric solution of the matrix equation $AX = B$ has been derived by the generalized singular value decomposition in [3]. The solvability conditions for the inverse Problem of the anti-bisymmetric matrix has been given in [4]. By the methods proposed in above references, it can solve the solution X of the matrix equation in $\mathbf{R}^{n \times n}$. However, it is difficult to solve the least squares anti-bisymmetric solution of the matrix equation $AXB + CXD = F$ by using these methods. Therefore, a new method is presented to solve Problem I in this paper.

Lemma 1. [1] Suppose that $A \in \mathbf{R}^{n \times n}$, the sufficient and necessary condition of $A \in$ **BASR**$^{n \times n}$ is $A = -A^T = SAS$, if $a, b \in \mathbf{R}$, $X, Y \in \mathbf{BASR}^{n \times n}$, then $aX + bY \in$ **BASR**$^{n \times n}$.

To introduce an iterative method for solving Problem I, we give the following theorem.

Theorem 1. Problem I is transformed the following the equation

$$
\begin{aligned}
A^{\mathrm{T}}(AXB + CXD)B^{\mathrm{T}} &+ C^{\mathrm{T}}(AXB + CXD)D^{\mathrm{T}} \\
&+ B(B^{\mathrm{T}}XA^{\mathrm{T}} + D^{\mathrm{T}}XC^{\mathrm{T}})A + D(B^{\mathrm{T}}XA^{\mathrm{T}} + D^{\mathrm{T}}XC^{\mathrm{T}})C \\
&+ SA^{\mathrm{T}}(ASXSB + CSXSD)B^{\mathrm{T}}S + SC^{\mathrm{T}}(ASXSB + CSXSD)D^{\mathrm{T}}S \\
&+ SB(B^{\mathrm{T}}SXSA^{\mathrm{T}} + D^{\mathrm{T}}XC^{\mathrm{T}})AS + SD(B^{\mathrm{T}}SXSA^{\mathrm{T}} + D^{\mathrm{T}}SXSC^{\mathrm{T}})CS = H
\end{aligned}
\tag{1}
$$

and it is always consistent, where

$$
\begin{aligned}
H = A^{\mathrm{T}}FB^{\mathrm{T}} &+ C^{\mathrm{T}}FD^{\mathrm{T}} - BF^{\mathrm{T}}A - DF^{\mathrm{T}}C \\
&+ SA^{\mathrm{T}}FB^{\mathrm{T}}S + SC^{\mathrm{T}}FD^{\mathrm{T}}S - SBF^{\mathrm{T}}AS - SDF^{\mathrm{T}}CS
\end{aligned}
$$

Proof. Finding $X \in \mathbf{BASR}^{n \times n}$ such that $\|AXB + CXD - F\| = \mathbf{\textit{min}}$ is equivalent to

$$
\begin{aligned}
&\|AXB + CXD - F\|^2 + \left\|B^{\mathrm{T}}XA^{\mathrm{T}} + D^{\mathrm{T}}XC^{\mathrm{T}} + F^{\mathrm{T}}\right\|^2 + \\
&\|ASXSB + CSXSD - F\|^2 + \left\|B^{\mathrm{T}}SXSA^{\mathrm{T}} + D^{\mathrm{T}}SXSC^{\mathrm{T}} + F^{\mathrm{T}}\right\|^2 = \mathbf{\textit{min}}
\end{aligned}
\tag{2}
$$

The following proof is given to solve the minimum problem (2) equivalent to solving the matrix Eq. (1). Considering matrix equation group

$$\begin{cases} AXB + CXD = F \\ B^\mathrm{T}XA^\mathrm{T} + D^\mathrm{T}XC^\mathrm{T} = -F^\mathrm{T} \\ ASXSB + CSXSD = F \\ B^\mathrm{T}SXSA^\mathrm{T} + D^\mathrm{T}SXSC^\mathrm{T} = -F^\mathrm{T} \end{cases} \tag{3}$$

The matrix equation group (3) can be straightened by rows of linear equations, that is

$$\begin{pmatrix} A\otimes B^\mathrm{T} + C\otimes D^\mathrm{T} \\ B^\mathrm{T}\otimes A + D^\mathrm{T}\otimes C \\ (AS)\otimes(B^\mathrm{T}S) + (CS)\otimes(D^\mathrm{T}S) \\ (B^\mathrm{T}S)\otimes(AS) + (D^\mathrm{T}S)\otimes(CS) \end{pmatrix} \cdot \overline{\mathrm{vec}}(X) = \begin{pmatrix} \overline{\mathrm{vec}}(F) \\ -\overline{\mathrm{vec}}(F^\mathrm{T}) \\ \overline{\mathrm{vec}}(F) \\ -\overline{\mathrm{vec}}(F^\mathrm{T}) \end{pmatrix} \tag{4}$$

Since the linear Eq. (4) is equivalent to the matrix equation group (3) for the least squares anti-bisymmetric solution. That is, the solution of the Eq. (2). The normal equations of the system of linear Eq. (4) is

$$U\begin{pmatrix} A\otimes B^\mathrm{T} + C\otimes D^\mathrm{T} \\ B^\mathrm{T}\otimes A + D^\mathrm{T}\otimes C \\ (AS)\otimes(B^\mathrm{T}S) + (CS)\otimes(D^\mathrm{T}S) \\ (B^\mathrm{T}S)\otimes(AS) + (D^\mathrm{T}S)\otimes(CS) \end{pmatrix} \cdot \overline{\mathrm{vec}}(X) = U\begin{pmatrix} \overline{\mathrm{vec}}(F) \\ -\overline{\mathrm{vec}}(F^\mathrm{T}) \\ \overline{\mathrm{vec}}(F) \\ -\overline{\mathrm{vec}}(F^\mathrm{T}) \end{pmatrix} \tag{5}$$

That is $W \cdot \overline{\mathrm{vec}}(X) = U(\overline{\mathrm{vec}}(F), -\overline{\mathrm{vec}}(F^\mathrm{T}), \overline{\mathrm{vec}}(F), -\overline{\mathrm{vec}}(F^\mathrm{T}))^\mathrm{T}$.
Where

$$\begin{aligned} W =\ & (A^\mathrm{T}A) \otimes (BB^\mathrm{T}) + (A^\mathrm{T}C)\otimes(BD^\mathrm{T}) + (C^\mathrm{T}A)\otimes(DB^\mathrm{T}) + (C^\mathrm{T}C)\otimes(DD^\mathrm{T}) \\ & + (BB^\mathrm{T})\otimes(A^\mathrm{T}A) + (BD^\mathrm{T})\otimes(A^\mathrm{T}C) + (DB^\mathrm{T})\otimes(C^\mathrm{T}A) + (DD^\mathrm{T})\otimes(C^\mathrm{T}C) \\ & + (SA^\mathrm{T}AS)\otimes(SBB^\mathrm{T}S) + (SA^\mathrm{T}CS)\otimes(SBD^\mathrm{T}S) \\ & + (SC^\mathrm{T}AS)\otimes(SDB^\mathrm{T}S) + (SC^\mathrm{T}CS)\otimes(SDD^\mathrm{T}S) \\ & + (SBB^\mathrm{T}S)\otimes(SA^\mathrm{T}AS) + (SBD^\mathrm{T}S)\otimes(SA^\mathrm{T}CS) \\ & (SDB^\mathrm{T}S)\otimes(SC^\mathrm{T}AS) + (SDD^\mathrm{T}S)\otimes(SC^\mathrm{T}CS) \end{aligned}$$

$$\begin{aligned} U =\ & [A^\mathrm{T}\otimes B + C^\mathrm{T}\otimes D, B\otimes A^\mathrm{T} + D\otimes C^\mathrm{T}, (SA^\mathrm{T})\otimes(SB) + (SC^\mathrm{T})\otimes(SD), \\ & (SB)\otimes(SA^\mathrm{T}) + (SD)\otimes(SC^\mathrm{T})] \end{aligned}$$

The linear equation group (5) is reduced to a matrix equation, namely the matrix Eq. (1).

For convenience of discussion, we introduce a matrix function

$$
\begin{aligned}
M(X) = M(X) = {} & A^{\mathrm{T}}(AXB+CXD)B^{\mathrm{T}}+C^{\mathrm{T}}(AXB+CXD)D^{\mathrm{T}} \\
& +B(B^{\mathrm{T}}XA^{\mathrm{T}}+D^{\mathrm{T}}XC^{\mathrm{T}})A+D(B^{\mathrm{T}}XA^{\mathrm{T}}+D^{\mathrm{T}}XC^{\mathrm{T}})C \\
& +SA^{\mathrm{T}}(ASXSB+CSXSD)B^{\mathrm{T}}S+SC^{\mathrm{T}}(ASXSB+CSXSD)D^{\mathrm{T}}S \\
& +SB(B^{\mathrm{T}}SXSA^{\mathrm{T}}+D^{\mathrm{T}}XC^{\mathrm{T}})AS+SD(B^{\mathrm{T}}SXSA^{\mathrm{T}}+D^{\mathrm{T}}SXSC^{\mathrm{T}})CS
\end{aligned}
$$

Therefore, the Eq. (2) is changed to $M(X) = H, X \in \mathbf{BASR}^{n\times n}$. It can be proved by Lemma 1, then $M(X) \in \mathbf{BASR}^{n\times n}$

In summary, the solution of the problem I is the anti-symmetric solution of the matrix Eq. (1)

The following proof of matrix Eq. (1) must have an anti-symmetric solution. Because the normal equation group (5) has a solution, the Eq. (1) has a solution. Let $\tilde{X}$ be a solution (not necessarily anti-symmetric solution), then

$$
M(\tilde{X}) = H \tag{6}
$$

Ordering $f(\tilde{X}) = [\tilde{X} - \tilde{X}^{T} + S(\tilde{X} - \tilde{X}^{T})S]/4$, By the Lemma 1, $f(\tilde{X})$ is the anti-symmetric solution that is proved by the Eq. (6), which is a solution of the Eq. (1).

2 An Iterative Method for Solving Problem I

Lemma 2. Suppose that $X, Y \in \mathbf{R}^{n\times n}$, then (1) $M(X \pm Y) = M(X) \pm M(Y)$ (2) $(M(X), Y) = (X, M(Y))$ Now, we introduce an iterative method for solving the linear matrix Eq. (1), or equivalently a method for solving Problem I.

(1). Input the initial matrix $X_0 \in \mathbf{BASR}^{n\times n}$, and then compute

$$
R_0 = H - M(X_0), \quad Q_0 = M(R_0), \quad k := 0
$$

(2). If $R_k = o$, then stop; else, $k : = k + 1$
(3). Compute

$$
\begin{aligned}
X_k &= X_{k-1} + \left(\|R_{k-1}\|^2 \big/ \|Q_{k-1}\|^2 \right) Q_{k-1}, R_k = H - M(X_k) \\
&= R_{k-1} - \left(\|R_{k-1}\|^2 \big/ \|Q_{k-1}\|^2 \right) M(Q_{k-1})
\end{aligned}
$$

$$
Q_k = M(R_k) + \left(\|R_k\|^2 \big/ \|R_{k-1}\|^2 \right) Q_{k-1}
$$

(4). Go to step 3.

According to Lemma 1, the mathematical induction method can be used to prove, if $X_0 \in \mathbf{BASR}^{n\times n}$, then $X_k \in \mathbf{BASR}^{n\times n} (k = 1, 2, 3, \cdots)$.

Theorem 2. For an arbitrary initial anti-bisymmetric matrix X_0, X^* is a solution, the sequences X_i, R_i and Q_i satisfy

$$(Q_i, \ X^* - X_i) \ = \|R_i\|^2 \ (i = 0, 1, 2, \cdots) \tag{7}$$

Remark 1. Theorem 2 implies that, if $R_i \neq O$, then $Q_i \neq O$ $(i = 0, 1, 2, \cdots)$, then Algorithm cannot be terminated.

Theorem 3. For the sequences R_i and Q_i, if there is a positive number l such that $R_i \neq O$ for all $i = 0, 1, 2, \cdots l$, then

$$(R_i, \ R_j) \ = 0, (Q_i, \ Q_j) \ = 0 \ (i \neq j; \ i,j = 0, 1, \cdots, l) \tag{8}$$

Remark 2. By Theorem 3, we know that, for $\forall X_0 \in \mathbf{BASR}^{n \times n}$, the least squares anti-bisymmetric solution of the equation $AXB + CXD = F$ can be obtained in no greater than n^2 steps.

Since the sequences R_i satisfy $(R_i, \ R_j) \ = 0$, $(i \neq j; \ i,j = 0, 1, \cdots, l)$ in the finite dimension space $\mathbf{R}^{n \times n}$, then it must exist a number $0 < k \leq n^2$ such that $R_k = O$.

Lemma 3. [1] Suppose that the consistent system of linear equations $Ax = b$ has a solution $x^* = A^+ b \in R(A^T)$, then X^* is the unique least Frobenius norm solution of the system of linear equations.

For vector $x \in \mathbf{R}^{nm}$, we denote the following $m \times n$ matrix containing all the entries of vector x by mat(x): mat(x) = $[x(1 : m), \ x(m + 1 : 2m), \cdots, \ x((n - 1)m + 1 : nm)]^T$, where $x(i : j)$ denotes a vector containing the ith to jth elements of vector x.

By Algorithm and Remark 2, if the initial matrix is $X_0 = M(U_0)$, where U_0 is an arbitrary matrix in $\mathbf{BASR}^{n \times n}$, we can obtain $X_k, R_k, Q_k \in \mathbf{BASR}^{n \times n}$ $(k = 0, 1, 2, \cdots)$, and $Q_0 = M(V_0)(V_0 \in \mathbf{BASR}^{n \times n})$. Assume that $X_{k-1} = M(U_{k-1}), Q_{k-1} = M(V_{k-1})$ $(U_{k-1}, V_{k-1} \in \mathbf{BASR}^{n \times n})$. By the expression of X_k and Q_k in Algorithm

$$X_k = M(U_k), Q_k = M(V_k) \ (U_k, V_k \in \mathbf{BASR}^{n \times n}) \tag{9}$$

We introduce a symbol

$$\begin{aligned}
L = &(A^T A) \otimes (BB^T) + (A^T C) \otimes (BD^T) + (C^T A) \otimes (DB^T) + (C^T C) \otimes (DD^T) + \\
&+ (BB^T) \otimes (A^T A) + (BD^T) \otimes (A^T C) + (DB^T) \otimes (C^T A) + (DD^T) \otimes (C^T C) \\
&+ (SA^T AS) \otimes (SBB^T S) + (SA^T CS) \otimes (SBD^T S) + (SC^T AS) \otimes (SDB^T S) \\
&+ (SC^T CS) \otimes (SDD^T S) + (SBB^T S) \otimes (SA^T AS) + (SBD^T S) \otimes (SA^T CS) \\
&+ (SDB^T S) \otimes (SC^T AS) + (SDD^T S) \otimes (SC^T CS)
\end{aligned}$$

The matrix Eq. (1) by line straightening obtain the linear equations

$$L \cdot \overline{\text{vec}}(X) = \overline{\text{vec}}(H) \tag{10}$$

By $X_k = M(U_k)$, the rows can be straightened

$$\overline{\text{vec}}(X_k) = L \cdot \overline{\text{vec}}(U_k) \in R(L) = R(L^T).$$

By Remark 2, the existence of positive integer k_0 makes the matrix Eq. (1) of the anti-bisymmetric solution $X^* = X_{k_0}$, then $\overline{\text{vec}}(X^*) = \overline{\text{vec}}(X_{k_0}) \in R(L^T)$.

By Lemma 3, the $\overline{\text{vec}}(X^*)$ is the least Frobenius norm solution of the linear Eq. (6). Then X^* is the least norm solution of the Eq. (1), and X^* is the least norm solution of Problem I, X^* is

$$X^* = \text{mat}(L^+ \cdot \overline{\text{vec}}(H)) \tag{11}$$

Above conclusions on the solutions of Problem I can be expressed as following theorem.

Theorem 4. Problem I is always consistent, and for $\forall X_0 \in \mathbf{BASR}^{n \times n}$, the X_k gained by Algorithm converges to one of its solutions within limited steps. Furthermore, if the initial matrix $X_0 = M(U_0)$ (U_0 is an arbitrary anti-bisymmetric matrix) is choosed, the solution X^* obtained by Algorithm is the solution of Problem I. And X^* can be expressed as (11).

3   The Solution for Problem II

For Problem II, there certainly exists a unique solution since the solution set of Problem I is a nonempty closed convex cone. Noting that a symmetric matrix and an anti-bisymmetric matrix are orthogonal each other, for $X \in \mathbf{BASR}^{n \times n}$, $\bar{X} \in \mathbf{R}^{n \times n}$, then

$$\|X - \bar{X}\|^2 = \|X - ((\bar{X} - \bar{X}^T)/2 + (\bar{X} + \bar{X}^T)/2)\|^2 = \|X - (\bar{X} - \bar{X}^T)/2\|^2 + \|(\bar{X} + \bar{X}^T)/2\|^2$$
$$= \|X - [\bar{X} - \bar{X}^T + S(\bar{X} - \bar{X}^T)S]/4\|^2 + \|[\bar{X} - \bar{X}^T - S(\bar{X} - \bar{X}^T)S]/4\|^2 + \|(\bar{X} + \bar{X}^T)/2\|^2$$

And

$$\min_{X \in S_E} \|X - \bar{X}\|^2 = \min_{X \in S_E} \|X - [\bar{X} - \bar{X}^T + S(\bar{X} - \bar{X}^T)S]/4\|^2$$
$$+ \|[\bar{X} - \bar{X}^T - S(\bar{X} - \bar{X}^T)S]/4\|^2 + \|(\bar{X} + \bar{X}^T)/2\|^2$$

By the first equality of Lemma 2, we have

$$M\{X - [\bar{X} - \bar{X}^T + S(\bar{X} - \bar{X}^T)S]/4\} = H - M\{[\bar{X} - \bar{X}^T + S(\bar{X} - \bar{X}^T)S]/4\}$$

We introduce a symbol $\tilde{X} = X - [\bar{X} - \bar{X}^{\mathrm{T}} + S(\bar{X} - \bar{X}^{\mathrm{T}})S]/4$,

$$\tilde{F} = F - A\{[\bar{X} - \bar{X}^{\mathrm{T}} + S(\bar{X} - \bar{X}^{\mathrm{T}})S]/4\}B - C\{[\bar{X} - \bar{X}^{\mathrm{T}} + S(\bar{X} - \bar{X}^{\mathrm{T}})S]/4\}D$$

$$\tilde{H} = A^{\mathrm{T}}\tilde{F}B^{\mathrm{T}} + C^{\mathrm{T}}\tilde{F}D^{\mathrm{T}} - B\tilde{F}^{\mathrm{T}}A - D\tilde{F}^{\mathrm{T}}C$$
$$+ SA^{\mathrm{T}}\tilde{F}B^{\mathrm{T}}S + SC^{\mathrm{T}}\tilde{F}D^{\mathrm{T}}S - SB\tilde{F}^{\mathrm{T}}AS + SD\tilde{F}^{\mathrm{T}}CS$$

Then finding the unique solution of Problem II is equivalent to finding the anti-bisymmetric least norm solution of the matrix equation

$$M(\tilde{X}) = \tilde{H} \tag{12}$$

Using Algorithm with initial matrix $\tilde{X}_0 = M(U)$, where $\forall U \in \mathbf{BASR}^{n \times n}$, we can obtain the anti-bisymmetric least norm solution $\tilde{X}^*$ of the linear matrix Eq. (12).

References

1. Cheng, Y.-P., Zhang, K.-Y., Xu, Z.: Matrix Theory. Northwestern Polytechinical University Press, Xi'an (2004). China, pp. 337–340
2. Dai, H.: Using vibration experiment, the optimal correction rigidity, flexibility and mass matrix. J. Vib. Eng. **2**, 18–27 (1988)
3. Zhang, X.-D., Zhang, Z.-N.: The anti- bisymmetric optimal approximation solution of matrix equation $\Lambda X = B$. J. Appl. Math. **5**, 810–818 (2009)
4. Sheng, Y.P., Xie, D.X.: The solvability conditions for the inverse problem of anti-bisymmetric matrices. Numer. Calculation Comput. Appl. **2**, 111–120 (2002)
5. Jameson, A., Kreindler, E., Lancaster, P.: Symmetric positive semidefinite and positive definite real solutions of $AX = XA^{\mathrm{T}}$ and $AX = YB$. Linear Algebra Appl. **160**, 189–215 (1992)
6. Jiang, Z.-X., Qi-Chao, L.: Under the restriction of spectral matrix optimal approximation problem. Numer. Math. **1**, 47–52 (1986)
7. Chu, M.T., Funderlic, R.E., Golub, G.H.: On a variational formulation of the generalized singular value decomposition. SIAM J. Matrix Anal. Appl. **18**(4), 1082–1092 (1997)
8. Chu, K.E.: Symmetric solution of linear matrix equations by matrix decompositions. Linear Algebra Appl. **119**, 35–55 (1989)
9. Hu, X.Y., Zhang, L., Xie, D.X.: The solvability conditions for the inverse eigenvalue problem of anti-bisymmetric matrices. Math. Numer. Sin. **4**, 409–418 (1998)
10. Zhang, L., Xie, D.X., Hu, X.Y.: The inverse eigenvalue problem of bisymmetric matrices on the linear manifola. Math. Numer. Sin. **2**, 129–138 (2000)

Parallelization of the Conical Area Evolutionary Algorithm on Message-Passing Clusters

Weiqin Ying[1(✉)], Bingshen Wu[1], Yuehong Xie[1], Yuxiang Feng[1], and Yu Wu[2]

[1] School of Software Engineering, South China University of Technology, Guangzhou 510006, China
yingweiqin@scut.edu.cn
[2] School of Computer Science and Educational Software, Guangzhou University, Guangzhou 510006, China

Abstract. The Conical Area Evolutionary Algorithm (CAEA) has exhibited significant performance for bi-objective optimization problems. In this paper, a parallel partially evolved CAEA (peCAEA) on message-passing clusters is proposed to further reduce the runtime of the sequential CAEA for bi-objective optimization. Each island maintains an entire population but is responsible for evolution of only a portion of the population. Further, the elitist migration adopted in order to share information among islands and speed up the evolutionary process. Additionally, a dynamic directional migration topology presented here to obtain a satisfactory balance between convergence speed and communication costs. Experimental results on ZDT test problems indicate that the peCAEA obtains satisfactory solution quality with good speedup on clusters.

Keywords: Bi-objective optimization · Evolutionary algorithm · Migration topology · Parallelization · Message passing

1 Introduction

In many fields of engineering, there are plenty of problems involving simultaneous optimization of k objective functions, $k \geq 2$, known as Multi-Objective Optimization problems (MOPs). In the most elementary case of $k = 2$, an MOP becomes a bi-objective optimization problem (BOP), such as the wireless sensor network (WSN) layout problem [1]. Since these objectives possibly conflict with each other, there are multiple Pareto optimal solutions, which make a trade-off between two or more objectives [2].

Multi-Objective Evolutionary Algorithms (MOEAs) have the advantageous ability of being able to find multiple solutions in one simulation run against classical optimization methods. In the past decade, several MOEAs including the non-dominated sorting genetic algorithm II (NSGA-II) [2] were developed. However, the main disadvantage of the MOEAs based on dominance is that they generally spend lots of time checking for non-dominance in a population. In recent years, the multi-objective

F. Xhafa et al. (eds.), *Recent Developments in Intelligent Systems and Interactive Applications*, Advances in Intelligent Systems and Computing 541, DOI 10.1007/978-3-319-49568-2_6

evolutionary algorithm based on decomposition (MOEA/D) [3] exhibits its great advantage in performance over the dominance-based MOEAs. The MOEA/D decomposes an MOP into a number of scalar sub-problems, which helps it avoid checking for non-dominance. On this basis, a Conical Area Evolutionary Algorithm (CAEA) [4] using a conical partition strategy and a conical area indicator was proposed to further improve the efficiencies of decomposition-based MOEAs for bi-objective optimization. When updating the best solution found so far of a sub-problem, the CAEA does not need to check for update of neighbor sub-problems like MOEA/D. Therefore, if MOEA/D and CAEA use the same population size and the same stopping criterion, the ratio between their computational complexities is $O(T):O(1)$, where T means the size of neighbors.

Although decomposition-based MOEAs have made commendable progress, many MOPs typically involve objective functions with high computational costs. The desire to reduce the running time of MOEAs naturally leads to the use of parallel and distributed technologies. Some parallel paradigms could be applied to decompose tasks or data and in turn decrease runtime cost. In general, there are two major parallel MOEA (pMOEA) paradigms: master-slave model and coarse-grained model (called island model) [5]. Compared with the master-slave model, the island model is easier to implement, and more suitable for decomposition-based MOEAs due to the decomposition strategy. Besides, in terms of implementation for decomposition-based pMOEAs, the number of distributed-memory, process-based versions is far less than that of shared-memory, thread-based versions [6] up to now. The advantages of distributed-memory versions are that their scalabilities exceed those of shared-memory versions in most cases. So far, two distributed-memory parallel MOEA/Ds have been developed: the overlapped partitioning MOEA/D (opMOEA/D) [7] and the partially evolved MOEA/D (peMOEA/D) [8]

However, these parallelizations of MOEA/D are not yet directly available for parallel CAEAs on distributed-memory clusters. The overlapped partitioning island model of opMOEA/D could not be adopted by parallel CAEAs due to the globality of the CAEA's replacement procedure. In addition, the migration policy and the migration topology of peMOEA/D are not suitable for parallel CAEAs. In this paper, we propose a partially evolved parallelization of CAEA on message-passing cluster, referred to as peCAEA. The peCAEA utilizes the partially evolved island model to achieve significant time reductions. Furthermore, an elitist migration policy and a dynamic directional migration topology are adopted to obtain satisfactory solution quality in the peCAEA.

2 Partially Evolved Parallelization of CAEA

In the partially evolved parallel CAEA, the whole population exists on each separate node, but the evolutionary tasks are split evenly among processors. In other words, each node is responsible for the optimization of only a portion of sub-problems and the relevant individuals maintain the best solutions found so far of their corresponding sub-problems. After certain generations of evolution, the peCAEA propagates elitist individuals among processors by directional migration so that each node can share the evolutionary progresses with the other nodes.

2.1 Partially Evolved Island Model

In the sequential CAEA, a BOP is transformed to N scalar optimization sub-problems, and each individual keeps the best solution found so far of its corresponding sub-problem. Moreover, each sub-problem is assigned an exclusive decision subset and uses the conical area as its scalar objective is to find a local non-dominated solution in its own decision subset [4].

Each island in the partially evolved island model keeps the entire population but is only in charge of the evolution of a portion of the population. The partially evolved island model makes the peCAEA spend less time than the sequential CAEA. In the peCAEA, the partitions of population are simply obtained by linear division and the size of the r-th partition $P^{(r)}$ equals:

$$N^{(r)} = \begin{cases} N/q+1 & \textit{if } r< (N \mod q) \\ N/q & \textit{otherwise.} \end{cases} \tag{1}$$

In Eq. (1), q denotes the number of processors and $r\in[0..q-1]$. After the division, the k-th scalar sub-problem on the r-th island, referred to as $g^{(r,k)}$, uses the conical area as its scalar objective in the form:

$$\text{minimize } g^{(r,k)}(x) = S\left(F(x) - F^{\Delta}\left(\Omega^{(r)}\right)\right)$$
$$\textit{subject to } x \in \Omega^{(r,k)}. \tag{2}$$

In Eq. (2), $\Omega^{(r)}$ denotes the attainable current solution space of the r-th island. Since $\Omega^{(r)}$ varies during the evolution, the sub-problems are expected to change correspondingly. From the point of view of objective space, each island assigned to a set of sub-problems is in charge of the exploration of a segment of the Pareto front. Besides the individuals responsible for evolution, the remaining individuals on each node help the island complete the updating procedure. When generating a new offspring, any individual has a chance of being updated in the CAEA while only its neighbor individuals have that chance in the MOEA/D, which explains why the opMOEA/D [7] isn't suitable for parallel CAEAs.

2.2 Elitist Migration Among Islands

Although the running time of peCAEA decreases remarkably by introducing the partially evolved island model, the quality of the final Pareto solutions might descend sharply at the meantime. In order to maintain a satisfactory quality, an elitist migration policy is performed among islands to share critical evolutionary information in the peCAEA.

In the peMOEA/D [8], critical evolutionary information migrates according to a hybrid policy, which includes Utopian migration and Elitist migration. However, utopian migration has almost no influence on the final solutions quality of peCAEA, and instead increases the communication cost. A better utopian point could help to discover the more complete front for MOEA/D or CAEA. However, the difference lies in whether utopian points generated by an island can satisfy the demand of this island without the help of external utopian points or not. In the reproduction procedure, two similar neighbor individuals are selected as parents on each island in the peMOEA/D. Thus, the position in the objective space of the new offspring is more likely near the locations of its parents. In other words, the global search ability of an island in the peMOEA/D is relatively weak, which results in the difficulty to obtain a good utopian point by itself. As a result, without utopian points from outside, an island in the peMOEA/D only converges towards a narrow front and finally the peMOEA/D might only obtain some front segments. In contrast with the peMOEA/D, the location in the objective space of a new offspring generated by an island in the peCAEA has a chance to lie at any conical area, more likely to obtain a better utopian point. Therefore, utopian migration is not important for the peCAEA.

Algorithm 1. The main procedure of peCAEA

1: **for each** $r \in [0..q-1]$ **in parallel do**   ▷ q processors
2:      $N^{\{r\}} \leftarrow$ InitializePartialEvolvedSize(N, q, r)
3:      $\overline{V} \leftarrow$ InitializeObservationVector(r)
4:      $P^{\{r\}} \leftarrow$ InitializePopulation($\overline{V}$)
5:      $C^{(r)} \leftarrow$ InitialParticipatedSops($\overline{V}, N, N^{(r)}, r$)
6:      $z \leftarrow$ InitializeLocalUtopianPoint($P^{\{r\}}$)
7:      $gen \leftarrow 0$
8:      **while** $gen \leq MaxGen$ **do**   ▷ evolutionary loop
9:          $size \leftarrow$ the size of $N^{(r)}$
10:         $i \leftarrow 0$
11:         **while** $i < size$ **do**
12:             $parents \leftarrow$ AreaTourSelection($P^{(r)}$)
13:             $child \leftarrow$ Crossover($parents$)
14:             $child \leftarrow$ Mutation($child$)
15:             EvaluateFitness($child$)
16:             $z \leftarrow$ UtopianPointUpdate($child$)
17:             InternalSolutionsUpdate($child, P^{\{r\}}$)
18:             $i \leftarrow i + 1$
19:         **end while**
20:         DirectionalMigration()   ▷ directional migration
21:         $gen \leftarrow gen + 1$
22:     **end while**
23:     $P \leftarrow$ SynchronizePop($P^{(r)}$)
24: **end for**

In consideration of this reason, the peCAEA only adopts the elitist migration policy, which accelerates the evolution of each island. If an individual is updated on an island, it would be sent to certain islands in a later course of time. In addition, when receiving any elitist individual from outside, the island would calculate the conical area index of this individual and find whether the primary individual in this conical area should be replaced by the immigrator.

2.3 Dynamic Directional Migration Topology

Exchanging elitist individuals can speed up the evolutionary process. But the choice of migration topologies would have a great impact on how much it does. There are several existing simple migration topologies, like the bi-ring and the complete graph, for parallel MOEAs. In the bi-ring topology, the elitist individual $x^{(i,k)}$ of the i-th island will be sent to both the l-th island and the k-th one where $l = (i + 1+q) \% q$ and $k = (i-1 + q) \% q$ where $\%$ denotes the modulus operation, $j \neq l$ and $j \neq k$. In other words, fewer elitist individuals that could improve $P^{(j)}$ will be received by the j-th island. As a result, when the j-th island selects parents from $P^{(j)}$, the quality of offsprings would be worse. Conversely, an island in the complete topology sends any of its elitist individuals to all other islands and spends too much time on useless communication.

Algorithm 2. Directional migration

1: **function** DIRECTIONALMIGRATION()
2:    **if** gen mod $commI = 0$ **then**   ▷ elitist migration
3:       $i \leftarrow 0$
4:       **while** $i < N$ **do**
5:          **if** $updated[i] =$ true **then**
6:             $index \leftarrow$ ComputeMigrationDestination(i)
7:             SendElitistIndiv($P_i^{\{r\}}, index$)
8:             $updated[i] \leftarrow$ false
9:          **end if**
10:          $i \leftarrow i + 1$
11:       **end while**
12:       $j \leftarrow 0$
13:       **while** $j < q$ **do**
14:          $indivs^j \leftarrow$ ReceiveElitistIndiv(j)
15:          ExternalSolutionsUpdate($indivs^j, P^{\{r\}}$)
16:          $j \leftarrow j + 1$
17:       **end while**
18:    **end if**
19: **end function**

Therefore, a dynamic directional migration topology is introduced in the peCAEA to balance communication costs against convergence speed. In this topology, the i-th island would send the elitist individual $x^{(i,k)}$, $0 \leq k \leq N-1$, to the j-th island when, and only when, $x^{(i,k)}$ has been updated in the last θ generations and the index of its sub-problem is in $S^{(j)}$, where $S^{(j)}$ includes the indexes of sub-problems that the j-th island is responsible to optimize. Unlike the constant channels of the static migration topologies such as the ring topology or the complete one, the channels of the directional migration are dynamically constructed depending on which sub-problems the updated elitist individuals belong to. Thus, the elitist individuals received by an island are generally different from those received by another island.

2.4   Procedure of PeCAEA

The procedure of peCAEA is presented in Algorithm 1. Each island is initialized on each processor in Lines 2–6. In Line 4, the r-th island initializes its whole population. Then the indexes $C^{(r)}$ of partition $P^{(r)}$ are generated in Line 5. The code in Lines 11–19 represents one generation of evolution, which is the same as the sequential CAEA except Line 12. The function in Line 12 selects individuals from $P^{(r)}$ as parents. After one generation of evolution, each island exchanges information with other islands by function DirectionalMigration in Line 20. Finally, all the q partitions are collected to make up a whole front in Line 23.

Algorithm 2 describes the pseudocode of function DirectionalMigration. It first checks whether it's time to migrate in Line 2 of Algorithm 2. If so, each island sends its elitist individuals to the other islands in Lines 4–11 and receives immigrators from outside in Lines 13–17. If an individual is updated, function ComputeMigrationDestination would compute which island should receive it in Line 6. Then, this island would receive the elitist individuals from all the other islands. It's worth noting that $indivs^j$ could be *null* in Line 14.

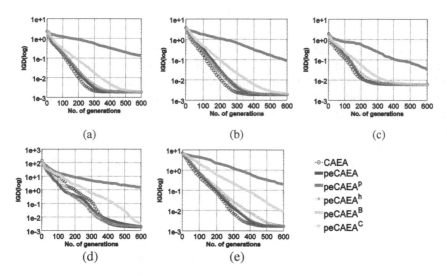

Fig. 1. Convergence curves of IGD values of each sequential and parallel CAEA for ZDT$_1$ (a), ZDT$_2$ (b), ZDT$_3$ (c), ZDT$_4$ (d), and ZDT$_6$ (e).

3   Experimental Results and Analysis

Several experiments are designed for the sequential CAEA, the peCAEA and its four variants, peCAEA^P, peCAEA^h, peCAEA^B and peCAEA^C in this section. The peCAEA^P represents the variant without any migration policy of peCAEA while the peCAEA^h denotes the variant with the hybrid migration policy of peCAEA. Similarly, the peCAEA^B uses the bi-ring migration topology while the peCAEA^C adopts the complete migration topology. All the above algorithms are written in C++ using MPI.

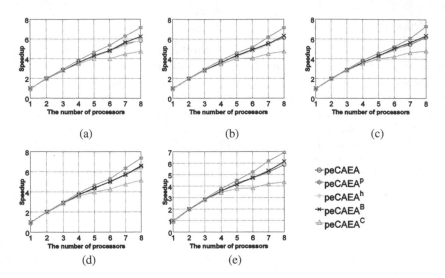

Fig. 2. Speedup curves achieved by each parallel CAEA with an increase in processors for ZDT_1 (a), ZDT_2 (b), ZDT_3 (c), ZDT_4 (d), and ZDT_6 (e).

The experiments are carried out on five widely used benchmark BOPs: ZDT_1 (convex), ZDT_2 (nonconvex) and ZDT_3 (disconnected, nonconvex) with 30 decision variables, ZDT_4 (multifrontality) and ZDT_6 (nonuniformity) with 10 decision variables. In order to simulate real world BOPs with complicated objective functions, we add an idle function into the objective functions of each ZDT test instance to increase the evaluation time.

All experiments are performed on a cluster composed of 20 interconnected computing nodes equipped with Intel Core I5 3.20 GHz CPU and 4 GB RAM. Some important parameters used in these algorithms are given as follows. The allowed generation *MaxGen* = 600 and the population size N = 200. Besides, the operator of crossover and mutation as well as their parameters are the same as used in [4, 9]. Each algorithm is applied 20 times independently for each test instance in our experiments.

The inverted generational distance (IGD) [4] metric is utilized to assess the quality of the obtained sets of solutions. The number of involved processors is set as q = 8 in our experiments. Figure 1 plots the convergence curve of average IGD values in logarithmic scalar over 30 runs of each sequential and parallel CAEA for each benchmark problem on the cluster. It is evident from Fig. 1 that the qualities of fronts discovered by the peCAEA are close to those by the sequential CAEA. Besides, the speedup metric is used to measure the effect of parallelization of algorithms. Figure 2 shows the speedup curves obtained by each parallel CAEA with the number of involved nodes increasing from 1 to 8. Figure 2 indicates that the peCAEA achieves significant time reductions. Although the peCAEA^P costs less time, not only it converges slower than the peCAEA obviously, but also the final IGD values achieved by it are far much higher than those by the peCAEA.

The experimental results suggest that the elitist migration accelerates the evolution and highly improves the solution quality. Figure 3 shows each front segment found by

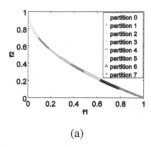
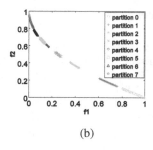

(a)   (b)

Fig. 3. Front segments discovered, respectively, by 8 partitions of peCAEA (a) and those by 8 partitions of peMOEA/D (b) for ZDT_1.

each processor in the peCAEA and peMOEAD for ZDT_1. The experiment results about IGD and speedup appear to make little difference between the peCAEA and $peCAEA^h$. Moreover, there is no disconnection between any neighbor partitions of front discovered by the peCAEA without any utopian migration while this kind of disconnection often occurs in the front by the peMOEA/D without any utopian migration [7], as shown in the case of ZDT_1 of Fig. 3. Consequently, it can be inferred that the utopian migration seems useless for the peCAEA. In Fig. 2, the speedup curves of the $peCAEA^B$ are similar to those of the peCAEA. However, the fronts found by the $peCAEA^B$ are much worse than those by the peCAEA. It suggests that the directional dynamic migration topology is more effective for the parallelization of CAEA than the bi-ring topology. Although the $peCAEA^C$ is also able to obtain the satisfactory solution quality, it consumes much more time due to the use of the complete topology.

4 Conclusions

The peCAEA achieves higher performance in terms of both convergence performance and speedup for bi-objective ZDT test instances on clusters. On one hand, the partially evolved island model highly improves the effect of parallelization of CAEA. On the other hand, the elitist migration with the dynamic directional migration topology speeds up the evolutionary process at the meantime. Our future work will focus on applications of the peCAEA for some complicated bi-objective problems from the real world and the parallelization of several decomposition-based MOEAs in higher dimensional objective spaces.

Acknowledgments. This work is supported by the National Natural Science Foundation of China (No. 61203310, 61503087), the Natural Science Foundation of Guangdong Province, China (No. 2015A030313204, 2015A030310446), the China Scholarship Council (CSC) (No. 201406155076, 201408440193), the Pearl River S&T Nova Program of Guangzhou (No. 2014J2200052), the Fundamental Research Funds for the Central Universities, SCUT (No. 2013ZZ0048, 2013ZM0104), and the Science and Technology Planning Project of Guangdong Province, China (No. 2013B010401003).

References

1. Molina, G., Alba, E., Talbi, E.G.: Optimal sensor network layout using multi-objective metaheuristics. J. Univ. Comput. Sci. **14**(15), 2549–2565 (2008)
2. Deb, K., Pratap, A., Agarwal, S., Meyarivan, T.: A fast and elitist multiobjective genetic algorithm: NSGA-II. IEEE Trans. Evol. Comput. **6**(2), 182–197 (2002)
3. Zhang, Q., Li, H.: MOEA/D: A multiobjective evolutionary algorithm based on decomposition. IEEE Trans. Evol. Comput. **11**(6), 712–731 (2007)
4. Ying, W., Xu, X., Feng, Y., Wu, Y.: An efficient conical area evolutionary algorithm for bi-objective optimization. IEICE Trans. Fundam. Electron. Commun. Comput. Sci. **E95-A**(8), 1420–1425 (2012)
5. van Veldhuizen, D.A., Zydallis, J.B., Lamont, G.B.: Considerations in engineering parallel multiobjective evolutionary algorithms. IEEE Trans. Evol. Comput. **7**(2), 144–173 (2003)
6. Durillo, J.J., Zhang, Q., Nebro, A.J., Alba, E.: Distribution of computational effort in parallel MOEA/D. In: Coello, C.A.C. (ed.) LION 2011. LNCS, vol. 6683, pp. 488–502. Springer, Heidelberg (2011). doi:10.1007/978-3-642-25566-3_38
7. Wu, Y., Xie, Y., Ying, W., Xu, X., Liu, Z.: A partitioning parallelization with hybrid migration of MOEA/D for bi-objective optimization on message-passing clusters. IEICE Trans. Fundam. Electron. Commun. Comput. Sci. **E99-A**(4), 843–848 (2016)
8. Ying, W., Xie, Y., Wu, Y., Wu, B., Chen, S., He, W.: Universal partially evolved parallelization of moea/d for multi-objective optimization on message-passing clusters. Soft Comput., pp. 1–14 (2016). http://dx.doi.org/10.1007/s00500-016-2125-y
9. Ying, W., Xie, Y., Xu, X., Wu, Y., Xu, A., Wang, Z.: An efficient and universal conical hypervolume evolutionary algorithm in three or higher dimensional objective space. IEICE Trans. Fundam. Electron. Commun. Comput. Sci. **E98-A**(11), 2330–2335 (2015)

Design of Heat Preservation Bathtub with Water Saving

Zaiqiang Ku and Li Cheng[✉]

College of Mathematics and Physics, Huanggang Normal University,
Hubei 438000, China
kzqhgnu@163.com, chenglialbert@163.com

Abstract. This paper studies the relationship between the quantity of consuming water and the time of bath. First, we construct a bi-objective optimization model to determine the optimal strategy that a person in the bathtub can keep the initial temperature. By analyzing the change of temperature of water through given parameter, we find that it is the saving water mode if the temperature of bathing water is 55 and the time of spending in bathing is less than 25 min. Meanwhile, influx rate of hot water should be controlled at 0.216 m^3/h to guarantee that minimal amount of water is consumed. The authors in this paper also consider the sensitivity and fitness of the model depending upon the shape and volume of the tub.

Keywords: Double layer bathtub · Bi-objective optimization model · Hot bath

1 Introduction

The process of bathing in bathtub is comfortable and relaxing. Unfortunately, the bath water would be cool after a while. The principal objective of this research work is about maintaining the temperature of the bathtub. The objectives set in this paper have been pursued in a two step approach. In the first instance for the purpose of minimizing the quantity of water being wasted, a bi-objective optimization model has been constructed to determine the mechanism that the person in the bathtub can adopt to keep the temperature and making it resemblance with the initial temperature. In the second step looking at the model, we consider the sensitivity and fitness of the model depending upon the shape and volume of the tub. We also examine the impact of bubble bath additive in the bathtub to assist in cleansing on the results obtained from this model.

We use the theory of heat exchange to analyze the process of heating system and establish the model of temperature. Based on the analysis of the model, we use MATLAB soft to solve the problem, and the result shows that it is the saving water mode if the temperature of bathing water is 55 and the time of spending in bathing is less than 25 min. However, if it takes longer than 25 min in bathing, then the influx rate of hot water should be controlled at 0.216 m^3/h to guarantee that minimal amount of water is consumed.

© Springer International Publishing AG 2017
F. Xhafa et al. (eds.), *Recent Developments in Intelligent Systems and Interactive Applications*,
Advances in Intelligent Systems and Computing 541, DOI 10.1007/978-3-319-49568-2_7

2 Problem Description

Sometimes, we may run into the water become cold. So it is very important to keep the range of people acceptable. This article begins with the analysis of bubble bath composition and infinitesimal analysis unit on each face of the in and out flow of time interface temperature gradient. So we can draw the conclusion that the bubble bath due to the loss of heat and with no addition of bubbles can get the effect of bubble bath model. To the identity by the seller to provide the non each face on and out of the flow in the explanation and introduce the advantages and disadvantages of the product. Firstly, we design PMMA materials bathtub shape is a cuboids in the middle, both ends of which are semi cylindrical (length, width and height are a, b and c, respectively) the bottom surface of the radius is $r(2r = b)$. Secondly we design a folding lid of bath crock above, covers the back surface only one end of the semicircular parts are not covered. Thirdly, bathtub designed for double, the thickness is l.

To be more specific, the following assumptions are made throughout the paper.

(1) Assuming the bath room temperature is 25, and the most suitable bath temperature range is 37～40.
(2) The temperature of the inner wall of the bathtub and the outer wall temperature of the bath tub are kept constant, and the heat conduction process is in a steady state.
(3) The heat conduction is uniform, and the heat conductivity coefficient is the constant.

3 The Model

3.1 Analysis of Bath Process

Bath to maintain the temperature in between, so the heat loss requires timely heating water supply heat energy. The bath process divided into three processes: feasible bath, bath temperature, safety bath, which are shown as follows (Fig. 1).

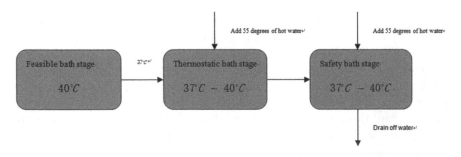

Fig. 1. A process flow chart

3.2   Analysis of Heat Gain and Loss

We analyse the whole process of heat balance of bath so that we can know heat loss and the heat should be balanced.

Where Q_I is heat for the injection of hot water. The formula is:

$$Q_I = cm\Delta T \tag{1}$$

Where c is the specific heat $(J/kg\,^\circ C)$, m is the quality (kg), ΔT is the temperature difference. There are three main types of heat loss: convection heat dissipation, radiation heat radiation, and heat conduction.

(1) Convection heat dissipation [1]

Unit of time, through the water surface (dF), the surface of the heat transfer to the air is dQ_{O1}, can be expressed as:

$$dQ_{O1} = \alpha(T - \theta)dF$$

(2) Radiation heat radiation [2]

Radiation heat sink calculation:

$$dQ_{O2} = \varepsilon\sigma(T + 273)^4 dF$$

Where ε is a blackness, σ is a constant take value 5.6×10^{-8} W/m$^2 \cdot$ hPa, and dF is surface area (m^2).

(3) Heat conduction [3, 4]

Heat conduction's formula is:

$$dQ_{O3} = tqdF$$

(4) The total thermal resistance of series process which is equal to the sum of its thermal resistance, and the principle of superposition of the so-called series resistance, the thermal resistance of each layer is superimposed on the total thermal resistance of the wall

$$\frac{T_1 - T_2}{q} = \frac{\delta_1}{\lambda_1} + \frac{\delta_2}{\lambda_2} + \frac{\delta_3}{\lambda_3}$$

3.3   A Double Multi-objective Optimization Model

On the basis of the above analysis, the following multi-objective optimization model for double bath objective function can be established:

$$\begin{cases} \max(t_1 + t_2), \\ \min(v) \end{cases},$$

Where the constraints are as follows:

$$
\begin{cases}
P_{O1} = \left(\int_0^{S_1} \alpha(T - \theta)dF + \int_0^{S_2} \beta(p_v'' - p_v)dF + \int_0^{S_2} \varepsilon\sigma(T + 273)^4 dF) \right) \Big/ 3.6 + \int_0^{S_3} qdF \\
P_{O2} = \left(\int_0^{S_1'} \alpha(T - \theta)dF + \int_0^{S_2} \beta(p_v'' - p_v)dF + \int_0^{S_2} \varepsilon\sigma(T + 273)^4 dF) \right) \Big/ 3.6 + \int_0^{S_3'} qdF \\
\quad\quad P_{O1}t_1 = h_1 S_1 \rho c(T_{max} - T_{min}) \\
\quad\quad P_{O2} = v\rho c(T_0 - T) \\
\quad\quad vt_2 = S_1(H - h_1) - V_0 \\
\quad\quad S_3 = S_2 h_1 \\
\quad\quad S_3' = S^2 H
\end{cases}
$$

Here S_1 is the opened area of bathtub cover ($S_1 = 0.3964$ m^2);
S_1' is the bathtub closed area of bathtub cover ($S_1' = 0.0966$ m^2);
S_2 is the surface area of bathtub ($S_2 = 0.8827$ m^2);
S_3 is the side of the bathtub with the ground and the area;
V_0 is volume in water ($V_0 = 0.045$ m^3);
T_{max} is the suitable temperature maximum;
T_{min} is the suitable minimum temperature;
T_0 is the add hot water temperature.

Combined with the real life situations the determination of the optimal water saving scheme by the model optimization calculation reveals that $h_1 = 0.3$, and $T_0 = 55$. Model solution result is as follows: $\max(t_1 + t_2) = 25$, and $\min(v) = 0.216$.

If people spend less than 25 min for having a bath, then there is no wastage of water. If someone spends more than 25 min for having bath according to flow rate, then the wastage of water is (0.216 m^3/h).

3.4   Effect of Bubble Bath Agent

The main components of bubble bath are sodium bicarbonate, sodium carbonate and tartaric acid. Bubble film is a layer of film formed by the cleaning agent molecules surrounded by water. Bubble can achieve the basic thermal insulation effect whose thickness is 0.5 m. It will have uniform dispersion in the surface of liquid because of the effect of hydrophobic bonds, forming very thin covering. It can hinder hydrant pass through skin layer. Suppress the evaporation of the liquid.

By Fourier transformation, the vector expression is:

$$ q = -\lambda \cdot gradT \tag{2} $$

Where q is heat flux density ($Kcal/m^2h$), $gradT$ is temperature gradient, and λ is thermal conductivity ($Kcal/mh°C$).

To analyze the temperature gradient of the inlet and outlet flow rate of the interface of the differential element in the T interface at t time, and the temperature gradient of the e interface at the time.

Unit of time in the X direction of the outflow of hexahedral heat:

$$\lambda \frac{\partial^2 T}{\partial x^2} dxdydz \tag{3}$$

Unit of time in the Y direction of the outflow of hexahedral heat:

$$\lambda \frac{\partial^2 T}{\partial y^2} dxdydz \tag{4}$$

Unit of time in the Z direction of the outflow of hexahedral heat

$$\lambda \frac{\partial^2 T}{\partial z^2} dxdydz \tag{5}$$

Then the total calories per unit time for the outflow of hexahedron:

$$Q_{total} = \lambda \left[\frac{\partial^2 T}{\partial x^2} + \frac{\partial^2 T}{\partial y^2} + \frac{\partial^2 T}{\partial z^2} \right] dxdydz \tag{6}$$

The thermal conductivity of water is 0.599 W/m · K, its thermal conductivity is close to the air approximately, take 0.025 W/m · K (ignore film thickness). Hence available heat loss:

$$Q_{water} > Q_{bubble} \tag{7}$$

Therefore, bubble bath can reduce the heat loss of the heat conduction process, and can also reduce the added heat.

4 Sensitivity Analysis

In this section, we focus on the analysis of 30 min under the premise of bath time.

According to the Figs. 2 and 3, we can draw the following results. The initial water level and water temperature have a significant impact on the time of water saving

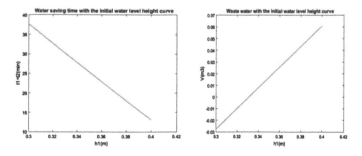

Fig. 2. Sensitivity analysis with respect to initial water level

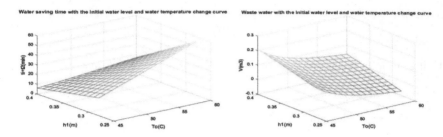

Fig. 3. Sensitivity analysis with respect to water temperature change

minimises wastage of water, and the optimization scheme is more reasonable according to the actual situation.

5   Conclusion

In this paper, we have investigated the relationship between the quantity of consuming water and the time of bath. For the purpose of minimizing the quantity of wasting water, we construct a bi-objective optimization model to determine the strategy. We have also considered the sensitivity and fitness of the model depending upon the shape and volume of the tub.

References

1. Zhao, Z.: Discussion on water surface evaporation coefficient formula. Shuili Xuebao **40**(12), 1440–1443 (2009)
2. Zhao, Z.: Water cooling hojicha formula and its application. Shuili Xuebao **35**(5), 34–38 (2004)
3. Yang, S.: Heat Transfer. Higher Education Press, Beijing (2006)
4. Boween, L.: The ratio of heat losses by conduction and by evaporation from any water surface. Phys. Rev. **27**(6), 779–787 (1926)

A Label-Correlated Multi-label Classification Algorithm Based on Spearman Rank Correlation Coefficient

Zhiqiang Li[1], Shihui Wang[1], and Hongchen Guo[2(✉)]

[1] School of Software, Beijing Institute of Technology, Beijing, China
lizq@bit.edu.cn, sunflowervva@aliyun.com
[2] Network Information Technology Centre,
Beijing Institute of Technology, Beijing, China
guohongchen@bit.edu.cn

Abstract. This paper proposes an improved multi-label text classification model based on label-correlation called LC-ASVM and it confirms frequent label pairs by the degree of support among feature labels. The model also calculates Spearman rank correlation coefficient of label-pairs to build the label correlation matrix and then measures confidence of SVM matching each category by calculating the projection distance of one point to the hyperplane. Finally the proposed model updates the label correlation matrix through the iterations layer by layer of Adaboost.

Keywords: Multi-label text classification · Label-Correlation · Spearman

1 Introduction

It is difficult to describe accurately the semantic information of the real objects by assuming that they only have one single category label. In this research work the assumption is that multi-label classification can better meet the characteristics and laws of objects in real world. The mainstream label-correlated classifier algorithms including CBA, CMAR, ML-kNN all have defects. CBA has restriction on the size of frequent item set, which cause its performance for high-dimensional data declining. CMAR always have the situation of over-pruning caused by the greatly increased memory overhead on big sample sets. ML-kNN has many improvements doing associative classification with big sample sets. But it also has a significantly low level of accuracy coming across a smaller or unbalanced sample set [1]. Due to the defection of some current algorithms of label-correlated classifier, there is evidences of inadequacies in some aspects existing in the performance of some classifiers. Hence they cannot meet the classification demand of increasing accuracy.

This paper proposes an improved multi-label text classification model based on label-correlation called LC-ASVM. It builds a label correlation matrix and a confidence matrix through the relevance and the confidence among the tags of feature, thereby improving the accuracy level of multi-label classification.

© Springer International Publishing AG 2017
F. Xhafa et al. (eds.), *Recent Developments in Intelligent Systems and Interactive Applications*,
Advances in Intelligent Systems and Computing 541, DOI 10.1007/978-3-319-49568-2_8

2 An Introduction to Related Algorithms

2.1 Spearman Rank Correlation Coefficient

Spearman correlation coefficient estimates the correlation between two variables with monotonic function [2]. Assuming X and Y as two random variables (or two sets) with N elements, and use X_i and Y_i to respectively represent their i-th ($1 <= i <= N$) value, sort X and Y, with the same order, getting x and y. Do the corresponding subtractions on the elements of set x and set y to get a ranking difference set d. Equations (1) is calculated with ranking sets x and y.

$$\rho = \frac{\sum_{i=1}^{N} (x_i - \bar{x})(y_i - \bar{y})}{\sqrt{\sum_{i=1}^{N} (x_i - \bar{x})^2 \sum_{i=1}^{N} (y_i - \bar{y})^2}} \tag{1}$$

2.2 Confidence and Confidence Interval

Confidence interval is the length or the distance of the area in which the general parameters are located at a certain confidence level. Confidence is defined as the probability of making mistakes when the estimated general parameters fall in a certain range [3, 4].

2.3 Multi-label Classification Strategy of Adaboost-SVM

The similarity threshold of a category can be set by the results of training process marking it with the categories exceeding the threshold, hence it is reasonable to transform the multi-label classification as multi-class classification based-on Adaboost-SVM [5–7].

Multi-label classification strategy of Adaboost-SVM works in the following way. Establish SVM classifier between any two categories, and suppose the sample divided into N classes, and then N (N−1)/2 classifiers need to be designed. [8]. While doing classification on a test sample, it calculates the matching degree between each category and it is done for classification prediction.

3 Label-Correlated Multi-label Classification Model Algorithm (LC-ASVM)

3.1 General Design

This model has four parts: data preprocessing, correlation analysis of characteristics, multi-label weight adjustment, and build combined classifier. As shown in Fig. 1.

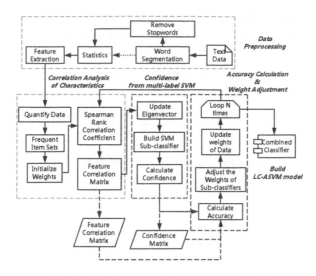

Fig. 1. The framework of LC-ASVM model

3.2 Correlation Analysis of Characteristics

The resulting feature label matrices are sparse and the classification weights are probably similar [8]. We denote the raw data after vectorizable processing as X, and Y is the label-correlation vector after operations of label correlation matrix. Through label correlation analysis, the remaining labels of the category get weights, so that the share of this category weight in all data weights is are improved, and can be more prominent in the multi-label classification [9].

In the first round of the iteration, all data are given the same weight of 1/N. Yet starting from the second round, build the accuracy vectors of the last data Accuracy = {Acc_1, Acc_2,...Acc_n} and all the vectors contained represent whether the classification results are correct in each dimension. 1 if correct, −1 if containing misclassified data. Then the Spearman correlation coefficient is updated in line with the analysis to the last round of data. The lower the case of data being misclassified, the higher the credibility of label associated matrices is considered, and vice versa. The introduction of label correlation vectors increases the dimensional weight whose matching degree is low while retaining the possibility of classification result.

3.3 Confidence Form Multi-label SVM

K(K−1)/2 two-class classifiers are constructed and their confidence is measured according to the projection of the distance between point and the hyperplane in sub-classification.

In SVM, make let ρ_i represents the probability of the points of the same category included in the neighborhood of test point. X_i. If ρ_i is more than a certain threshold, we consider this sample as being classified correctly. And if ρ_i is less than a certain threshold, we consider this sample as misclassified. Hence the samples between the two categories are fuzzy classification results. The confidence is calculated as Eqs. (2) and (3).

$$f(x_1) = \exp\left(\frac{-1}{|d(x_i)| * \rho_i}\right) \tag{2}$$

$$d(X_1) = \frac{\text{sgn}(\omega x + b)}{|\omega|} = \frac{\text{sgn}\left(\sum_{i=1}^{i} y_j \alpha_j k(x_j x_i) + b\right)}{|\omega|} \tag{3}$$

y_j presents the corresponding hyperplane determined by α_j and its kernel function. K (x_j, x_i) Assuming j adjacent sample points are in the neighborhood scope of test samples x_i, and j_s sample points belong to the same category with it. Then $\rho_i = j_s / j$.

For the data with n categories we build n $(n-1)/2$ sub-classifiers and for the corresponding category i and j of each sub-category, the confidence equals to the sum of all the confidence of this sample in the training set. Then the result of each sub-category is credited into the confidence matrix. The confidence of each category of the test sample x_i shows as Eq. (4):

$$y = [f_1, f_2, \ldots, f_n]^T \tag{4}$$

Each classifier has its confidence vector y, and the total confidence vector formed based-on   iteration   of   AdaBoost   is   confidence   of   each   classifier, $Confidence = \sum_{i=1}^{k} \alpha_i y / K$.

3.4   Build LC-ASVM Model

The model firstly quantifies the traineds data, getting frequent 2-item sets, and get the label associated matrix through Spearman Correlation, and build sub-categories on the basis of the label associated vectors, and then calculate the matching degree between data and each category.

During Adaboost-SVM, the weight of misclassified contents is adjusted to higher values in the next round of training. The weight of weak classifier is calculated as Eqs. (5)

$$\alpha_m = \frac{1}{2}\ln\{\frac{1 - \varepsilon_m}{\varepsilon_m}\} \tag{5}$$

Among them, the accuracy rate is calculated as Eqs. (6)

$$\varepsilon t = \sum_{i=1}^{N} W_i^{(m)} I(h_t(X)) \tag{6}$$

h_t is a confidence matrix built on the SVM sub-classifiers, as shown in Eqs. (7)

$$h_t = f \cdot X^T \tag{7}$$

AdaBoost updates the weights through Eqs. (8) and (9).

$$W_{m+1,i} = \frac{W_{mi}}{Z_m} \exp(-\alpha_m t_i I(h_t(X)) \cdot Acc_i), i = 1, 2 \ldots N \tag{8}$$

$$Z_m = \sum_{i=1}^{N} W_{mi} \exp(-a_m t_i I(h_t(X)) \cdot Acc_i) \tag{9}$$

And make $Z_m = 1$. After adjustment, the weight ratio of misclassified data increases, thus affects the associated degree of frequent feature pair in the next training round. Then label correlation matrix adjusted as AdaBoost iterating and converging, label correlation matrix tends to be stable. The classification process of LC-ASVM model is shown in Fig. 2.

1	Input: Training set S={$(x_{1,1},x_{1,2} \ldots,x_{1,n}) \ldots (x_{m,1},x_{m,2} \ldots,x_{m,n})$},$x_{i,j} \in$ X
2	Initialization: translate S to the data vector V
3	For t=1,2, ...,T do
4	For i∈ {1,2,3 ... p} do
5	For j∈ {1,2,3 ... q}do
6	Calculate the support : Support=P($V_i^{(t)} \cap V_j^{(t)}$)
7	If (Support > Support_Threshold) do
8	Calculate Spearman Rank Correlation Coefficient ρ <- Eqs. (1)
9	Make the corresponding value of label associated matrix A_{ij}=ρ
10	End
11	End
12	End
13	For classifier=1,2, ... , k(k-1)/2 do
14	Calculate the distance between point and hyperplane d($x_{classifier}$) <- Eqs. (3)
15	Calculate confidence of classification f($x_{classifier}$)<- Eqs. (6)
16	End
17	Calculate accuracy ε_t <-Eqs. (9),among which h_t<-Eqs.(10). if ε_t=0 or $\varepsilon_t \geq$1/2, T=t-1; or else $V^{(T)}=V^{(t)}$, jump to step21
18	Adjust the weight of sub-classifiers α_t <-Eqs. (8)
19	Update the weight of data $w_{m+1,i}$<-Eqs. (11),among which i=1,2,...,n
20	Update data vector $V^{(t+1)}=V^{(t)}$
21	Output: the updated data vector $V^{(T)}$

Fig. 2. Pseudo code of LC-ASVM model classification algorithm

4   Experiment and Analysis

4.1   Experimental Datasets

The experimental data has totally about 300,000 pieces of data consists of Chinese text classification corpus TanCorpV1.0 [10] and data of the campus network collected from school server which is to retain data after segmentation, feature words extraction, getting rid of stop words and other steps. Extract the feature words of 14,150 texts in TanCorpV1.0, and each forming a piece of test data.

4.2   Experimental Results and Analysis

Effect Verification of Ensemble Learning Classification Model. Set the main parameters of LC-ASVM: SVM adopts compound kennel function made up of RBF and polynomial; 20 sub-classifiers of AdaBoost, and 12 categories of SVM.

Do cross-validation on the 30,000 pieces of data selected randomly, recording their Precision, Accuracy, Recall, false alarm rate and other performance information, and then calculate F1-score of each classifier, and make the comparison, as shown in Table 1.

Table 1. Classification performance values of each classifier

Algorithm	Precision	Recall	Accuracy	F1-score
SVM	83.44 %	78.15 %	78.81 %	0.807084
AdaBoostMH	84.69 %	77.69 %	81.92 %	0.810391
Non-LC-ASVM	84.84 %	79.71 %	82.76 %	0.82195
LC-ASVM	83.17 %	82.63 %	85.03 %	0.828991

The time for completion of error classification significantly reduced, but data misclassified increases. Thus LC-ASVM has a slightly lower accuracy rate, but an obvious advantage on recall. Model LC-ASVM performs better than others in the experiments about overall performance—accuracy and F1-score.

Effect Verification of Classification Model with Association Rules. Randomly select 1000, 3000, 5000, 10000, 30000, 50000 pieces of data from data set, and respectively train the four classifiers CBA, CMAR, S-ML-kNN, and LC-ASVM. Randomly select 1000 pieces of data as test data, and record their Accuracy and Recall, as well as F1-score to measure their comprehensive classification performance.

Figure 3 shows the results of the above classification performance comparison experiments.

CBA and CMAR perform effectively in small sample sets. Yet as the size of data set increasing, their accuracy are declining.LC-ASVM. S-ML-kNN algorithm have the property of cold start in small sample set, causing its low accuracy, with data set increasing, its accuracy improves a lot. With small sample sets, LC-ASVM has the

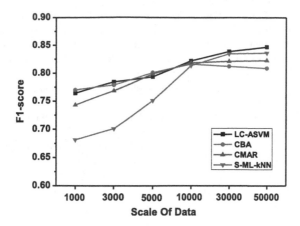

Fig. 3. Classification performance F1-score comparison of each classifier

equivalent classification ability with CBA. Being able to fully tap the relevance among labels in the case of serious lack of data, and when it comes to the data set increasing, LC-ASVM shows more obvious advantage. While CBA and CMAR had stabilized, the S-ML-kNN also shows basically stable classification performance. However in the context of big data era such scale of training sets are still limited in some aspects. By contrast, LC-ASVM model still has space to improve. As Compared to the above three current multi-label algorithms with association rules, LC-ASVM model has a rather high comprehensive classification capability and higher stability under various conditions.

5 Conclusions

This paper proposes an improved multi-label text classification model based on label-correlation called LC-ASVM, which performs well whether for big sample sets or small. The practicability and feasibility of LC-ASVM is also proved though the experiment using data of campus network. Hence it can play a role in the network behavior analysis with its stable multi-label associated classification.

References

1. Yan, Z., Bin, Z., Ruhua, Z.: Research on text classification algorithm. Softw. Guide **10**, 54–56 (2013)
2. Xiang, W., Yan, J., Bin, Z., Zhaoyun, D., Zheng, L.: Computing semantic relatedness using Chinese wikipedia links and taxonomy. J. Chin. Comput. Syst. **11**, 2237–2242 (2011)
3. Xiaofan, L., Xiaoqing, D., Youshou, W.: Theoretical analysis of confidence estimation of nearest neighbor classifier. Sci. Bull. **03**, 322–325 (1998)
4. Jielong, L., Yanshan, X., Zhifeng, H., Yibang, R., Liyang, Z.: Active learning for multi-instance multi-label classification based on SVM. Comput. Eng. Des. **01**, 254–258 (2016)

5. Li, W., Han, J., Pei, J.: CMAR: accurate and efficient classification based on multiple class-association rules. In: IEEE International Conference on Data Mining, pp. 19–21 (2001)
6. Fu, Z., Wang, L., Zhang, D.: An improved multi-label classification ensemble learning algorithm. In: Li, S., Liu, C., Wang, Y. (eds.) CCPR 2014. CCIS, vol. 483, pp. 243–252. Springer, Heidelberg (2014). doi:10.1007/978-3-662-45646-0_25
7. Esuli, A., Fagni, T., Sebastiani, F.: Boosting multi-label hierarchical text categorization. Inf. Retrieval **11**(4), 287–313 (2008)
8. Zhen Wang. Multi-label classification algorithm based on correlation of learning Tags. University of Science and Technology of China (2015)
9. Bi, W., Kwok, J.T.: Multilabel classification with label correlations and missing labels. In: Twenty-Eighth AAAI Conference on Artificial Intelligence. AAAI Press (2014)
10. Tan, S., Cheng, X., Ghanem, M.M., et al.: A novel refinement approach for text categorization. In: ACM International Conference on Information and Knowledge Management, pp. 469–476 (2005)

Effect of Film Thickness on Load-Carrying Property of Seawater Dynamic-Hydrostatic Hybrid Thrust Bearing

Zhao Jianhua$^{(\boxtimes)}$, Gao Dianrong, and Wang Qiang

Key Laboratory of Advanced Forging and Stamping Technology and Science,
Hebei Provincial Key Laboratory of Heavy Machinery Fluid Power Transmission
and Control, Yanshan University, Qinhuangdao 066004, China
zhaojianhua@ysu.edu.cn

Abstract. Since the viscosity of sea water is low, it is difficult to form the lubrication film, the overload and the "liner burn" phenomenon of the seawater. Lubricated hydrostatic-dynamic hybrid thrust bearing are happening so frequently that the operation stability and service life are reduced drastically. Therefore, it is necessary to analyze the maximum bearing capacity of the thrust bearing in order to prevent the phenomenon of "liner burn". The film thickness have the great influence on the load performance of seawater thrust bearing so that the rules of film thickness on static, dynamic and vibration performance of the bearing is quantitatively studied in the paper. First, the structural features, the flow channel, the bearing mechanism and primarily parameters of seawater thrust bearing are derived. Subsequently calculating equations of hydrodynamic, hydrostatic and the total capacity are deduced. Second, the carrying capacity, static stiffness and power loss are selected as the static property indexes. The adjustment time, dynamic stiffness and phase margin are selected as dynamic property indexes. The natural frequency, active and passive amplitude coefficients are selected as vibration property indexes. Finally, under constant-flow and PM-Controller supply model, the change rules between the performance indexes and film thickness are quantitatively analyzed. The results show that as film thickness increases carrying capacity, static stiffness, power loss, dynamic stiffness and natural frequency decreased conversely, and phase margin and passive amplitude coefficients remain the same, and then adjustment time, active amplitude coefficients increased in constant-flow supply model. The proposed research analyzed the influence of film thickness on static, dynamic and vibration performance of seawater thrust bearing and provided a basis for the design of the bearing.

Keywords: Seawater Hydrostatic-Dynamic thrust bearing · Static property index · Dynamic property index · Vibration property index · Film thickness

1 Introduction

The seawater-lubricated hydrostatic-dynamic hybrid thrust bearing which uses the seawater as lubricated medium is widely applied in hydropower generation, deep-sea exploration, food industry and so on. Because it enjoys the advantage of not polluting

© Springer International Publishing AG 2017
F. Xhafa et al. (eds.), *Recent Developments in Intelligent Systems and Interactive Applications*,
Advances in Intelligent Systems and Computing 541, DOI 10.1007/978-3-319-49568-2_9

the environment, extensive sources, low prices, and the bearing thrust plate with small thermal deformation for the large heat exchange coefficient of seawater. Therefore many scholars have been attracted to study it and have made great achievement [1, 2].

By using Gambit grid division of the ladder structure bearing film and establish a three-dimensional finite element and the research shows that the carrying capacity is increased with the increase of rotating speed and eccentricity ratio.

The literature [4] different bearing models were established with the Fluent, the influence of the thickness of water film was analyzed, the results indicate that with the increase of the water cavity thickness or inlet opening diameter, load capacity increases, then basically remains unchanged.

As the seawater viscocity is low, it is difficult to form lubricate film between thrust plate and shaft shoulder and liquid resistant decreased gradually with the increase in thickness of the film which leads to reduce carrying capacity of the thrust bearing opposed support system and static stiffness decreased too. As the fit clearance between thrust plate and shaft shoulder (about 10 µm), easy to overload, resulting in direct contact with the "liner burn" phenomenon [5]. Therefore this article intends to explore the influence of film thickness on the thrust carrying capacity–static performance, dynamic performance and vibration performance–under the constant flow supply and PM controller in order to prevent overload and extend the service life.

2   Brief of Seawater-Lubricated Hydrostatic-Dynamic Thrust Bearing

The seawater-lubricated hydrostatic-dynamic hybrid thrust bearing has six rectangular radial bearing cavity and six fan-shaped thrust bearing cavity respectively, which limits radial and axial movement of the rotation axis. Between adjacent supports separated by back to the sink to prevent internal stream of freshwater and marine. The fan bearing cavity consists of the central pressure area, water sealing, wedge-shaped tank, inlet holes. Wedge-shaped Groove in the bearing rotates to provide dynamic pressure bearing effect, improving static, dynamics and vibration resistance performance.

2.1   Lubricate Flow Through the Channel in the Sector Opposed Supporting Ladder-Cavity

Select a pair of opposed supporting ladder-cavity (Fig. 1) for the study.

2.2   Supporting Mechanism for Opposed Supporting Ladder-Cavity

When the lubricating fluid flows through the sealing water with the action of fluid resistance it makes the water's edge position have a great pressure forming a hydro-static bearing effect. When the lubricating fluid flows through the ladder position with the double functions of the rotational speed difference between the upper and lower bearing and of the ladder wedge cavity space. in this way, which produces a dynamic

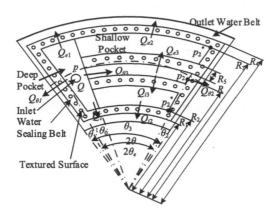

Fig. 1. Ladder-shallow sector cavity support pads schematic

pressure bearing effect. As a result, the mechanism of thrust bearing to seal the water is formed by double suspension bearing—hydrostatic pressure bearing and ladder cavity founded through the dynamic pressure bearing.

2.3 Initial Design Parameters of the Thrust Bearing

The initial parameters for seawater thrust bearing are shown in Table 1.

Table 1. Initial parameters for thrust bearing

Index	Pump pressure p_s/MPa	Flow Q/(L/min)	Speed ω(r/min)	$\theta_e/°$	$\theta_0/°$	$\theta_1/°$
Value	10	1.44	2700	55	10	5
Index	Lubricant	Temperature $t/°C$	Dynamic viscosity η/Pa·s	$\theta_2/°$	$\theta_3/°$	R_1/mm
Value	water	20	1.005×10^{-3}	5	40	45
Index	Sealing gap h/μm	Depth of shallow cavity h_p/μm	$\theta/°$	R_2/mm	R_3/mm	R_4/mm
Value	30	15	30	50	95	100

3 Property Index of the Thrust Bearing

3.1 Static Property Index

(1) Carrying capacity: Carrying capacity represents the film can withstand maximum external load when it is the theoretical limit, which is expressed as:

$$F_0 = \frac{2p_0 A_{e,j} h_0}{\dfrac{A^2 \Omega_{26,0}^2}{\rho p_0} + 1} \frac{1}{\Omega_{26,0}} \left. \frac{\partial \Omega_{26}}{\partial h} \right|_{h=h_0} \tag{1}$$

(2) Static stiffness Static stiffness j_0 represents the film resist external load caused by deformation capacity, its mathematical expression:

$$\dot{j}_0 = -\frac{2p_0 A_{e,j}}{\left(\frac{A^2 \Omega_{26,0}^2}{\rho p_0} + 1\right)\Omega_{26,0}} \left.\frac{\partial \Omega_{26}}{\partial h}\right|_{h=h_0} \tag{2}$$

(3) Power loss W represent the sum of the friction power of static pressure cavity and pump power, and its mathematical expression is:

$$W = \mu v^2 \left(\frac{A_s}{h} + \frac{A_t}{h + h_p}\right) + p_s q \tag{3}$$

3.2 Dynamic Property Index

(1) Adjustment time: Adjustment time represent how fast system response, generally related to the time constant control system. In this paper the allowable error is 2 %.

$$t_s = 4T = \frac{4m}{R_0 A_b A_e}\left(1 + \frac{R_0^2 A^2}{\rho p_0}\right) \tag{4}$$

(2) Dynamic stiffness: Dynamic stiffness represent the film can resist external load of water lubricated thrust bearing in frequency range, and its mathematical expression is:

$$J = \sqrt{\left(2\frac{\frac{p_0 A_e}{R_0}}{1 + \frac{R_0^2 A^2}{\rho p_0}}\left.\frac{\partial R}{\partial h}\right|_{h_0} + m\omega^2\right)^2 + \left(\frac{2R_0 A_b A_e \omega}{1 + \frac{R_0^2 A^2}{\rho p_0}}\right)^2} \tag{5}$$

The load is set to $\omega = 10$ Hz.

(3) Phase margin: During design not only the system is stable, but also the system requirements from the critical point has a certain stability margin, that have the appropriate relative stability.

$$\gamma = 180° + \arctan\frac{2R_0 A_b A_e \omega}{2\frac{p_0 A_e}{R_0}\left.\frac{\partial R}{\partial h}\right|_{h_0} + \left(1 + \frac{R_0^2 A^2}{\rho p_0}\right)m\omega^2} \tag{6}$$

3.3   Vibration Property Index

(1) Natural frequency The natural frequency is only related to the physical properties
of the thrust bearing system itself, its mathematical expression:

$$f_n = \frac{1}{2\pi}\sqrt{\frac{k}{m}} \tag{7}$$

(2) Amplitude magnification coefficients Amplitude magnification coefficient is
characterization the degree of vibration performance of bearing system.

$$K_b = \sqrt{\frac{k^2+(c\omega)^2}{(k-m\omega^2)^2+(c\omega)^2}}(\text{Passive}) \quad K_a = \frac{1}{\sqrt{(k-m\omega^2)^2+(c\omega)^2}}(\text{Active}) \tag{8}$$

4   Influence of Shallow Cavity Depth on Static Performance

Adjustable restrictor regulate bearing cavity flow using is constant-flow supply. Under
constant-flow supply model the carrying capacity of supporting cavities decreased with
the film thickness increases sequentially, as shown in Fig. 2.

With increasing film thickness the opposed support cavity static stiffness decreases,
as shown in Fig. 3.

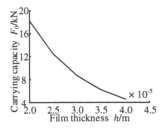
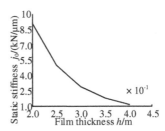

Fig. 2. Curve between carrying capacity and film thickness

Fig. 3. Curve of static stiffness and film thickness

As film thickness sequentially increases, power loss of the opposed support cavity
also decreases, But the decrease in range is limited and can be ignored.

5   Influence of Film Thickness on Dynamic Property Index

With increase in film thickness, the opposed support cavity adjustment time also increases, as shown in Fig. 4.

With the film thickness increases, the opposed support dynamic stiffness decreases, as shown in Fig. 5.

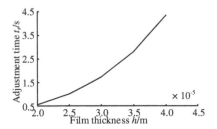   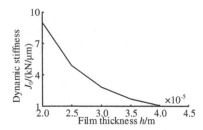

Fig. 4. Curve of adjustment time and film thickness

Fig. 5. Curve of Dynamic stiffness and film thickness

With the film thickness increases the opposed support phase margin decreases. But the change in magnitude is very small, and it can be ignored.

6   Influence of Film Thickness on Vibration Property Index

With the film thickness increases the opposed support natural frequency decreases, as shown in Fig. 6.

With the film thickness increases the opposed support active magnification amplitude coefficient increases as shown in Fig. 7.

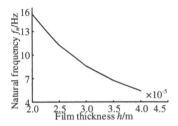   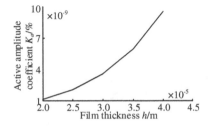

Fig. 6. Curve of natural frequency and film thickness

Fig. 7. Curve of active amplitude coefficient and film thickness

As the film thickness increases opposed support passive magnification amplitude coefficient increases, but change magnitude is very small and can be ignored.

7 Conclusion

As film thickness increases carrying capacity, static stiffness, power loss decreases conversely. As film thickness increases, adjustment time also increases, dynamic stiffness decreased, and then phase margin remain the same. As the film thickness increases, natural frequency decreases, passive amplitude coefficients remain the same and active amplitude coefficients also increases.

Acknowledgement. The project was financially supported by Younger Teachers Independent Research Program Science B Series of Yanshan University (13LGB003) and Doctoral Fund Project of Yanshan University (B815).

References

1. Ginzburg, B.M., Bakhareva, V.E., Anisimov, A.V.: Polymeric materials for water-lubricated plain bearings. Russ. J. Appl. Chem. **79**(5), 695 (2006)
2. Chen, W., Fan, H.J., Wu, L.J.: Review of water lubricated bearing for high speed spindle. Eng. Sci. **15**(1), 21–27 (2010)
3. Liu, H.J., Guo, H., Zhang, S.L.: Research on static characteristics of Deep/Shallow pockets hybrid bearing based on FLUENT. Lubr. Eng. **38**(10), 35–38 (2013)
4. Wang, Y.Z., Jiang, D.: Load capacity analysis of water lubricated hydrostatic thrust bearing based on CFD. J. Donghua Univ. **41**(4), 428–432 (2015)
5. Meng, F.M.: Three-dimensional finite element analysis for thermo-elasto-hydrodynamic performances for a water lubricated bearing system. J. Chongqing Univ. **36**(2), 121–126 (2013)

Research on Static Performance of Water-Lubricated Hybrid Bearing with Constant Flow Supply

Zhao Jianhua[⊠], Gao Dianrong, and Wang Qiang

Key Laboratory of Advanced Forging and Stamping Technology and Science,
Hebei Provincial Key Laboratory of Heavy Machinery Fluid Power Transmission
and Control, Yanshan University, Qinhuangdao 066004, China
zhaojianhua@ysu.edu.cn

Abstract. The article focuses on the thrust bearing with freshwater and marine surface tecter of Water-lubricated Hybrid and established a three dimensional mathematical model. Based on one of opposing ladder-cavity supporting systems, initial and working state equation under constant flow supply model, carrying capacity of thrust bearing with orifices throttling supply is derived. The mathematical expressions of carrying capacity, static rigidity, and total power loss are established and the static performance indexes of opposing cavity are analyzed. The influence of flow supply, pumping system, shallow cavity depth, film thickness, rotational speed on static indexes of thrust bearing is quantitatively analyzed.

Keywords: Water-lubricated hybrid bearing · Mathematical model · Orifices throttling · Carrying capacity · Static rigidity · Power loss

1 Introduction

Water embodies ceratin special physical properties. Due these properties the water-lubricated hybrid bearing with small viscosity and large leakage it is difficult to establish the hydrostatic oil film, predisposed to tolerate overload, burning of bearing liner and so on. The author of the article focuses on a thrust bearing with fresh water and marine surface tecter of Water-lubricated Hybrid, compared with the traditional oil-lubricated bearing. This new bearing have numerous advantages such as zero-pollution, small thermal deformation, anti-corrosion and small friction loss. Hence it has wide array of applications such as sailing and hydroelectric generation. Moreover, the performance indexes of the bearing relate to many structures and technological parameters where each parameter is intercoupling. So we have to analyze the static performance indexes to find some rules between each index with structures and technological parameters so that we can do more things for researching the thrust bearing with freshwater and marine surface tecter of Water-lubricated Hybrid.

In the recent years a number of scholars and practitioners have conducted seminal works on the structures and performance of the water-lubricated bearing all over the world.

© Springer International Publishing AG 2017
F. Xhafa et al. (eds.), *Recent Developments in Intelligent Systems and Interactive Applications*,
Advances in Intelligent Systems and Computing 541, DOI 10.1007/978-3-319-49568-2_10

Lin [1] designed the water-lubricated hybrid sliding bearings based on porous restrictor. Guo [2] analyzed implementation of inside pressure feedback in water-lubricated hydrodynamic journal bearing. Yang [3] established experimental modeling of new long orifice-type restrictor of high speed turbine hybrid bearing. The journey of exploring literature reveals that there is a dearth of earlier research work on the static performance of water-lubricated hybrid bearing with constant flow supply. Since the parameters of bearing are inter-coupling each other it is necessary to research on the influence of design parameters on static performance of the bearing.

Design a style of thrust bearing with freshwater and marine surface tecter of Water-lubricated Hybrid. The radial support of the upper and lower ends and the closed hydrostatic bearing of thrust in between constitute the bearing block. The radical movement and swing of the rotation axis was limited by the radial support of the upper and lower ends. the axial movement was limited by the closed hydrostatic bearing of thrust in between, which is the main portion to load bearing of the new style of the water-lubricated hybrid bearing. There are 6 Rectangular oil chambers of static pressure uniformed in radical supporting and each of them is separated by black tank to prevent freshwater and marine internal steam. There are 6 sector oil chambers comprise the bearing. There is a 1/4 runner of the new thrust bearing with freshwater and marine surface tecter of Water-lubricated Hybrid.

2 Static Performance Analysis

2.1 Flow Condition

In order to analyze easy thrust bearing capacity, the assumption can be made as follows.

- Flow state of lubricant is laminar and inertia force can be ignored.
- Liquid viscous pressure characteristics in low pressure can be ignored.
- Deform of rigid body of bearing surface can be ignored.

2.2 Bearing Force Calculation

(1) Initial equation of opposing ladder-cavity in initial state, gravity of spindles is balanced by ladder-cavity of bearing capacity, equilibrium equation is as follows [7]:

$$\begin{cases} p_{2,0}A_{e,2} = p_{1,0}A_{e,1} \\ p_{i,0} = q_{i,0}\Omega_{26,0} \end{cases} (i = 1, 2) \tag{2.1}$$

In initial state, mass of the spindle can be ignored, and each parameter of opposing ladder-cavity supporting system is equal.

(2) Working state equation in working state, the weight, supporting, external load of spindle & deduced equilibrium equation of the systems:

$$\begin{cases} p_1 A_{e,1} = f + p_2 A_{e,2} \\ p_i = q_{i,0} \Omega_{26} \end{cases} (i = 1, 2) \tag{2.2}$$

(3) Bearing capacity of the contraposition ladder-cavity

On the basis of the initial and working state equation of the thrust bearing systems, deduced the equilibrium equation of the bearing capacity during the constant flow:

$$f = 2q_0 A_e \frac{\partial R}{\partial h} \Delta h \tag{2.3}$$

The capacity of the contraposition ladder-cavity of thrust hybrid bearing is comprised by the pressure of hydrostatic and dynamic. Rotate speed of the bearing is invariant, so there is no dynamic pressure effect of the contraposition ladder-cavity, only about the hydrostatic pressure.

According the (2.3):

$$f = 2q_0 A_e \frac{\partial \Omega_{26}}{\partial h}\bigg|_{h=h_0} \Delta h \tag{2.4}$$

2.3   Static Performance Indexes

Carrying capacity, static rigidity and the total power loss have content to the static performance of contraposition ladder-cavity supporting systems.

(1) Carrying capacity: Carrying capacity (F_0) is represent maximum external load of when film reach cortical boundary(zero-thickness),the formula is as follows:

$$F_0 = \frac{2p_0 A_{e,j} h_0}{\frac{A^2 \Omega_{26,0}^2}{\rho p_0} + 1} \frac{1}{\Omega_{26,0}} \frac{\partial \Omega_{26}}{\partial h}\bigg|_{h=h_0} \tag{2.5}$$

(2) Static rigidity static rigidity (j_0) is represent the capacity of the film resist distortion caused by the external load is one of the important index of the freshwater and marine face tecter of Water-lubricated Hybrid. the formula is as follows:

$$j_0 = -\frac{\partial F}{\partial h}\bigg|_{h=h_F} = -\frac{2p_0 A_{e,j}}{\left(\frac{A^2 \Omega_{26,0}^2}{\rho p_0} + 1\right) \Omega_{26,0}} \frac{\partial \Omega_{26}}{\partial h}\bigg|_{h=h_0} \tag{2.6}$$

(3) Power loss Power loss W(without the overflow loss) is represent the power of pump and the power of friction of hydrostatic cavitary, the formula is as follows:

$$W = \mu v^2 \left(\frac{A_s}{h} + \frac{A_t}{h+h_p} \right) + p_s q \qquad (2.7)$$

During design and the thrust bearing working, which have connected with static performance of the bearing(carrying capacity, static rigidity, the power loss) is rotational speed, pump systems, flow supply, the deep of cavity and space of the film.

2.4 Influence of Parameter on Static Index

(1) Initial design parameter The indexes of the bearing of Water-lubricated Hybrid(as shown in Table 1), the indexes of the ladder-cavity(as is shown in Table 2):

Table 1. Initial design parameter of bearing

Parameter	Pumping pressure p_s/MPa	Bearing pressure p_0/MPa	Film interval h_0/μm
Value	10	3	30
Parameter	Orifice area(A/m^2)	Rotate speed ω(r/min)	Flow Q/(L/min)
Value	1.5501×10^{-6}	2700	11.0045
Parameter	Lubricants	Temperature t/°C	Dynamic viscosity η/Pa·s
Value	Pure water	20	1.005 × 10-3

Table 2. Initial design parameter of ladder-cavity

Parameter	Film thickenss h/μm	Shallow cavity Deep h_p/μm	R_1/ mm	R_2/ mm	R_3/ mm	R_4/ mm
Value	30	15	45	50	95	100
Parameter	θ/°	θ_e/°	θ_0/°	θ_1/°	θ_2/°	θ_3/°
Value	30	55	10	5	5	40

(2) The influence of the static performance by the flow supply, carrying capacity, static rigidity and total power loss of the contraposition ladder-cavity would enlargement with it. As is shown in the Figs. 1 and 2.

With the increasing rotation speed of the bearing, there will be no change of carrying capacity, static rigidity and total power loss, as is shown in Fig. 1.

(3) The influence of the static performance on the pumping pressure is that as the pumping pressure increase there will be no change in the carrying capacity and static rigidity.

When the pumping pressure increases, the overflow of the oil systems and throttling loss enhances, and the power loss of the contraposition ladder-cavity would enhance

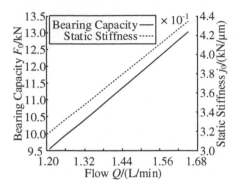

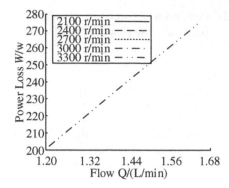

Fig. 1. Relationship between Flow, Carrying Capacity and Static Rigidity

Fig. 2. Relationship between Power Loss and Flow

too. As the rotation speed increases, the power loss of the contraposition ladder-cavity changes very little, and to that extent that we can ignore it. As is shown in Fig. 3.

(4) The influence of static by deep of shallow ladder-cavity is that with the ladder-cavity becoming deeper, difference between deep and shallow cavity will reduce and hydrodynamic effect will become lower which caused by the lubricants flow into the ladder-cavity. Hence when the deep of the ladder-cavity increases, carrying capacity and static rigidity of the contraposition ladder-cavity gets reduced, as is shown in Fig. 4.

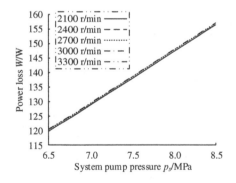

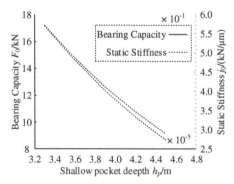

Fig. 3. Relationship between power loss and pumping pressure

Fig. 4. Relationship between shallow cavity deep and carrying capacity, static rigidity

When the deep of the shallow ladder-cavity increases, its fluidic resistor will decline. So the power loss of the throttle will become lower which caused by the lubricants flow the ladder-cavity. But it is not sharp, we can ignore it. With the rotation speed of the bearing increases, the hydrodynamic effect of the cavity also increases, the wedge-shape extrusion effect of the lubricants will increase, the calorific is magnify and the power loss to the cavity will increase too. as is shown in Fig. 5.

(5) The influence of the static performance by the film thickness when the anti-thrust pad of the spindle reach the threshold limit of thickness. When the thickness of the film increases, the thickness also increases too which is in the other side. Fluidic resistor, supporting capacity will reduce. Results in the supporting capacity become magnified following the thickness of the film. while the static rigidity of the ladder-cavity supporting systems is formed by the static rigidity of the two supporting systems. So with the thickness of the film increase, the rigidity of the anti-thrust pad will reduce which lead to the static rigidity of the contraposition ladder-cavity supporting systems reduced followed the thickness of the film, as is shown in Fig. 6.

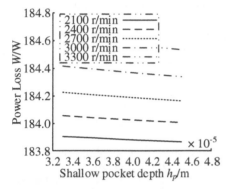

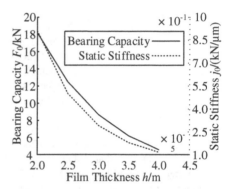

Fig. 5. Relationship between power loss and shallow cavity deep

Fig. 6. Relationship between film thickness and carrying capacity, static rigidity

As the interval between films increases, fluidic resistor of the contraposition cavity will also increase. There will be more power loss of the lubricants. With the increase in rotation speed, power loss of contraposition will increase, but range is limited and can be ignored.

3   Conclusion

1. Static and dynamic supporting comprise to the thrust bearing of Water-lubricated Hybrid, and the supporting static pressure is dominant, the supporting of the dynamic pressure have nothing to do with the static performance of the contraposition ladder-cavity.
2. Carrying capacity will enlargement followed with the flow rate increase, thickness decrease, the depth reduce of the shallow cavity.
3. The static rigidity will enlargement followed with the flow rate increase, thickness of the film decrease, the depth reduce of the shallow cavity.

4. The power loss will enlargement followed with the flow rate, pumping pressure and rotation speed of bearing increase, the depth reduce of the shallow cavity, thickness of the film decrease.
5. Carrying capacity and the static rigidity of the ladder-cavity remain unchanged, which is basically impregnable.

Acknowledgement. The project was financially supported by Younger Teachers Independent Research Program Science B Series of Yanshan University (13LGB003) and Doctoral Fund Project of Yanshan University (B815).

References

1. Bin, L., Feng, L.: Design of water-lubricated hybrid sliding bearings based on porous restrictor. Bearing (2), 4–8 (2007)
2. Zhangjie, G.H.: Implement of inside pressure feedback in water-lubricated hydrodynamic journal bearing. Bearing (9), 11–14 (2007)
3. Peiji, Y., Xiaoyang, Y.: Experimental modeling of new long orifice-type restrictor of high speed turbine hybrid bearing. Mech. Sci. Technol. **31**(11), 1831–1840 (2012)
4. Pan, D., Yabin, Z., Hua, X.U.: The structure of new journal hybrid bearing for high-speed machine spindle and its performance. Lubr. Eng. **34**(2), 11–24 (2009)
5. Guoyuan, Z., Xiaoyang, Y.: Experiment for water-lubricated high-speed hydrostatic journal bearings. Tribology **26**(3), 38–240 (2006)
6. Shengxian, D., Qiushan, Ma.: Basic performance analysis of high-speed spindle system supported by hybrid bearings. Manuf. Technol. Mach. Tool **9**, 136–139 (2011)
7. Fengming, S.G.: Design of the Hybrid Bearing of the Flow Supply. World Publishing Corporation, Beijing (1993)

A New Method for Desiccant Package Quality Check

Yingchun Fan[1,2(✉)], Zuofeng Zhou[1,2], Hongtao Yang[1,2],
and Jianzhong Cao[1,2]

[1] Xi'an institute of Optics and Precision Mechanics of CAS,
Xi'an 710119, Shaanxi, People's Republic of China
fyc@snnu.edu.cn
[2] Shaanxi Normal University, Xi'an 710119, Shaanxi
People's Republic of China

Abstract. The desiccant is widely used and plays an important role in the daily life and industrial manufacturing. Ironically its package quality check is still a challenging issue. In this paper we present a new method for desiccant package quality check. First, the package lie on the assembly line in arbitrarily direction corrected to vertical direction. Then the colicor information is used to distinguish the contaminated package. Finally, we combine the texture analysis and region growth to obtain the detection result. The experimental results show the effectiveness of the proposed method and the detection accuracy can achieve 92.85 %.

Keywords: Desiccant package · Quality check · Region growth · Texture analysis

1 Introduction

The quality of the packaging is becoming more and more important especially in the medicine, aviation and space, food, transportation and storage fields. And a good quality packaging can make sure that customers are satisfied with the products' quality. Packaging is done across wide spectrum of products and services with diverse size and appearances. It's impossible to find an all-purpose method to detect all packaging. Therefore most of researchers only research a sort of packaging.

The measures of checking the packaging are radically changing with the development of machine vision and digital image processing. Qi [1] presented an algorithm based on machine vision which applied to the cigarette packing line. This paper focused on inferior product in plastic blister tablets' packing through an inspection method based on support vector Machine is proposed [2]. A set of solution about defect detection of the aluminum-plastic blister package based on machine vision is proposed in seminal work [3]. The texture information of the object plays an important role in these papers. And the textural features are extensively used in the image classification, pattern recognition and intelligent system. In fact the detection of packaging can be supposed as a process of classifying the bad samples from the qualified samples. Zhai [4] use LAWS texture and uncertainty texture spectrum to describe how to use texture analysis method to achieve the classification of rice paper. The author proposed

© Springer International Publishing AG 2017
F. Xhafa et al. (eds.), *Recent Developments in Intelligent Systems and Interactive Applications*,
Advances in Intelligent Systems and Computing 541, DOI 10.1007/978-3-319-49568-2_11

a texture-based method for classifying cracked concrete surfaces from digital images using neural networks [5]. Motonori Doi [6] proposed a skin color texture analysis method based on wavelet transform for detecting different texture patterns including local non-periodic features. Cui and Zhou [7] achieve characteristic of leather texture classify the face and the inverse of leather. And some papers are presented to evaluate the texture feature extraction methods [8].

In this paper we present a new method for desiccant packaging integrity detection and this method is transplantable for these packaging that have the similar texture features. This paper is organized as follows. First, correcting the image captured on the assembly line. Second, extracting the parts of the image that we are interested in and wiping out contaminated samples. Finally, using the method of image region growth to the integrity detect of the desiccant packaging.

2 The Proposed Method

In this section we give the detailed descriptions of the proposed method. First, the captured image is needed to be corrected. In this step we need to find the rotation direction of the sample, calculate the rotation angle and analyze the texture feature of the packaging. Second, we extract the four corners and four sides without corners in the meanwhile wipe out contaminated samples. Finally, the integrity of the packaging will be detected using the method of image region growth, in this step we consider the corner and the side texture information of the packaging respectively. Specific steps are shown in the Fig. 1.

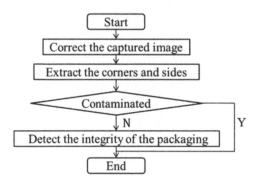

Fig. 1. Flow diagram of the proposed method

2.1 Correct the Captured Image

The packaging on the assembly line is in optional position. Before dealing with these samples, it is necessary to correct the captured image.

Find the Rotation Direction. In this step we apply the geometric method to find the rotation direction of the packaging, as it is shown in Fig. 2.

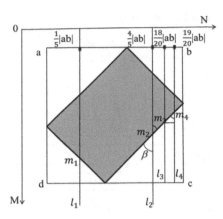

Fig. 2. The geometric approximation of the captured image

In the coordinate system O-MN, the packaging of desiccant is the rectangle filled with shadow. Finding the area of the packaging and lock it with the rectangle frame abcd. Then choosing two reference points $\frac{1}{5}|ab|, \frac{4}{5}|ab|$, construct two perpendicular lines l_1, l_2, through the two points, the crossover points of perpendicular lines and the side of the shadow area are m_1, m_2. Supposing m_1, m_2 coordinates are $(m_1^l, n_1), (m_2^l, n_2)$, we can get such a rule, when $m_1^l > m_2^l$, the direction of rotation is clockwise, the included angle β of l_2 and the side of packaging is the rotation angle, otherwise anti-clockwise, and the rotation angle is the included angle of l_1 and the side.

Calculate the Rotation Angle. When calculate the angle we need to choose other points, if the rotation direction is clockwise, choose two points on the right of l_2, otherwise, choose two point on the left of l_1. In order to understand the procedure of calculation we choose the situation of clockwise to elaborate this question as shown above.

Choosing $\frac{18}{20}|ab|, \frac{19}{20}|ab|$ as the reference points, construct two perpendicular lines l_3, l_4, the crossover points are m_3, m_4. The m_3, m_4 coordinates are $(m_3^l, n_3), (m_4^l, n_4)$. According to the geometrical relationship, we can get,

$$\beta = \arctan \frac{|m_3^l - m_4^l|}{|n_4 - n_3|} \tag{1}$$

The Texture Feature Analysis of the Packaging. The good samples used in our experiment are satisfied with three features:

1. Three sides of the packaging are filled with lattices on the surface; one side is smooth except the regions of the two corners.
2. The central region does not contain the texture information that can help us detect the integrity of the packaging. It is not necessary to consider the region.
3. The textures of the corner and the side that removed the corners are different. It is better to analyze them respectively.

2.2   Wipe Out the Contaminated Samples

The polluted samples are needed to be wiped out before next step.

We can attain this goal follow the steps of the flow diagram shown as the Fig. 3.

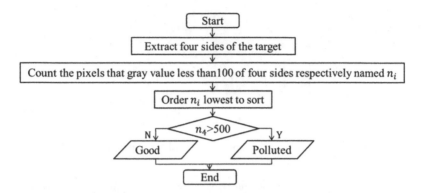

Fig. 3. The flow diagram that discard the polluted samples

The formula $n_4 > 500$ is changeable according to the actual demand.

2.3   Detect the Integrity of the Packaging

The steps of dealing corner and side are same except the chosen of some parameters. So we just discuss the detecting method of corner.

The Image region growth can be done as the follow steps:

1. Extracting four corners of the target.
2. Getting the edges of the four corners through the Canny-operator.
3. Acquiring new images through the Image region growth. The way as follows:

   Assuming the parts of edge value 1, other part value 0.
   If $img(i, j) = 1 \&\& img(i, j + \tau) = 1$, then $img(i, j + k) = 1$ $(k = 1{:}\tau)$.
   If $img(i, j) = 1 \&\& img(i + \tau, j) = 1$, then $img(i + k, j) = 1$ $(k = 1{:}\tau)$.
   The τ in the expression is an empirical value. It can be changed according to different type of the desiccant packaging. In our experiment the value of τ is set to 10.

4. Counting the total 1 in the four images N_i then calculate the number of 1 in the unit area. The computational formula is,

$$n_i = \frac{N_i}{m * n} * 100, \quad (m,n) = size(img) \tag{2}$$

5. Ordering n_i lowest to sort. If $n_3 > 2 * n_1$, the packaging is side missing or corner missing.

3   Experiment Result

The bad samples include: side missing, corner missing, the side or corner be polluted. As shown in the Fig. 4. There are four kinds of desiccant packaging which includes one qualified and three unqualified. And the parts lead to unqualified surrounded by white frames.

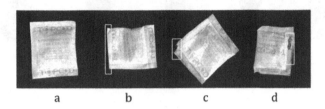

Fig. 4.   Samples of the packaging. From left to right: a. qualified packaging, b. side missing, c. corner missing, d. the polluted packaging.

We use the sample b in Fig. 4 to show the processes of Image region growth. The corresponding pictures of each steps is shown in the Fig. 5.

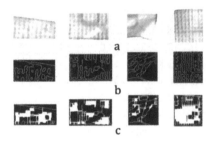

Fig. 5.   From top to bottom: a. four corners of the target, b. the edges of the four corners got through the Canny-operator, c. new images through the Image region growth.

The values of n_i as follows:

$$n_1 = 17.0015, \quad n_2 = 34.1461, \quad n_3 = 36.0071, \quad n_4 = 54.5026$$

From the values we can see that the region of missing embossing has less number of 1 than other region. However the computational formula $n_3 > 2 * n_1$ is an empirical formula, it did a good job on our experimental subject.

Results: to measure the quality of the proposed algorithm, we tested a group of picture. Figure 6 shows 12 pictures of our test 1,2,3,4,5,10,11 are qualified packaging, 6,7,9 are unqualified packaging that missing one of sides, 8,12 are unqualified packaging that missing one of corners. The corresponding experimental data and the observed result of the 12 pictures shown in the Table 1.

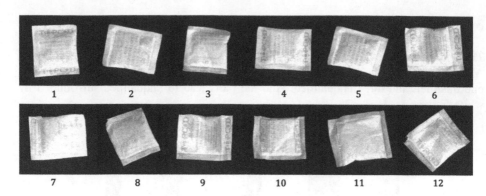

Fig. 6. Some of our test pictures

Table 1. The experimental data and the judge result

Number	n_1	n_3	Result	Actual	Validity
1	16.9553	19.7815	1	1	T
2	26.3989	31.726	1	1	T
3	26.4436	38.9982	1	1	T
4	34.7755	39.8944	1	1	T
5	23.4783	31.4851	1	1	T
6	21.5848	44.799	0	0	T
7	9.6189	31.6958	0	0	T
8	16.2172	53.1253	0	0	T
9	17.0015	36.0071	0	0	T
10	26.0163	34.7043	1	1	T
11	31.7424	39.675	1	1	T
12	9.9145	30.0706	0	0	T

qualified: 1, unqualified: 0, correct: T, false: F.

4   Conclusion

The proposed algorithm can distinguish the unqualified packaging from the samples effectively. However the proposed method has suffers from limitations such as if the packaging has large fold the reflections on the surface occurs severely and sides have the embossing but when the packaging is opened, this method will fail. At present the proposed method was only running successfully on the software and did not implement in hardware, and it can only detect one sample at a time. Our future work is to implement the proposed on the hardware platform.

Acknowledgments. This work is supported by Foundation of West Light of CAS (No. Y42961 1213) and Distinguished Young Technology of Shaanxi province (No. 2016-KJXX01).

References

1. Qi, R.: Research on machine vision based package inspection system. Master Dissertation, Dalian University of Technology (2007)
2. Peng, Z.: The research of pharmaceutical packaging inspection technology based on machine vision. Master Dissertation, Hunan University (2009)
3. Lv, Z.: Research of defect detection about aluminum-plastic blister package based on machine vision. Master Dissertation, Huazhong University of Science & Technology (2011)
4. Zhai, H., Huang, H., He, S., et al.: Rice paper classification study based on signal processing and statistical methods in image texture analysis. In: IEEE/ACIS, International Conference on Computer and Information Science, pp. 189–194 (2014)
5. Chen, Z., Derakhshani, R.R., Halmen, C., et al.: A texture-based method for classifying cracked concrete surfaces from digital images using neural networks. Thromb. Res. 3(14), 2632–2637 (2011)
6. Doi, N.M., Tominaga, S.: Image analysis and synthesis of skin color textures by wavelet transform. In: IEEE Southwest Symposium on Image Analysis and Interpretation, pp. 193–197. IEEE Computer Society (2006)
7. Cui, Y., Zhou, Z.: Application of pattern recognition on classifying texture images. In: Fifth World Congress on Intelligent Control and Automation, WCICA 2004, vol. 5, pp. 4124–4126 (2004)
8. Sharma, M., Singh, S.: Evaluation of texture methods for image analysis. In: Intelligent Information Systems Conference, the Seventh Australian and New Zealand, pp. 117–121 (2010)

A Novel Algorithm for Laser Spot Detection Based on Quaternion Discrete Cosine Transform

Xiaodong Miao[1,2(✉)], Xiang Shi[1], Zhixing Xu[1], Aiqing Wang[1], and Wei Wang[1,2]

[1] Kangni Mechanical & Electrical Co., Ltd., Nanjing 210009, China
nuaaxiaodong@163.com
[2] School of Mechanical and Power Engineering,
Nanjing University of Technology, Nanjing 211800, China

Abstract. Laser spot detection is an important problem in optical measurement. The precision and speed of the detection algorithm influence the optical measurement system directly. The traditional algorithms such as Hough Transform and Gravity model are unsatisfactory in complex conditions. The laser spot detection algorithm referred in this paper is based on the quaternion discrete cosine transform and moment and a method is adopted to approximate the edge of the laser spot. Not only the center and edge can be detected simultaneously but also the robustness of noise is better than others. The algorithm is suitable for the real-time optical measurement.

Keywords: Laser spot · Image processing · Quaternion discrete cosine transform

1 Introduction

Laser spot detection is an important technique, which can be used widely in optical measurement, gaming, laser triangulation method and even in military applications such as target detection or tracking. However, the environment is complex, the noise, uneven exposure, beam divergence. So the algorithms need to be robust, highly efficient and satisfy the specific application [1–3].

Normally spot detection methods are based on by circular or ellipsoidal like shapes recognizing combined with color feature, and followed by the mathematical computation of the spot center. Traditional algorithms such as Hough transform, gravity, have been used in some special environment. However, only the center is considered in the method such as gravity and some important parameters should be pre-adjusted in Hough transform, which is not suitable for dynamic environment [4, 5].

A novel algorithm for laser spot detection based on quaternion discrete cosine transform is introduced in this paper. The organization is as follows. The next section provides a brief introduction about the quaternion discrete cosine transform. Section 3 introduces some conventional algorithm for spot detection such as Hough transform and image moment method. The experimental results obtained by our new algorithm as

F. Xhafa et al. (eds.), *Recent Developments in Intelligent Systems and Interactive Applications*,
Advances in Intelligent Systems and Computing 541, DOI 10.1007/978-3-319-49568-2_12

well as brief comparisons, are presented in Sect. 4. References shows a summary of this paper's contributions.

2 The Quaternion Discrete Cosine Transform Based Laser

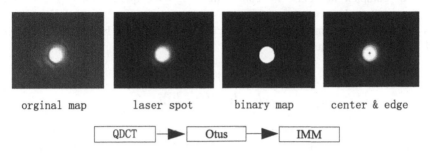

orginal map   laser spot   binary map   center & edge

Fig. 1. Diagram of spot detection algorithm

Spot Detection

The algorithm for laser spot detection based on quaternion discrete cosine transform is shown in Fig. 1.

Firstly, for the original laser map, the QDCT algorithm is used to change its form, in the QDCT space, the filter is used to reduce the high frequency, which means noise and some other unmoral things, and then we use IQDCT to get the laser spot.

Secondly, the Otus algorithm is used to segment the laser spot to binary map, the threshold t $t = Max[w_0(t) \cdot (u_0(t) - u)^2 + w_1(t) \cdot (u_1(t) - u)^2]$, w_0 is property of background, u_0 is the equal value of background, w_1 is property of front ground, u_1 is the equal value of front ground, u is the equal value of the whole image;

Thirdly, the image moment method (IMM) algorithm is used to compute the centers.

2.1 Quaternion Discrete Cosine Transform

The concept of Quaternion was formally introduced by Hamilton, which is denoted by H, a 4-dimensional algebra [6]. A quaternion q is defined as $q = a + bi + cj + dk \in H$ with $a, b, c, d \in R$ where i,j and $k(i^2 = j^2 = k^2 = ijk = -1)$ provide the necessary basis to define a product in H. The corresponding Hamilton product of two quaternion q_1 and q_2 defined as:

$$q_1 q_2 = (a_1 + b_1 i + c_1 j + d_1 k)(a_2 + b_2 i + c_2 j + d_2 k) \tag{1}$$

It must be noted that, the Hamilton product is not commutative (e.g., $ij = k$ while $ji = -k$).Thus, the left-side and right-sided operations should be paid attention in the following. A quaternion q is called real, if $q = a+0i + 0\ k$, and pure imaginary, if $q = 0 +bi + cj + dk$. If we multiply an arbitrary quaternion with a real quaternion $q_3 = y$, we obtain a simple element, i.e.:

$$q_1 q_3 = a_1 y + b_1 yi + c_1 yj + d_1 yk \tag{2}$$

As for complex numbers, we can define conjugate quaternion $\bar{q} = a - b - i - cj - dk$ as well as the norm $|q| = \sqrt{q \cdot \bar{q}}$.

And then, we can transform every $M \times N \times C$ image with less than 4 channels I_c $c \leq 4$ as a quaternion equation:

$$
\begin{aligned}
I_Q &= I_4 + I_1 i + I_2 j + I_3 k \\
&= I_4 + I_1 i + (I_2 + I_3 i)j
\end{aligned}
\tag{3}
$$

We represent the 4th channel as the scalar part, because then we obtain a pure imaginary quaternion matrix for color spaces with 3 channels, such as RGB, HIS.

Following the definition of the quaternion DCT in [7], we can transform the M*N quaternion matrix:

$$QDCT^L(p, q) = \alpha_p^M \alpha_q^M \sum_{m=0}^{M-1} \sum_{n=0}^{N-1} u_Q I_Q(m, n) \beta_{p,m}^M \beta_{q,n}^N \tag{4}$$

$$QDCT^R(p, q) = \alpha_p^M \alpha_q^M \sum_{m=0}^{M-1} \sum_{n=0}^{N-1} I_Q(m, n) \beta_{p,m}^M \beta_{q,n}^N \mu_Q \tag{5}$$

Where μ_Q is a unit quaternion, i.e. $\mu_Q^2 = -1$, that serves as DCT axis. In accordance with the definition of the traditional DCT. We can define α and β as:

$$\alpha_p^M = \begin{cases} \sqrt{\frac{1}{M}} for & p = 0 \\ \sqrt{\frac{1}{M}} for & p \neq 0 \end{cases}, \quad \beta_{p,m}^M = \cos\left[\frac{\pi}{M}(m+0.5)p\right] \tag{6}$$

Consequently, the corresponding inverse quaternion DCT is defined as follows:

$$IQDCT^L(m, n) = \sum_{p=0}^{M-1} \sum_{q=0}^{N-1} \alpha_p^M \alpha_q^N u_Q C_Q(p, q) \beta_{p,q}^M \beta_{m,n}^N \tag{8}$$

$$IQDCT^R(m, n) = \sum_{p=0}^{M-1} \sum_{q=0}^{N-1} \alpha_p^M \alpha_q^N C_Q(p, q) \beta_{p,q}^M \beta_{m,n}^N \mu_Q \tag{9}$$

2.2 Image Moment Method (IMM)

The image moment method (IMM) is used to marking the centers after our proposed QDCT-based algorithm. Although some new methods may get better accuracy for center location, the IMM is simple and easy to be used in field-programmable gate arrays (FPGA). We can calculate the center using the definition of the centroid [8]:

$$
\begin{cases}
x_0 = \sum_{i=1}^{M}\sum_{j=1}^{N} j \cdot g(i,j) \Big/ \sum_{i=1}^{M}\sum_{j=1}^{N} g(i,j) \\
y_0 = \sum_{i=1}^{M}\sum_{j=1}^{N} i \cdot g(i,j) \Big/ \sum_{i=1}^{M}\sum_{j=1}^{N} g(i,j)
\end{cases}
\tag{10}
$$

3 Experiment

In order to test the above algorithms, an artificial image is made firstly in Fig. 2(a), Gaussian noise is added, the original center is (344, 288), as shown in Fig. 2(b), the

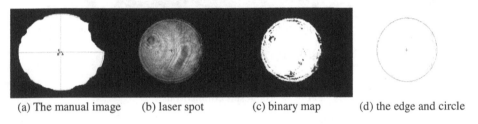

(a) The manual image   (b) laser spot   (c) binary map   (d) the edge and circle

Fig. 2. The analysis of laser spot

radius r is 199. The comparison is made between our algorithm and the traditional on the Intel CPU core 6500, the central point by Hough is (336.7,289.6), the computational time is 11 ms,The result using our method is shown in Fig. 2(d), the center is (345.15,289.2), the computational time is 8 ms.

Some typical images are chose to test the algorithms as show in Fig. 3, the original maps are in the first column, the results by traditional algorithm are in the second column, and the results by our algorithm are in the third column.

In the experimental results in Table 1, the right two columns are a center offset we try to prove the out come of our QDCT-based algorithm. The results are normalized with pixel and length.

The Hough algorithm can show the outer contour, however the detection is based on assume that the spot is a circle or a ellipse. In our paper, we prove the QDCT is better in accuracy and efficiency, no matter the spot is which shape.

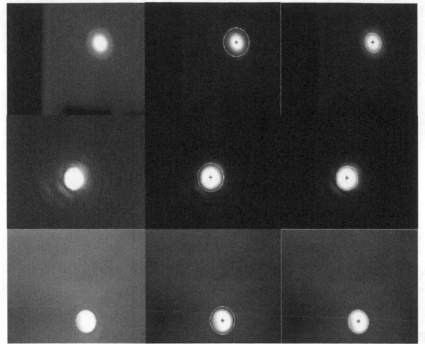

(1)original map   (2) result of traditional algorithm (3) result of our algorithm

Fig. 3. The detection of laser spot

Table 1. The result comparisons

		Centroid Coordinates					
		Hough-x	Hough-y	QDCT -x	QDCT -y	Offset x(10^{-3})	Offset y(10^{-3})
Image	1	302.330	544.317	302.452	544.337	−0.095	−0.021
	2	363.200	600.375	363.304	600.349	−0.081	0.027
	3	131.899	460.435	132.180	460.296	−0.220	0.145
	4	223.383	487.399	226.049	490.340	−2.083	−3.064
	5	578.307	531.479	586.126	523.829	−6.109	7.969
	6	625.730	463.174	654.045	520.999	1.186	−10.331

4   Conclusions

This paper proposed a QDCT-based method for laser spot detection, regardless its shape, color, and noise.

As shown in the experimental results, this algorithm may be useful in the fields such as laser measurement, motion sensing games, space detection.

References

1. Dong, H., Wang, L.: Non-iterative spot center location algorithm based on gaussian for fish-eye imaging laser warning system. Optik **123**, 2148–2153 (2012)
2. Li, Y., Lei, S., Yunshan, C.: Design and implementation of high-speed laser spot detection systems. Laser Technol. **39**(4), 533–536 (2015)
3. Liu, H.L., Hou, W., Fan, Y.L., et al.: An improved algorithm of laser spot center location. Comput. Meas. Control **22**(1), 1655–1660 (2014)
4. Wang, L.L., Hu, Z.W., Ji, H.X.: Laser spot center location algorithm based on gaussian fitting. J. Appl. Opt. **33**(5), 985–990 (2012)
5. Yang, P., Xie, L., Liu, J.L.: Zernike moment based high-accuracy sun image centroid algorithm. J. Astronaut. **34**(9), 1963–1970 (2011)
6. Schauerte, B., Stiefelhagen, R.: Predicting human gaze using quaternion DCT image signature saliency and face detection. In: IEEE Workshop on, Applications of Computer Vision, pp. 137–144 (2012)
7. Vázquez-Otero, A., Khikhlukha, D., Solano-Altamirano, J.M., Dormido, R., Duro, N.: Laser spot detection based on reaction diffusion. Sensors **16**(3), 315–326 (2016)
8. Krstinić, D., Skelin, A.K., Milatić, I.: Laser spot tracking based on modified circular hough transform and motion pattern analysis. Sensors **14**(11), 20112–20133 (2014)

Natural Frequency Modification Using Frequency-Shift Combined Approximations Algorithm

Xue Liu, Qing Shao, Tao Xu, and Guikai Guo[✉]

School of Mechanical Science and Engineering,
Jilin University, Changchun, People's Republic of China
181231735@qq.com, shaoqing14@mails.jlu.edu.cn,
{xutao,ggk}@jlu.edu.cn

Abstract. This paper proposes an algorithm for determining structural changes for modifying natural frequencies of a structure in a prescribed manner using Frequency Shift Combined Approximations (FSCA) approach. This algorithm is based on the reduced basis method, natural frequency sensitivity analysis and Taylor series expansion. The application of this algorithm to a truck body finite element analysis is described. Natural frequency modification of the finite element model is shown to converge to satisfied results. The proposed algorithm is easy to implement and is suitable for structural dynamic modification and optimization in many engineering applications.

Keywords: Structural modification · Sensitivity analysis · Vibration reanalysis frequency shift · Combined approximations

1 Introduction

In order to make a design structure to satisfy the predetermined demands such as the demands of the natural frequencies the designer will modify the structure repeatedly. Due to modification the responses get changed on the structure. Appropriate structural changes must then be determined to meet the design objectives. Structural dynamic modification and reanalysis approach are concerned with finding the parameters changes in an analytical estimate.

Reanalysis technology was established to evaluate responses of changed structures in process of design and optimization [1]. Reanalysis of structure for displacements and stresses have been discussed since the 20th century [2]. Combined Approximations (CA) approach is one of the most effective methods for solving static displacement equations [3]. After CA method was founded, extended CA methods were proposed [4].

Reanalysis methods for vibration problems have been presented since the early 21st century [5]. Kirsch grafted the CA approach to solve eigenproblems [6]. Combining CA and Rayleigh quotient, an extended CA method of eigenproblem for large changes was presented by Chen [7]. A Modified Combined Approximations (MCA) method for solving large-scale structure dynamic problem was discussed [8]. With a suitable frequency shift coefficient, FSCA approach allowed to calculate higher modes accurately [9].

F. Xhafa et al. (eds.), *Recent Developments in Intelligent Systems and Interactive Applications*,
Advances in Intelligent Systems and Computing 541, DOI 10.1007/978-3-319-49568-2_13

In the structural modification procedures, sensitivities are usually needed. Several methods for accurate sensitivity calculation were discussed [10], but the low computational efficiency for large scale structures becomes a main obstacle. Based on CA, sensitivities of displacements were presented [11]. Reanalysis sensitivities with CA method when the eigenvalues are repeated was discussed [12].

In this study, a sensitivity analysis using FSCA method is proposed to acquire the derivatives of the natural frequencies. The formulations of natural frequency sensitivity reanalysis and modification based on Taylor series expansion are expressed in Sect. 2 and then the application of this algorithm to a truck body finite element analysis is described in Sect. 3. Efficiency consideration and conclusion are discussed in Sect. 4.

2 Formulation and Solution

2.1 Natural Frequency Sensitivity Analysis Using FSCA Method

Natural frequency sensitivity is aimed to find which parameter is most important for the response of structure, which is indispensable to structural modification. In practice, natural frequency has the same mean with eigenvalue in mathematics. Given an initial structure with stiffness matrix $\mathbf{K}^{(0)}$ and mass matrix $\mathbf{M}^{(0)}$, the equation of the first m eigenvalues and eigenvectors can be expressed:

$$\mathbf{K}^{(0)}_{n \times n} \mathbf{\Phi}^{(0)}_{n \times m} = \mathbf{M}^{(0)}_{n \times n} \mathbf{\Phi}^{(0)}_{n \times m} \mathbf{\Lambda}^{(0)}_{m \times m} \tag{1}$$

Where $\mathbf{\Lambda}^{(0)}$ denotes the matrix of the first m eigenvalues and $\mathbf{\Phi}^{(0)}$ is the corresponding matrix of first m eigenvectors, n is DoFs for the initial system. Assuming there are changes in the stiffness and mass matrices, respectively.

$$\mathbf{K} = \mathbf{K}^{(0)} + \Delta\mathbf{K} \qquad \mathbf{M} = \mathbf{M}^{(0)} + \Delta\mathbf{M} \tag{2}$$

The eigenproblem of the changed structure can be rearranged:

$$\underset{n \times n}{\mathbf{K}} \; \underset{n \times m}{\mathbf{\Phi}} \; \underset{m \times m}{\mathbf{\Lambda}^{-1}} = \underset{n \times n}{\mathbf{M}} \; \underset{n \times m}{\mathbf{\Phi}} \tag{3}$$

Where $\mathbf{\Lambda}$ denotes the matrix of the first m eigenvalues and $\mathbf{\Phi}$ is the corresponding matrix of first m eigenvectors for the changed structure.

Equation 3 is rearranged using a frequency-shift factor:

$$\underset{n \times m}{\mathbf{\Phi}} = [(\mathbf{M} - \mu^{-1}\mathbf{K})^{-1} \; \underset{n \times n}{\mathbf{K}}] \; \underset{n \times m}{\mathbf{\Phi}} [\mathbf{\Lambda}^{-1} - \mu^{-1} \underset{m \times m}{\mathbf{I}}] \tag{4}$$

Given an initial $\mathbf{\Phi}^{(i)}$, we can compute $\mathbf{\Phi}^{(i+1)}$ by solving iterative formula as Eq. 5.

$$\underset{n \times m}{\mathbf{\Phi}^{(i+1)}} = (\mathbf{M} - \mu^{-1}\mathbf{K})^{-1} \; \underset{n \times n}{\mathbf{K}} \; \underset{n \times m}{\mathbf{\Phi}^{(i)}} (\mathbf{\Lambda}^{-1} - \mu^{-1}\mathbf{I}) \tag{5}$$

Assuming that a linear expression of $\Phi^{(i)}$, where $i = 0, 1, \cdots, s - 1$, can be close to the exact solutions, the linear expression is given:

$$
\begin{aligned}
\Phi^c_{n\times m} &= a_0 \Phi^{(0)}_{n\times m} + a_1 \Phi^{(1)}_{n\times m} + a_2 \Phi^{(2)}_{n\times m} + \cdots + a_{s-1} \Phi^{(s-1)}_{n\times m} \\
&= a_0 \Phi^{(0)}_{n\times m} + a_1 (\mathbf{M} - \mu^{-1}\mathbf{K})^{-1} \underset{n\times n}{\mathbf{K}} \, \Phi^{(0)}_{n\times m} (\mathbf{\Lambda}^{-1} - \mu^{-1}\mathbf{I}) + a_2 ((\mathbf{M} - \mu^{-1}\mathbf{K})^{-1} \underset{n\times n}{\mathbf{K}})^2 \Phi^{(0)}_{n\times m} (\mathbf{\Lambda}^{-1} - \mu^{-1}\mathbf{I})^2 \\
&\quad + \cdots + a_{s-1} ((\mathbf{M} - \mu^{-1}\mathbf{K})^{-1} \underset{n\times n}{\mathbf{K}})^{s-1} \Phi^{(0)}_{n\times m} (\mathbf{\Lambda}^{-1} - \mu^{-1}\mathbf{I})^{s-1} \\
&= [\Phi^{(0)}_{n\times m}, (\mathbf{M} - \mu^{-1}\mathbf{K})^{-1} \underset{n\times n}{\mathbf{K}} \, \Phi^{(0)}_{n\times m}, ((\mathbf{M} - \mu^{-1}\mathbf{K})^{-1} \underset{n\times n}{\mathbf{K}})^2 \Phi^{(0)}_{n\times m}, \cdots, ((\mathbf{M} - \mu^{-1}\mathbf{K})^{-1} \underset{n\times n}{\mathbf{K}})^{s-1} \Phi^{(0)}_{n\times m}] \\
&\quad \cdot [a_0\mathbf{I}, \; a_1(\mathbf{\Lambda}^{-1} - \mu^{-1}\mathbf{I}), \quad a_2(\mathbf{\Lambda}^{-1} - \mu^{-1}\mathbf{I})^2 \quad \cdots \quad a_{s-1}(\mathbf{\Lambda}^{-1} - \mu^{-1}\mathbf{I})^{s-1}]^T \\
&= \underset{n\times ms}{\mathbf{R}} \; \underset{ms\times m}{\mathbf{X}}
\end{aligned}
$$

$$(6)$$

Premultiplying Eq. 3 by $\mathbf{R}^T$, a condensed equation is got and expressed in the following form:

$$
[\underset{ms\times ms}{\mathbf{R}^T \; \mathbf{K} \; \mathbf{R}}] \; \underset{ms\times m}{\mathbf{X}} = [\underset{ms\times ms}{\mathbf{R}^T \; \mathbf{M} \; \mathbf{R}}] \; \underset{ms\times m}{\mathbf{X}} \; \underset{m\times m}{\mathbf{\Lambda}}
$$

$$(7)$$

The matrices $[\mathbf{R}^T\mathbf{K}\mathbf{R}]$ and $[\mathbf{R}^T\mathbf{M}\mathbf{R}]$ of the condensed system are much smaller than those in the initial system. So we can calculate a new $ms \times ms$ system in Eq. 7 instead. The computing time can be greatly reduced.

Aim to get more precise high frequency results highest mode vector of interest is chosen to calculate the factor.

$$
\mu = \frac{\underset{n\times 1}{\varphi^{(0)T}_m} \; \underset{n\times n}{\mathbf{K}} \; \underset{n\times 1}{\varphi^{(0)}_m}}{\underset{n\times 1}{\varphi^{(0)T}_m} \; \underset{n\times n}{\mathbf{M}} \; \underset{n\times 1}{\varphi^{(0)}_m}}
$$

$$(8)$$

Considering a problem of calculating $\frac{\partial \lambda^{(0)}_j}{\partial x}$ of the eigenvector λ_j with respect to a design variable x at $x^{(0)}$, the central-difference approximation is used:

$$
\frac{\partial \lambda^{(0)}_j}{\partial x} = \frac{\lambda_j(x^{(0)} + \delta x) - \lambda_j(x^{(0)} - \delta x)}{2\delta x}
$$

$$(9)$$

Where δx is a given small step. In process of design, repetition finite difference calculations are required. This formula is not very efficient by exact eigenvalue analysis. But by solving a much smaller eigenproblem using FSCA method, the efficiency can be improved.

2.2 Natural Frequency Modification Based on Taylor Series Expansion

The sensitivities obtained by FSCA method are used to predict the natural frequencies of the modified structure. Taylor series expansion is adopted here, which is used to find how eigenvalue change with respect to variations in the properties of the structure.

Assume the design variables are x_i, $i = 1, \cdots, t$, respectively. In order to get expected frequencies, a mathematical optimization model is formed using least squares estimation:

$$\min \sum_{j=1}^{r} [\lambda_j(\Delta x_1, \Delta x_1, \cdots, \Delta x_t) - \lambda_j^{OBJ}]^2 \tag{10}$$

Where λ_j^{OBJ} is defined for the objective of jth eigenvalue, and λ_j can be expressed with one order Taylor expansion:

$$\lambda_j(\Delta x_1, \Delta x_1, \cdots, \Delta x_t) = \lambda_j^{(0)} + \sum_{i=1}^{t} \frac{\partial \lambda_j}{\partial x_i} \Delta x_i \tag{11}$$

Where $\frac{\partial \lambda_j}{\partial x_i}$ is the sensitivity respect to the design variable x_i. Or else we can express the eigenvalue with a two order Taylor expansion like this:

$$\lambda_j(\Delta x_1, \Delta x_1, \cdots, \Delta x_t) = \lambda_j^{(0)} + \sum_{i=1}^{t} \frac{\partial \lambda_j}{\partial x_i} \Delta x_i + \left(\sum_{i=1}^{t} \Delta x_i \frac{\partial}{\partial x_i} \right)^2 \lambda_j \tag{12}$$

3 Numerical Example

To demonstrate the efficient of the proposed algorithm, a finite element model of truck body is modified with 3 thickness parameters, which are hood, frame and carriage respectively, as an example. The model contains 11664 degrees of freedom. The finite element model of truck body, design variables which are to be changed, original values and variation ranges are signed in Fig. 1. The first 3 natural frequencies of initial structure are 9.697 Hz, 9.778 Hz and 11.177 Hz, respectively. After the sensitivities were got, Taylor series expansions are formed in Table 1.

Considering the first 3 natural frequencies are changed by prescribed amounts, two cases are taken for example. The first 3 frequencies are changed to 9.9 Hz, 10.0 Hz and 11.4 Hz respectively for a small modification case 1, and changed to 10.0 Hz, 11 Hz and 12 Hz respectively for a large modification case 2. Results are listed in Table 2.

The reanalysis of the modified structure shows that the frequency constraints of first 3 orders are satisfied using Taylor expansions of both one and two orders in small modification case 1. When the require of frequency modification is large in case 2, the results using two order expansion are better than that using one order expansion in the variance analysis.

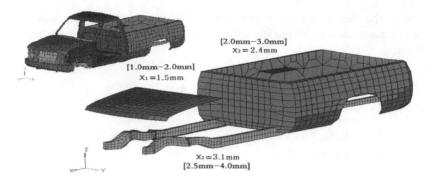

Fig. 1. Finite element model of truck body and design variables

Table 1. Taylor series expansions with the sensitivity information

Order of expansion	Mode	Taylor series expansions
1st	1	$\omega_1 = 9.697 + 0.00003 \times (x_1 - 1.5) + 0.00217 \times (x_2 - 3.1) + 2.94953 \times (x_3 - 2.4)$
	2	$\omega_2 = 9.778 + 0.00119 \times (x_1 - 1.5) + 0.01747 \times (x_2 - 3.1) + 3.00946 \times (x_3 - 2.4)$
	3	$\omega_3 = 11.177 + 1.01384 \times (x_1 - 1.5) + 0.16558 \times (x_2 - 3.1) + 0.23546 \times (x_3 - 2.4)$
2nd	1	$\omega_1 = 9.697 + 0.00003 \times (x_1 - 1.5) + 0.00217 \times (x_2 - 3.1) + 2.94953 \times (x_3 - 2.4)$ $- 0.00009(x_1 - 1.5)^2 - 0.00049(x_2 - 3.1)^2 - 0.31357(x_3 - 2.4)^2$ $+ 0.00001(x_1 - 1.5)(x_2 - 3.1) + 0.45026(x_1 - 1.5)(x_3 - 2.4)$ $+ 0.33270(x_2 - 3.1)(x_3 - 2.4)$
	2	$\omega_2 = 9.778 + 0.00119 \times (x_1 - 1.5) + 0.01747 \times (x_2 - 3.1) + 3.00946 \times (x_3 - 2.4)$ $- 0.00148(x_1 - 1.5)^2 - 0.00673(x_2 - 3.1)^2 - 0.14121(x_3 - 2.4)^2$ $- 0.00022(x_1 - 1.5)(x_2 - 3.1) + 0.08986(x_1 - 1.5)(x_3 - 2.4)$ $- 0.02183(x_2 - 3.1)(x_3 - 2.4)$
	3	$\omega_3 = 11.177 + 1.01384 \times (x_1 - 1.5) + 0.16558 \times (x_2 - 3.1) + 0.23546 \times (x_3 - 2.4)$ $- 1.43520(x_1 - 1.5)^2 - 0.25548(x_2 - 3.1)^2 - 0.85532(x_3 - 2.4)^2$ $+ 1.65848(x_1 - 1.5)(x_2 - 3.1) + 0.37832(x_1 - 1.5)(x_3 - 2.4)$ $- 0.24772(x_2 - 3.1)(x_3 - 2.4)$

Table 2. Results of the modified parameters and natural frequencies

Case	Order of expansion	x_1 mm	x_2 mm	x_3 mm	ω_1 Hz	ω_2 Hz	ω_3 Hz	Variance
1	1st	1.8	3.0	2.5	10.018	10.087	11.547	0.043
	2nd	1.5	3.3	2.5	10.019	10.093	11.213	0.058
2	1st	1.2	4.0	2.7	10.611	10.658	10.718	2.135
	2nd	1.8	3.5	2.6	10.339	10.405	11.839	0.494

4 Conclusion

The paper presents a natural frequency modification method and its application for increasing the natural frequencies of a truck body. Taylor series expansions made up of the sensitivities with respect to the structural changes and least squares estimations are used to get the modified design parameters. Sensitivities used here are got from FSCA method. The numerical example demonstrates the efficient and accuracy for the proposed algorithm. It is expected that natural frequency modification using proposed method could reduce the computing time for problems where better frequency properties are needed in structural dynamic design and optimization.

Acknowledgements. This work is supported by Project of National Science Foundation of China (NSFC) (Grant No. 11502092), and Plan for Scientific and Technological Development of Jilin Province (Grant Nos. 20140520111JH and 20160520064JH).

References

1. Kirsch, U.: Reanalysis and sensitivity reanalysis by combined approximations. Struct. Multi. Optim. **40**, 1–15 (2010)
2. Phansalkar, S.R.: Matrix iterative methods for structural reanalysis. Comput. Struct. **4**, 779–800 (1974)
3. Kirsch, U.: A unified reanalysis approach for structural analysis, design, and optimization. Struct. Multi. Optim. **2003**, 67–85 (2003)
4. Zuo, W., Yu, Z., Zhao, S., Zhang, W.: A hybrid fox and kirsch's reduced basis method for structural static reanalysis. Struct. Multi. Optim. **46**, 261–272 (2012)
5. Chen, S.H., Yang, X.W., Lian, H.D.: Comparison of several eigenvalue reanalysis methods for modified structures. Struct. Multi. Optim. **20**, 253–259 (2000)
6. Kirsch, U.: Approximate vibration reanalysis of structures. AIAA J. **41**, 504–511 (2003)
7. Chen, S.H., Yang, X.W.: Extended kirsch combined method for eigenvalue reanalysis. AIAA J. **38**, 927–930 (2000)
8. Zhang, G., Nikolaidis, E., Mourelatos, Z.P.: An efficient re-analysis methodology for probabilistic vibration of large-scale structures. J. Mech. Des. **131**, 051007 (2009)
9. Xu, T., Guo, G., Zhang, H.: Vibration reanalysis using frequency-shift combined approximations. Struct. Multi. Optim. **44**, 235–246 (2011)
10. Fox, R.L., Kappor, M.P.: Rates of change of eigenvalues and eigenvectors. AIAA J. **6**, 2426–2429 (1968)
11. Zuo, W., Bai, J., Yu, J.: Sensitivity reanalysis of static displacement using taylor series expansion and combined approximate method. Struct. Multi. Optim. **53**, 953–959 (2016)
12. Zhao, S., Guo, G., Zhang, W., Liu, D.: Efficient procedures of sensitivity analysis for structural vibration systems with repeated frequencies. J. Appl. Math. **2013**, 1–7 (2013)

Adaptive Stochastic Ranking Schemes for Constrained Evolutionary Optimization

Yu Wu[1]([✉]), Weiqin Ying[2], Bingshen Wu[2], and Dongxin Peng[2]

[1] School of Computer Science and Educational Software,
Guangzhou University, Guangzhou 510006, China
wuyu@gzhu.edu.cn
[2] School of Software Engineering, South China University of Technology,
Guangzhou 510006, China
yingweiqin@scut.edu.cn

Abstract. Stochastic ranking (SR) is a popular constraint handling method for constrained evolutionary optimization. A constant probability parameter P_f in the classic SR is applied to choose one of constraint violation and fitness as the basis for comparison of a pair of individuals. This paper presents two adaptive SR schemes, referred to as a linear decline SR and a reciprocal decline SR respectively, which are inspired by cooling schedules for simulated annealing. Both adaptive SRs aim to achieve different optimization goals at the various stages of constrained evolutionary optimization by adaptively adjusting the probability P_f according to the current generation. Experimental results on 13 benchmarks problems show that both adaptive SRs are more competitive for local search at the terminal stage than the classic SR, and the reciprocal decline SR obtains the best performance in terms of solution accuracy and convergence speed.

Keywords: Constrained optimization · Evolutionary strategy · Constraint handling · Stochastic ranking · Simulated annealing

1 Introduction

Many optimization problems with various types of constraints could be modeled as constrained optimization problems (COPs) in a wide range of real-world applications [1]. Without loss of generality, COPs in the minimization sense can be described as follows:

$$minimize \ f(x) \quad x = \{x_1, x_2, \ldots, x_n\} \in S$$
$$subject \ to \ g_j(x) \le 0, \quad j = 1, \ldots, l$$
$$h_j(x) = 0, \quad j = l+1, \ldots, m$$

where $f(x)$ represents the objective to be minimized, and S denotes an n-dimensional decision space bounded by the following parametric constraints:

© Springer International Publishing AG 2017
F. Xhafa et al. (eds.), *Recent Developments in Intelligent Systems and Interactive Applications*,
Advances in Intelligent Systems and Computing 541, DOI 10.1007/978-3-319-49568-2_14

$$\underline{x_i} \le x_i \le \bar{x_i}, \quad i = 1, \ldots, n$$

$gj\,(x)$ and $hj(x)$ are, separately, the j-th inequality constraint and the $(j–l)$-th equality constraint.

Unlike unconstraint problems, the degree of constraint violation becomes an additional evaluation criterion for any solution of constrained problems. It can be calculated as follows:

$$G(x) = \sum_{j=1}^{l} \max\{0, g_j(x)\} + \sum_{j=l+1}^{m} \max\{0, |h_j(x)| - \delta\}$$

where the tolerance parameter δ is generally extremely small. The algorithms [2–4] or optimization mechanisms [5] should compare or select solutions based on their objective values and violation degrees of constraints [6] for constrained optimization. Consequently, the additional constraint handling techniques about how to deal with the constraints and evaluate individuals play an important role in constrained evolutionary optimization. Although Michalewicz divided the existing constraint handling techniques into five categories, there has been a great deal of interest in the research community in the primary methodologies of penalty function, stochastic ranking and multi-objective optimization methods in recent years. Penalty function methods are easy to implement but hard to tune some effective penalty factors. The multi-objective techniques are employed to solve COPs in diverse optimization algorithms [7, 8]. In multi-objective techniques, the concept of Pareto dominance is a crucial relation and the optimization goal is to find a uniformly spread of non-dominated solutions. The main difficulty of the multi-objective techniques is how to reduce the time of non-dominance checking. The stochastic ranking method with simple preference rules [9, 10] not only has fewer parameters than the penalty function methods but also spends less time checking for non-dominance than the multi-objective techniques.

The classic SR algorithm introduced a random probability to keep a balance between constraint violation and fitness during optimization. The ranking sequence of each population depends on the values of a random number. Apart from the case that both of two compared adjacent individuals are feasible, if the random value is less than or equal to the probability parameter Pf, the comparison between the adjacent individuals is decided by their values of objective function; otherwise, their degrees of constraint violation are utilized to determine the comparison result. Runarsson and Yao [10] proposed an improved SR evolutionary algorithm based on evolution strategy and differential variation operator. The parameter of Pf is usually set as a constant value in both optimization process of the classic SR and improved SR methods. However, there are several stages with different optimization goals in the whole evolutionary process. It is blind to apply an invariable biases search at different optimization stages. The invariable biases search may lead to the fact that the population converges slowly or even converges to infeasible regions at one certain specific stage of the evolutionary process.

In this paper, we analyze the features of biases search at different evolutionary stages. Further, two adaptive SR schemes, respectively referred to as a linear decline SR and a reciprocal decline SR, are presented to enhance the performance at variable evolutionary stages by borrowing the idea of cooling schedules of simulated annealing. This paper is organized as follows. After the introduction, Sect. 2 discusses characteristic of various optimization stages for solving COPs, analyses the roles of the probability parameter Pf during the whole evolutionary process, and presents our proposed methods. Section 3 gives the experimental results and analysis. Finally, Sect. 4 concludes the paper.

2 Theoretical Analysis and Proposed Approach

There are usually some difficulties about the search space when solving COPs. The searching space is constructed by a variety of feasible and infeasible regions. For some particular COPs, there are several disjoined feasible regions in the search space. Sometimes, these feasible zones may be small size, or the proportion of feasible regions in the search space is extremely low. Even the global feasible optima for these particular COPs maybe locate on the boundaries between feasible and infeasible regions. Therefore, some additional techniques is supposed to be adopted to keep diversity of population in the beginning of evolution or to improve the ability of local search at the late stage of evolution since a lot of local feasible optima for COPs make the search difficult. In order to satisfy the above optimization goals of two stages in some degree, the value of Pf in SR can be used to adjust the ratio between infeasible and feasible solutions and determine the comparison preference on constraint violation or fitness.

The classic SR algorithm is sensitive to the parameter value of Pf which is normally set in an interval [0, 1]. The most recommended value of Pf is set as a constant value of 0.45 in the classic SR. And it would not be changed in the whole evolutionary process. The comparison rules in SR is described as follows: two adjacent feasible individuals are surely compared based on their values of objective function; if there are less than two feasible individuals, the comparison preference between the adjacent two individuals depends on the value of random number. If the random value is less than or equal to Pf, the fitness is chosen as a basis for comparison of the adjacent individuals; otherwise, they are compared according to the violation degree. It means the parameter Pf in the above rules is used to control how much proportion of individuals in the population to be compared based on the objective function values, no matter they are feasible or infeasible.

When the value of Pf is set large, some solutions with the better values of objective function, which may be feasible or infeasible, have a larger chance to lie at the top of the ranking sequence. When Pf becomes low, the randomness degree of ranking would decrease and the top of the sequence would include more feasible solutions and a few of infeasible solutions with a small value of constraint violation. It is worth noting that, the selection of parent individuals which are used to generate new offspring candidates and the generation of the next population, are closely related to the individuals at the top of the ranking sequence. Consequently, it is not necessary to keep the value of Pf constant at various evolutionary stages with different optimization goals.

A relatively large value of Pf in the early evolution is beneficial to keep high degree of randomness and to maintain population diversity. It results in that many solutions with good values of objective function, regardless of feasible or infeasible individuals, would be ranked at the top of the ranking sequence and would be retained to the next generation with a high probability. The individuals at the top of the ranking sequence play a great role in improving the capability of exploration to search more extensive areas. As the SR algorithm evolves, the decrease of Pf could reduce the search randomness and could more strictly compare the individuals according to their degrees of constraint violation. It leads to that many individuals with the smaller values of constraint violation would be ranked in the top order. This means that infeasible individuals with small degrees of constraint violation and feasible individuals would be sorted at the top of the ranking sequence and would be allowed to survive to the next generation with a high probability. In other words, the individuals at the top are focusing on searching feasible regions and a part of infeasible regions which is close to the feasible regions. It is beneficial to search the global optima located on the feasible boundaries and to escape from the local optima. As a result, the main goal of improving the ability of local search could be achieved.

According to the above analysis, the parameter Pf is expected to be set as a relatively high initial value to make searching more stochastic and to explore as wide regions as possible at the beginning of optimization. As the SR algorithm evolves, Pf should be gradually decreased. With a small value of Pf, more individuals are ranked in order of constraint violation value with a high probability. It helps the algorithm enhance the performance of local search. In simulated annealing, the temperature cools slowly so that the system finally converges to a state of minimum energy steadily. The standard cooling schedule is normally described as $Tt = T0 * \alpha^t$ where $T0$ is the initial temperature of the system, the value of cooling parameter α is usually set between 0 and 1, and t denotes the cooling time for the system. As the cooling time goes on, the temperature exhibits an exponential decline. Inspired by cooling schedules of simulated annealing, two schemes to decrease the value of parameter Pf adaptively in the SR algorithm are designed in this paper. The first scheme reduces the value of Pf at a linear rate, in which Pf changes dynamically in accordance with the ratio between the current generation and the maximum generation $maxGen$ in the following form:

$$P_f = P_{f_0} * \left(1 - \frac{t}{maxGen}\right), \quad t = 0, 1, \ldots maxGen$$

Taking into consideration the initial population at the current generation $t = 0$, parameter Pf is equal to the initial value Pf_0. When an modified SR algorithm with the adaptive linear decline parameter, referred to as SR-LD, reaches one half of the maximum generation, the value of Pf would decline to one half of the initial value Pf_0. The value would become much smaller as the current generation increases to the maximum generation. The details of the adaptive SR scheme in one generation of the algorithm SR-LD are presented in the pseudocode of Algorithm 1.

The experimental results on 13 benchmark problems show that the classic SR algorithm find lots of feasible solutions, respectively, within less than 100 generations for problems g01, g02, g04, g06, g08, g09, g10, and g12, and after more than 300 generations for problems g03, g05, g11 and g13. This is because the feasible regions are extremely small and the estimated ratios between the feasible regions and the whole search space roughly equal to 0 for problems g03, g05, g11 and g13. It suggests that the optimization should be concentrated on improving the ability of local search when many feasible solutions had been found during the evolutionary process. If the declining speed is too fast, the likelihood of converging in local optima is very large and there exists a high risk of not taking full advantage of valuable infeasible solutions. As the value of *Pf* decreases quickly, infeasible solutions are sorted at the end with a high probability and survive to the next generation of population with an extremely small probability.

Algorithm 1. Adaptive SR with the linear decline P_f

Input:
- λ: the population size ;
- t: the current generation ;

```
1  for i ← 1 to λ do
2      for j ← 1 to λ − 1 do
3          P_{f_t} = P_{f_0} * (1 − t/maxGen);    // maxGen is the maximum generation
4          if (G(x_j) == G(x_{j+1}) == 0)or(U(0,1) < P_{f_t}) then
5              if f(x_j) > f(x_{j+1}) then
6                  Exchange the locations of these two adjacent individuals;
7              end
8          end
9          else
10             if G(x_j) > G(x_{j+1}) then
11                 Exchange the locations of these two adjacent individuals;
12             end
13         end
14     end
15 end
```

The second scheme is aiming to adjust the balance between convergence speed and ability of local search. As shown in the experimental results of the classic SR algorithm, the appropriate declining speed of *Pf* is supposed to be slower and smoother after several hundred generations than that in the linear decline SR. Therefore, the second scheme is designed with a reciprocal decline rate. The adaptive decline value of *Pf* in the modified SR algorithm with a reciprocal decline parameter, referred as SR-RD, is calculated as follows:

$$P_f = P_{f_0} * \frac{t}{1 + t/maxGen}, \quad t = 0, 1, \ldots maxGen$$

The only difference between SR-LD and SR-RD is that the value of *Pf* in SR-RD is updated in the form of reciprocal decline.

3   Experimental Results

13 benchmark problems [10] were used to evaluate the performance of two proposed adaptive SR algorithms and the classic SR algorithm. The properties of these test instances were reported in [8, 10]. All experiments for SR, SR-LD and SR-RD were carried out in the same experimental conditions. The population size is set as $\lambda = 200$ and the parent population size $\mu=30$ in these algorithms. The probability parameter Pf in SR is set as a constant value of 0.45, which is also the value of Pf_0 in SR-LD and SR-RD. The tolerance parameter $\delta = 0.0001$, and the other parameters in these algorithms are set as the same as those in [10]. In our experiments, every algorithm performed 30 times independently for each among 13 instances. The statistical results of the best objective value, median, mean, standard deviation are recorded, respectively, in Table 1 for g01-g06 and Table 2 for g07-g13. The "optimal" means the objective value of the best known solution.

For problems g01, g03, g04, g05, g06, g08, g11 and g12, the three algorithms always find the optimal solutions in each run. For problem g09, the optimal solutions are found by all algorithms, but the SR-LD and SR-RD are performed better than the classic SR. The results achieved by SR-RD are consistently with the higher precision. The estimated feasible region proportion of problem g02 is more than 90%, which means most of the search regions are feasible [8, 10]. SR-LD and SR-RD obtain extremely approximate optimal results and they discover the better "best" result than SR. While SR finds the better "worst" result than SR-LD, SR-RD outperforms SR in terms of the criteria of "best", "median", "mean" and "worst". Most of the search regions are infeasible and the estimated feasible region ratio is very small for problem g07, g10 and g13. All the three algorithms find extremely approximate optimal results for the above three problems. SR-LD performs better in terms of the "mean" results as well as the "best" results and worse than "worst" results than SR for problems g07 and g10. SR-RD achieves the best performance in terms of all criteria among three algorithms. Especially, SR-RD can consistently find the optimal solution in all 30 runs while SR-LD and SR have the similar performance and sometimes get trapped into a local optimal solution of 0.438851219909 for problem g13.

Figure 1 plots the convergence curves of average objective errors in logarithmic scalar in 30 independent runs of each algorithm for problems g02, g07, g09 and g10, respectively. These problems are more difficult to be solved than the others. Most of the search regions for problem g02 are feasible and one half of the search regions for problem g09 is also feasible. Conversely, the estimated feasible region ratio in the whole search space for problem g07 and g10 nearly equal to 0 which means most of their search regions are infeasible.

As shown in Fig. 1(a) and (b), the convergence curves of SR and SR-LD are nearly the same for problem g02 and g07. After about 200 generations, the two algorithms converge to an approximate optimal solution and fail to improve the solution towards higher accuracy. However, SR-RD continues to converge towards the optimal solution. It obtains the solutions with a much higher accuracy after about 200 generations than SR and SR-LD. It is evident from Fig. 1(c) that the performance of SR-RD is the best among the three algorithms for problem g09 while SR-LD performs better than the

Table 1. Statistical results achieved by SR, SR-LD and SR-RD for problems g01-g06

fcn	Optimal	Algorithms	Best	Median	Mean	Worst
g01	−15.000	SR	−15.000000000077	−15.000000000077	−15.000000000077	−15.000000000077
		SR-LD	−15.000000000077	−15.000000000077	−15.000000000077	−15.000000000077
		SR-RD	−15.000000000077	−15.000000000077	−15.000000000077	−15.000000000077
g02	−0.803619	SR	−0.794893529982	−0.7688114946785	−0.762187345231	−0.712966318613
		SR-LD	−0.803618720880	−0.774232012796	−0.763792085590	−0.69533654415
		SR-RD	−0.803618869695	−0.78625227168	−0.781790366895	−0.74920376566
g03	−1.000	SR	−1.00000000	−1.00000000	−1.00000000	−1.00000000
		SR-LD	−1.00000000	−1.00000000	−1.00000000	−1.00000000
		SR-RD	−1.00000000	−1.00000000	−1.00000000	−1.00000000
g04	−30665.539	SR	−30665.5386718738	−30665.5386718738	−30665.5386718738	−30665.5386718738
		SR-LD	−30665.5386718738	−30665.5386718738	−30665.5386718738	−30665.5386718738
		SR-RD	−30665.5386718738	−30665.5386718738	−30665.5386718738	−30665.5386718738
g05	5126.498	SR	5126.49810959446	5126.49810959446	5126.49810959446	5126.49810959446
		SR-LD	5126.49810959446	5126.49810959446	5126.49810959446	5126.49810959446
		SR-RD	5126.49810959446	5126.49810959446	5126.49810959446	5126.49810959446
g06	−6961.814	SR	−6961.81387574495	−6961.81387574495	−6961.81387574495	−6961.81387574495
		SR-LD	−6961.81387574495	−6961.81387574495	−6961.81387574495	−6961.81387574495
		SR-RD	−6961.81387574495	−6961.81387574495	−6961.81387574495	−6961.81387574495

Table 2. Statistical results achieved by SR, SR-LD and SR-RD for problems g07-g13

fcn	Optimal	Algorithms	Best	Median	Mean	Worst
g07	24.306	SR	24.30633493397	24.307412018056	24.311407554677	24.331907900479
		SR-LD	24.306315108618	24.307510380312	24.311273894600	24.340158033618
		SR-RD	24.306212171191	24.306670052014	24.306880268432	24.308189745893
g08	-0.095825	SR	-0.095825041163	-0.095825041163	-0.095825041163	-0.095825041163
		SR-LD	-0.095825041163	-0.095825041163	-0.095825041163	-0.095825041163
		SR-RD	-0.095825041163	-0.095825041163	-0.095825041163	-0.095825041163
g09	680.630	SR	680.630057374282	680.630057379555	680.630059198452	680.630075101729
		SR-LD	680.630057374282	680.630057374640	680.630057453157	680.630058100362
		SR-RD	680.630057374282	680.630057374376	680.630057405856	680.630057557687
g10	7049.248	SR	7049.248045134460	7050.221621230270	050.726890999790	7053.619632893530
		SR-LD	7049.248528325060	7049.309994193530	7049.659070972610	7050.8114676 0483
		SR-RD	7049.248035867630	7049.261097251370	7049.367593354660	7049.955532128970
g11	0.75	SR	0.75	0.75	0.75	0.75
		SR-LD	0.75	0.75	0.75	0.75
		SR-RD	0.75	0.75	0.75	0.75
g12	-1.0000000	SR	-1.00000000	-1.00000000	-1.00000000	-1.00000000
		SR-LD	-1.00000000	-1.00000000	-1.00000000	-1.00000000
		SR-RD	-1.00000000	-1.00000000	-1.00000000	-1.00000000
g13	0.053950	SR	0.053949847765	0.053949847765	0.130930122194	0.438851219909
		SR-LD	0.053949847765	0.053949847765	0.130930122194	0.438851219909
		SR-RD	0.053949847765	0.053949847765	0.053949847765	0.053949847765

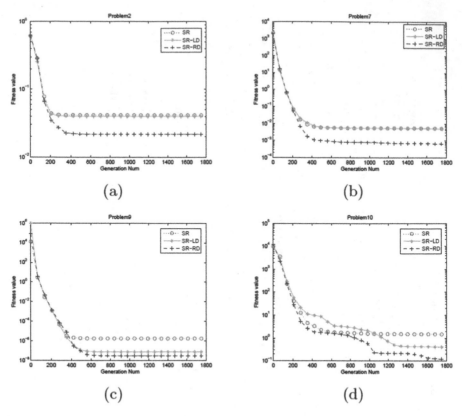

Fig. 1. Convergence curves of average objective errors in 30 runs of each algorithm for problems g02 (a), g07 (b), g09 (c) and g10 (d)

classic SR. For problem g10, the convergence speeds of the three algorithms are similar during the first 200 generations while the SR-LD and SR-RD perform better than the SR in terms of convergence speed as well as solution accuracy at the terminal stage of evolution, as shown in Fig. 1(d). In other words, two adaptive SR algorithms could explore much more extensive areas at the beginning stage of evolution and could enhance local search capacity to improve solution quality at the terminal stage.

4   Conclusion

On the basis of the classic SR algorithm, this paper proposes an adaptive linear decline SR and an adaptive reciprocal decline SR to enhance the performance at various evolutionary stages by decreasing the value of parameter Pf. They are able to maintain population diversity at the early stage of evolution and improve local search capacity and solution quality at the later stage. Experimental results on 13 benchmark problems show that SR-LD and SR-RD are capable of converging to highly accurate solutions and jumping out of local optima by adaptively reducing the value of Pf when solving

COPs. Meanwhile, SR-RD is the most competitive among all the three algorithms. Further research focuses on extending the adaptive SR schemes to solve constraint multi-objective optimization problems.

Acknowledgments. This work was supported in part by the National Natural Science Foundation of China (Nos. 61203310 and 61503087), the Natural Science Foundation of Guangdong Province, China (No. 2015A030313204), the China Scholarship Council (CSC) (Nos. 201406155076 and 201408440193), the Pearl River Science & Technology Nova Program of Guangzhou (No. 2014J2200052), and the Fundamental Research Funds for the Central Universities, SCUT (No. 2013ZZ0048).

References

1. Mezura-Montes, E., Velzquez-Reyes, J., Coello, C.A.C.: Modified differential evolution for constrained optimization. In: Proceedings of IEEE Congress on Evolutionary Computation, pp. 25–32 (2006)
2. Hamza, N.M., Essam, D.L., Sarker, R.A.: Differential evolution with a constraint consensus mutation for solving optimization problems. In: Proceedings of IEEE Congress on Evolutionary Computation, pp. 991–997. IEEE (2014)
3. Takahama, T., Sakai, S.: Constrained optimization by the epsilon constrained differential evolution with an archive and gradient-based mutation. In: Proceedings of IEEE Congress on Evolutionary Computation, pp. 1–9. IEEE (2010)
4. Elsayed, S.M., Sarker, R.A., Essam, D.L.: Multi-operator based evolutionary algorithms for solving constrained optimization problems. Comput. Oper. Res. **38**(12), 1877–1896 (2011)
5. Venkatraman, S., Yen, G.G.: A genetic framework for constrained optimization using genetic algorithms. IEEE Trans. Evol. Comput. **9**(4), 424–435 (2005)
6. Mezura-Montes, E., Coello, C.A.C.: Constraint-handling in nature-inspired numerical optimization: Past, present and future. Swarm Evol. Comput. **1**(4), 173–194 (2011)
7. Cai, Z., Wang, Y.: A multiobjective optimization-based evolutionary algorithm for constrained optimization. IEEE Trans. Evol. Comput. **10**(6), 658–675 (2006)
8. Wang, Y., Cai, Z.: Combining multiobjective optimization with differential evolution to solve constrained optimization problems. IEEE Trans. Evol. Comput. **16**(1), 117–134 (2012)
9. Runarsson, T.P., Yao, X.: Stochastic ranking for constrained evolutionary optimization. IEEE Trans. Evol. Comput. **4**(3), 284–294 (2000)
10. Runarsson, T.P., Yao, X.: Search biases in constrained evolutionary optimization. IEEE Trans. Syst. Man Cybern. Part C **35**(2), 233–243 (2005)

A Case-Based Reasoning Method with Relative Entropy and TOPSIS Integration

Jian Hu[(⊠)] and Jinhua Sun

School of Management, Chongqing University of Technology,
Chongqing 400054, China
jianhu-hit@163.com, sjhl009@163.com

Abstract. This paper addresses a new method of case-based reasoning (CBR). The aim of this work presented here is to provide effective warning knowledge for decision-makers. At first we design the similarity calculation methods according to the different case feature such as crisp number, interval number, crisp symbols and fuzzy linguistic variables. The similarity of each feature is calculated between target case and each historical case which step gets a similarity matrix. Then the CBR system employs a new ensemble measure for similarity matrix with two methods including relative entropy and the technique for order preference by similarity to an ideal solution (TOPSIS). On the basis, a new algorithm is designed, which is named as RTCBR. At the same time, RTCBR is tested on UCI data sets and compared with other two well-known CBR algorithms such as Euclidean distance CBR (ECBR) and Manhuttan distance CBR (MCBR). Empirical results indicate that RTCBR outperforms ECBR, MCBR, which can effectively improve the accuracy of CBR system.

Keywords: Case-Based reasoning · Relative entropy · TOPSIS · Integration

1 Introduction

Case-based reasoning method was first proposed by Professor Schank Roger of Yale University in 1982. It is an important problem solving and learning method based knowledge in the field of artificial intelligence. It solves the problem by reusing or modifying previous solutions to similar problems.

Case-based reasoning process consists of four main steps, which are similar case retrieval, case reuse, case revision and adjustment and case study. Among them, case retrieval based on similarity measure is the core of case reasoning method. The traditional similarity measure method uses the distance between the cases to get the most similar case. But there are two shortcomings in the traditional method. A disadvantage is the need to subjectively determine the weights of case features. Another weakness is that the data of cases can't be fully utilized.

In order to solve above problems, a new ensemble measure for similarity matrix is put forwarded with two methods integration, including relative entropy and the technique for order preference by similarity to an ideal solution (TOPSIS). On the basis, an algorithm, named as RTCBR is designed.

© Springer International Publishing AG 2017
F. Xhafa et al. (eds.), *Recent Developments in Intelligent Systems and Interactive Applications*,
Advances in Intelligent Systems and Computing 541, DOI 10.1007/978-3-319-49568-2_15

2 Related Work

There are always some arguments on limitations of traditional similarity measurement method of CBR, which can't improve the accuracy and efficiency of case reasoning. Thus, there are different multiple features ranking tools in the decision making literature including TOPSIS, ELECTRE, etc. TOPSIS is one of the most effective multiple criteria ranking methodologies, which stands for 'technique of order preference similarity to the ideal solution' which has been used widely for ranking purposes [1, 2].

In this research field some advanced techniques are integrated in CBR system. Many scholars began to attempts to use multi-CBR system to increase the performance of case retrieve than simply used CBR models. H. Li put forwarded a Multi-CBR–MV algorithm which is integrated majority voting method in case-based reasoning system and applied for financial distress prediction [3]. H. Li developed a new Multi-CBR-SVM model which is a Multiple CBR system with SVM method. In above model, k-nearest neighbor (KNN) algorithm is used to combine the classification result and SVM is utilized as the algorithm fulfilling combining-classifiers [4]. H. Jo put forward new structured model with multiple stages to apply for bankruptcy prediction. The integrated model consist of four phases (training, test, adjustment, and prediction), and three types of input data (training, testing, and generalization) [5]. Z.P. Fan gave a hybrid similarity measurement method to solve the similar measures with symbol attribute, interval attribute, numerical attribute and fuzzy attribute [6].

There are always some arguments on limitations of various predictive methods because of lacking of reliable basis, which can't provide more early warning knowledge for decision makers. Thus, there are different multiple criteria decision making tools in the decision making literature integrating with CBR, including TOPSIS, ELECTRE, TOPSIS. P. Chanvarasuth et al. proposed the integration method, which can utilize outranking relationship in the ELECTRE III with CBR. The new model can increase the ranking performance than the CBR model derived from Euclidean metric [7]. H. Li et al. used Electre-CBR-I and Electre-CBR-II two methods to construct a new CBR model by combining principles of the Electre decision-aiding, which could improve the accuracy and efficiency of case based reasoning [8]. H. Li et al. combined the technique for order performance by the similarity to ideal solution (TOPSIS) in CBR, which was a new type of multiple criteria CBR method with similarities to positive and negative ideal cases and was used in binary business failure prediction [9]. H. Malekpoora et al. proposed a novel case-based reasoning algorithm combining TOPSIS method, which was used to capture the past experience and expertise of oncologists [10].

3 CBR Framwork with Relative Entropy and TOPSIS

In CBR system, the historical case base is represented as $CB = \{c_i\}$ $(i = 1,2,...,M)$. Each case can be described through a set of features.

V is the feature number of case. The feature set can be expressed as $V = \{v_j\}$ $(j = 1,2,...,N)$.

Thus, the ith case c_i, can be represented as a vector $c_i = \{v_{i1}, v_{i2}, \ldots, v_{iN}\}$, which is N-dimensional. Let c_0 as the target case, which is also be represented as $c_0 = \{v_{01}, v_{02}, \ldots, v_{0N}\}$.

3.1 Similarity Measure

The degree of matching between two cases is calculated by similarity measures. We design four types of similarity measure of feature in oil price fluctuation case, such as crisp number, interval number, crisp symbol and fuzzy linguistic variable.

(1) Crisp Number

v_{0j} expresses the jth numeric feature value in target case and v_{ij} expresses jth feature value of the ith historical case. The distance between pairwise cases can be calculated as follows.

$$Sim_1(v_{0j}, v_{ij}) = [1 - d(v_{0j}, v_{ij})] \tag{1}$$

Where,

$$d(v_{0j}, v_{ij}) = \frac{|v_{ij} - v_{0j}|}{\max v_{ij} - \min v_{ij}} i \in M.$$

(2) Interval Number

$v_{0j} = [v_{0j}^L, v_{0j}^U]$ is named as an interval number in target case, which represents the jth feature value. $v_{ij} = [v_{ij}^L, v_{ij}^U]$ is jth feature value of the ith historical case. Then, the similarity measure between historical case and target case is calculated as follows.

$$Sim_2(v_{0j}, v_{ij}) = \exp[-d(v_{0j}, v_{ij})] \tag{2}$$

Where,

$$d(v_{0j}, v_{ij}) = \frac{\sqrt{(v_{ij}^L - v_{0j}^L)^2 + (v_{ij}^U - v_{0j}^U)^2}}{\max\left\{\sqrt{(v_{ij}^L - v_{0j}^L)^2 + (v_{ij}^U - v_{0j}^U)^2} \mid i \in M\right\}}.$$

(3) Crisp symbol

v_{0j} and v_{ij} are crisp symbol feature in target case and historical case. Then, a similarity measure between pairwise cases is calculated as follows.

$$Sim_3(v_{0j}, v_{ij}) = \begin{cases} 1 & if \quad v_{0j} = v_{ij} \\ 0 & if \quad v_{0j} \neq v_{ij} \end{cases} \tag{3}$$

3.2   A New Ensemble Measure

A similarity matrix S can be got by above similarity measures for each feature between target and each historical case. $S = (s_{ij})_{M \times N}$. The core of CBR is how to integrate the similarity in each feature between pairwise cases. A new ensemble measure is put forward with relative entropy and TOPSIS method integration.

Definition 1 Relative Entropy. c_i is the ith case, $c_i = \{v_{i1}, v_{i2}, \ldots, v_{iN}\}$. c_j is the jth case, $c_j = \{v_{j1}, v_{j2}, \ldots, v_{jN}\}$. Therein, $0 \leq v_{ik} \leq 1$, $0 \leq v_{jk} \leq 1$, $1 \leq j \leq N$. Then, the degree of deviation Can be expressed as Eq. (4).

$$D(c_i, c_j) = \sum_{k=1}^{N} [v_{ik} \log \frac{v_{ik}}{v_{jk}} + (1 - v_{ik}) \log \frac{1 - v_{ik}}{1 - v_{jk}}] \tag{4}$$

Relative entropy can represent the deviation, but not measure the similarity between pairwise cases. In order to make full use of the data information similarity matrix $S_{M \times N}$, relative entropy is integrated into the TOPSIS method. On the basis, a new ensemble measure on similarity matrix $S_{M \times N}$ is put forward.

Throgh TOPSIS method, the positive ideal solution S^+ and negative ideal S^- solution can be got from similarity matrix $S = (s_{ij})_{M \times N}$

$$S^+ = \{s_1^+, s_2^+, \cdots, s_N^+\} = \{(\max v_{ij} | j \in I), (\min v_{ij} | j \in J)\} \tag{5}$$

$$S^- = \{s_1^-, s_2^-, \cdots, s_N^-\} = \{(\min v_{ij} | j \in I), (\max v_{ij} | j \in J)\} \tag{6}$$

Wherein, I represents positive index and J represents negative index.

Definition 2 Relative Entropy Ideal Solution. ES^+ and ES^- are called relative entropy ideal solution, which is calculated by relative entropy between each element s_k of similarity matrix S and the positive ideal solution S^+ and negative ideal S^-.

$$ES^+ = \sum_{k=1}^{N} [s_k^+ \log \frac{s_k^+}{s_k} + (1 - s_k^+) \log \frac{1 - s_k^+}{1 - s_k}] \tag{7}$$

$$ES^- = \sum_{k=1}^{N} [s_k^- \log \frac{s_k^-}{s_k} + (1 - s_k^-) \log \frac{1 - s_k^-}{1 - s_k}] \tag{8}$$

Then, a new similarity measure is S^* constructed by relative entropy ideal solution ES + and ES-.

$$S_i^* = \frac{ES_i^-}{ES_i^+ + ES_i^-} \quad (i = 1, 2, \cdots, M) \tag{9}$$

The ensemble results of similarity matrix S can be got. According to S_i^* ranking result, we can select the most similar historical case.

4   RTCBR Algorithm

RTCBR algorithm procedure is shown as follow.

RTCBR algorithm
Input: target case c_0; historical case set $c_{i.}$
Output: case ranking result and the most similar case.
  Begin
      For each feature subspace v_j
   Calculating the similarity between target case and each historical case.
   Getting similarity matrix S.
   For Similarity matrix S
   Calculating S^+ and S^-.
Repeat
      Calculate the ES^+ , ES^- and S^* for each element of similarity matrix S.
   Until read each element of S.

Getting case ranking results and the most similar case.

5   Experiment Study and Analysis

5.1   Experiment Design

A five-fold cross validation method is used to test RTCBR algorithm. The testing data sets are from UCI machine learning database, which show as Table 1. Each data set is randomly and averagely divided into five parts. One fold data is selected as target case set and other fold data was historical case set. So we need to do five experiments on each data set. The experimental computer environment is Intel Core i5-4200, 8G memory. RTCBR algorithm use Matlab 7.0 to achieve.

Table 1. Testing data set

Data set	Feature		
	Number of feature	Record number	Class number
hepatitis	20	155	2
post-operative	8	90	3
adult	14	16281	2
car	6	1728	4

5.2   Result Analysis

In order to verify the accuracy of RTCBR algorithm, we select four data sets as test data, which is shown in Table 1. For the missing data in data sets, such as hepatitis and post-operative, We use the average value to fill the missing data. If the class feature of the target case and the most similar case is consistent, we consider that the result of case retrieval is correct. Otherwise, the case retrieval result is wrong. Then, five-fold cross validation is adopted. Each data set is randomly and averagely divided into five parts. One fold data is selected as target case set and other fold data was historical case set. So we need to do five experiments on each data set. The experimental results are shown in Table 2. The average accuracies on the four data sets are 83.23 %, 84.64 %, 89.70 % and 90.95 %. The accuracies of RTCBR algorithm for hepatitis and post-operative data sets are lower than adult and car data sets due to missing data of hepatitis and post-operative data sets. So we can get conclusion that the accuracy rate of RTCBR algorithm is higher.

Table 2. Testing accuracy rate %

| Data set | Experiment | | | | | |
	First	Second	Third	Fourth	Fifth	Average
hepatitis	80.64	77.42	87.10	83.87	87.10	83.23
post-operative	83.33	88.89	77.78	88.89	83.33	84.64
adult	89.50	90.50	89.0	91.0	88.50	89.70
car	91.30	89.86	91.88	91.01	90.72	90.95

In order to verify the relative effect of the RTCBR algorithm, we select two well-known CBR algorithms, Euclidean distance CBR (ECBR) and Manhuttan distance CBR (MCBR), to compare the retrieving accuracy rate of three algorithms. The experimental results are shown as Fig. 1. We could find that RTCBR has achieved the highest warning accuracy, which is better than ECBR and MCBR. The main reason of

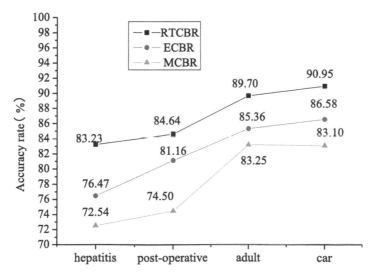

Fig. 1. Comparison experiment

this empirical result is that we design a new ensemble measure, which can increase the precision by using the each case data information.

6 Conclusion

To address the shortcomings of the traditional CBR method, a new ensemble measure of CBR is given in this paper. Relative entropy and relative entropy ideal solution are defined with the technique for order preference by similarity to an ideal solution (TOPSIS) to achieve above measure. Then, a new CBR algorithm, RTCBR, is designed. Through the real data from UCI testing show that RTCBR algorithm can solve weaknesses of the traditional CBR method, which accuracy is higher than the classical method, such as ECBR, MCBR. This paper adds up accuracy and efficiency from two aspects. First, we design a new ensemble measure, which can increase the precision by using the each case data information. Second, we put forward an ensemble measure which does not need to set up feature weights of case, which make full use of similarity matrix and avoid the influence of human factors.

In our future work we are going to further verify the validity of the algorithm in other data set.

Acknowledgment. This work is partially supported by the National Natural Science Foundation of China (Grant No. 71301181, Grant No. 71401021), by the Humanities and Social Science Project of Chongqing Municipal Education Commission (15SKG134), by the Science and Technology Project of Chongqing Municipal Education Commission (KJ1500911), and by National Statistical Science Research Project (2015 LY58).

References

1. Tavana, M., Li, Z., Mobin, M., Komaki, M., Teymourian, E.: Multi-objective control chart design optimization using NSGA-III and MOPSO enhanced with DEA and TOPSIS. Expert Syst. Appl. **50**, 1739 (2016)
2. Corrente, S., Grecoa, S., Słowińskib, R.: Multiple criteria hierarchy process for ELECTRE Tri methods. Eur. J. Oper. Res. **252**(1), 191–203 (2016)
3. Li, H., Sun, J.: Majority voting combination of multiple case-based reasoning for financial distress prediction. Expert Syst. Appl. **36**(3), 4363–4373 (2009)
4. Li, H., Sun, J.: Predicting business failure using multiple case-based reasoning combined with support vector machine. Expert Syst. Appl. **36**(6), 10085–10096 (2009)
5. Jo, H., Han, I.: Integration of case- based forecasting, neural network, and discriminant analysis for bankruptcy prediction. Expert Syst. Appl. **11**(4), 415–422 (1996)
6. Fan, Z.P., Li, Y.H., Wang, X.H., Liu, Y.: Hybrid similarity measure for case retrieval in CBR and its application to emergency response towards gas explosion. Expert Syst. Appl. **41**(5), 2526–2534 (2014)
7. Chanvarasuth, P., Boongasame, L.: Remove from marked Records Hybridizing principles of the ELECTRE III method with case-based reasoning for a travel advisory system: case study of Thailand. Asia Pacific J. Tourism Res. **20**(5), 585–598 (2015)

8. Li, H., Sun, J.: Hybridizing principles of the Electre method with case-based reasoning for data mining: Electre-CBR-I and Electre-CBR-II. Eur. J. Oper. Res. **197**(1), 214–224 (2009)
9. Li, H., Adelib, H., Sun, J., Han, J.G.: Hybridizing principles of TOPSIS with case-based reasoning for business failure prediction. Comput. Oper. Res. **38**(2), 409–419 (2011)
10. H. Malekpoora, Mishrab, N., Sumalyac, S., Kumarid, S.: An efficient approach to radiotherapy dose planning problem: a TOPSIS case-based reasoning approach. Int. J. Syst. Sci. Oper. Logistics (2016)

An Algorithm Design for Electrical Impedance Tomography Based on Levenberg Method

Mingyong Zhou[1(✉)] and Xin Lu[2]

[1] Guangxi University of Science and Technology, Liuzhou, China
Zed6641@hotmail.com
[2] Beihai College of Beihang University, Beihai 536000, China

Abstract. In this paper, a new method to construct an electrical impedance tomography algorithm based on Levenberg algorithm is proposed. First, Levenberg algorithm principle for nonlinear approximation is introduced and then the ideas to apply the Levenberg algorithm into our electrical impedance tomography outlined, and finally a detailed algorithm for electrical impedance tomography experiments is described. Initial simulations for the proposed algorithm are described and key issues are highlighted.

Keywords: Levenberg algorithm (L-algorithm) · Electrical impedance tomography (EIT)

1 Introduction

Levenberg-Marpuardt algorithm (L-algorithm) is a well- known iteration method to solve nonlinear inverse problems by transforming the nonlinear problems into linear ones first and then regularizing the problems [1–3]. As indicated in Refs. [1–3], the key point is to properly select the parameters in the L-algorithm for specific problems.

The most important issue in the L-algorithm is the estimation of the differential matrix that is used in the iteration steps as well as the avoidance of illness of the matrix related to the differential matrix.

In this paper, we first propose a design method to construct the differential matrix in a simplest way then we improve the algorithm in specific electrical impedance tomography issue and lastly, we apply the proposed algorithm into simulations.

2 Levenberg-Marpuardt Algorithm (L-Algorithm)

We outline the L-algorithm as follows, most parts are written in Refs. [1–3]. For nonlinear function F(x) = y as demonstrated in Fig. 1, one can transform it into linear problem as in Eq. (1).

$$F'(x_k^\delta)(x - x_k^\delta) = y^\delta - F(x_k^\delta) \tag{1}$$

© Springer International Publishing AG 2017
F. Xhafa et al. (eds.), *Recent Developments in Intelligent Systems and Interactive Applications*,
Advances in Intelligent Systems and Computing 541, DOI 10.1007/978-3-319-49568-2_16

$$\Phi_{\alpha_k}(x, x_n^\delta) = ||F'(x_n^\delta)(x - x_n^\delta) - (y^\delta - F(x_n^\delta))||^2 + \alpha_k||x - x_n^\delta||^2, \tag{2}$$

$$x_{k-1}^\delta = x_k^\delta + [F'(x_k^\delta)^* F'(x_k^\delta) + \alpha_k I]^{-1} F'(x_k^\delta)^* (y^\delta - F(x_k^\delta)), \tag{3}$$

Equation (2) is a general function proposed in Levenberg algorithm and Eq. (3) is Levenberg iteration steps starting with initial values. We intend to indicate that the key issues in the L-algorithm are the proper selection of parameters as well as the avoidance of the illness of the matrix that is related to the differential matrix constructed for specific problems.

3 Construct Differential Matrix in L-Algorithm

As indicated in Fig. 1, non-linear problem in Eq. (4) can be approximated by Eq. (5) via a linear transformation for nonlinear problem. Equation (6) is a matrix-vector representation for the transformed linear issue.

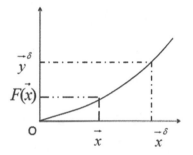

Fig. 1. Approximation of nonlinear problem

$$F(\vec{x}) = \vec{y} \tag{4}$$

$$F'(\vec{x}) \cdot (\vec{x} - \vec{x}^\delta) = \vec{y}^\delta - F(x) \tag{5}$$

$$F'(\vec{x})_{M \times N} \cdot \Delta\vec{\delta}_{N \times 1} = \vec{y}^\delta - F(\vec{x}) = \begin{pmatrix} \Delta y_1 \\ \Delta y_2 \\ \dots \\ \Delta y_M \end{pmatrix}_{M \times 1} \tag{6}$$

Without loss of generality, we assume the all delta to be equal to 1.0. Equation (7) then is the M x N differential matrix as defined in the L-algorithm.

$$F'_{M \times N} \triangleq \begin{pmatrix} \frac{1}{N}\Delta y_1 & \frac{1}{N}\Delta y_1 & \dots & \frac{1}{N}\Delta y_1 \\ \frac{1}{N}\Delta y_2 & \frac{1}{N}\Delta y_2 & \dots & \frac{1}{N}\Delta y_2 \\ \dots & \dots & \dots & \dots \\ \frac{1}{N}\Delta y_M & \frac{1}{N}\Delta y_M & \dots & \frac{1}{N}\Delta y_M \end{pmatrix}_{M \times N} \tag{7}$$

4   Construction of Improved Algorithm for Electrical Impedance Tomography

In the design of electrical impedance tomography, in Eq. (6) $\vec{x} \triangleq (\rho_0, \rho_1 \ldots \rho_{N-1})^T$ and $\Delta y_1, \Delta y_2 \ldots \Delta y_M$ are the surface measured voltages $v_o, v_1 \ldots v_{M-1}$, where positive integer number M is usually much larger than positive integer number N.

In electrical impedance tomography issue, the inner parts of the body has less impact on the measured voltages on the surface of the 2-D body see Fig. 2. In this paper, we proposed a weighted method to construct a differential matrix. On one hand, it can weigh the contribution of the inner impedance to the voltages measured on the 2-D surface. On the other hand, the weighing method can avoid the illness of the matrix calculation as required in L-algorithm. In particular, the revised differential matrix is constructed as follows.

$$F'_{M \times N} \triangleq \begin{pmatrix} \frac{w_1}{N} \Delta y_1 & \frac{w_2}{N} \Delta y_1 & \cdots & \frac{w_N}{N} \Delta y_1 \\ \frac{w_1}{N} \Delta y_2 & \frac{w_2}{N} \Delta y_2 & \cdots & \frac{w_N}{N} \Delta y_2 \\ \cdots & \cdots & \cdots & \cdots \\ \frac{w_1}{N} \Delta y_M & \frac{w_2}{N} \Delta y_M & \cdots & \frac{w_N}{N} \Delta y_M \end{pmatrix}_{M \times N} \tag{8}$$

Where, $w1 + w2 + \ldots + wN = N$, and N is the number of internal portioned impedance number.

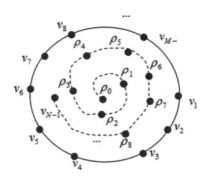

Fig. 2. 2-D inner impedance and surface measured voltages

5   Proposed Algorithm and Further Considerations

By adjusting the weights $w1, w2 \ldots wN$ in Eq. (8), the impact of the inner impedance on the measured voltages of 2-D body can be reflected. For example, we can increase $w1$ to "enlarge" the effect of the most inner impedance on the measured surface voltages. However, we can set $w1 < w2 < \ldots < wN$ in general and usual cases.

We now outline the proposed algorithm iterations that is based on L-algorithm:

1. We divide the 2-D body inner parts into N = 3, and we measure the M = 6 points of voltages v0, v1,...v5 on the surface.

2. According to w1 < w2 < w3 and w1 + w2 + w3 = 3, select the weight values w1, w2, w3.
3. According to Eq. (8) calculate the differential matrix F', the dimension is 6 × 3, and in Eq. (8) Δy1...Δy6 are the measured voltages.

Finally following the L-algorithm iterations [1], given the initial impedance values, the converged values of the 3 impedance values are the desired ones. The convergence is guaranteed by L-algorithm itself.

In fact, initial Matlab simulations indicated that through the weighting step, the illness of differential matrix F' can be reduced. Specifically

$$(F'^T \cdot F')^{-1}_{N \times N} \tag{9}$$

Equation (9) indicates that the matrix is N × N dimensions, and that the illness of differential matrix can be largely reduced by MATLAB simulations.

6 Conclusion

A new reconstruction method based on Levenberg algorithm (L-algorithm) is proposed in this paper. It is based on Levenberg method which solves inverse nonlinear issues. First we outlined the L-algorithm and then we revised the L-algorithm and showed how it can be applied in electrical impedance tomography issue. And lastly we proposed and discussed the proposed algorithm and related issues in real implementations. Initial simulations indicated that the proposed algorithm is feasible and potential issues were discussed.

References

1. Zhou, M.: Inverse problems in electrical engineering and several case study, pp. 25–26 (2014) (unpublished book)
2. Wang, Y.: Computation Methods and Applications of Inverse Problems. Higher Education Publishing Press, Beijing (2007) (in Chinese language)
3. Xiao, T., et al.: Numerical Solutions for Inverse Problems. Science Publishing Press, Beijing (2003) (in Chinese language)

A Return-Value-Unchecked Vulnerability Detection Method Based on Property Graph

Han Kun[1], Wu Bo[2(✉)], and Xin Dan[2]

[1] Xidian University, Xi'an, China
hankundsp@163.com
[2] Xi'an Communication Institute, Xi'an, China
cherry_wb@163.com, xindan625@126.com

Abstract. Traditional static analysis methods for binary software vulnerability detection are used only to make use of a single aspect of the target software, so it is difficult to obtain the hidden global properties and relationships which leads to low detection accuracy and high rate of false positives. To improve the effectiveness of the binary software static vulnerability detection, this paper proposes a fusion method for binary software vulnerability detection which first represents the binary software as a single property graph and then the vulnerability is modeled and detected based on this property graph. Because property graph includes integrated information such as the relations between function calls, control flow, data flow relationship and so on, researchers can model vulnerability more easily and accurately. It can detect unknown vulnerabilities accurately and effi-ciently. The experiments of prototype system show that this method can effectively detect Return-Value-Unchecked Vulnerability in binary software.

Keywords: Property graph · Vulnerability model · Vulnerability detection

1 Introduction

In earier practices people usually take into account economic losses caused by software vulnerabilities to emphasize the influence of software vulnerabilities. But since the appearance of "Stuxnet" and "Flame" it was found that software vulnerabilities are far from economic loss. Software vulnerability is the one of the most important strategic resources of network space competition. So all countries are actively developing techniques and methods to discover software vulnerabilities. Automatic vulnerability detection related theories and techniques have become important and herculean research pursuits in security community.

There are many methods to detect vulnerabilities, basically divided into two categories static or dynamic. In this paper we mainly focus on static analysis approach where a method of automatic derivation of program operation behavior originated from the optimization requirements of the early high-level language. The basic idea of program analysis is widely used in the field of software engineering, program safety analysis and so on. For example, buffer overflow detection, not initialized variable detection, and variable release, reuse detection and other typical security analysis

© Springer International Publishing AG 2017
F. Xhafa et al. (eds.), *Recent Developments in Intelligent Systems and Interactive Applications*,
Advances in Intelligent Systems and Computing 541, DOI 10.1007/978-3-319-49568-2_17

application scenarios. At present, there are a variety of static analysis methods and tools for detection software vulnerabilities, but these methods and tools are usually not modeling and detecting vulnerabilities [1] from multi perspective, leading to the low detection accuracy rate and the existence of a large number of false positives.

In order to comprehensively utilize of all kinds of static analysis methods, this paper proposes a new method for static detection of binary software vulnerabilities based on property graph. In this method the binary software is first disassembled, followed by intermediate language conversion and intermediate language based control flow data flow analysis, and all the analysis results expressed as a fusion of a property graph. Based on the property graph, vulnerabilities are modeled and detected. The experimental results of the prototype system show that the proposed method can effectively detect vulnerabilities in binary software.

The rest of this paper is organized as follows. In Sect. 2, we illustrate the framework of the vulnerability detection approach. In Sect. 3, how a binary software can be represented as a property graph is elaborated. In Sect. 4, how vulnerabilities can be modeled and detected is elaborated. The experimental results evaluated on real-world software of our prototype system have been described in Sect. 5. Finally Sect. 6 concludes this paper.

2 System Design

In a nutshell the process and algorithm of our vulnerability discovery method is shown in Fig. 1. Firstly, our approach processes the target binary with disassemble engines such as IDA, capstone which converts the binary code to assembly code, and then translates code into the SSA form of intermediate code. Secondly, in the intermediate representation form control flow analysis, data flow analysis, reach-constant routine analysis and other analysis are performed and then all the analysis results are converted to nodes and edges in the unified property graph. Finally based on the property graph representation, potential vulnerabilities are modeled and detected.

The proposed detection method uses binary software disassemble as the foundation, while the disassembly is a mature technology, there are a lot of multi platform supporting and multi instruction set supporting engines, therefore it will not be discussed here. Intermediate representation conversion method and intermediate representation

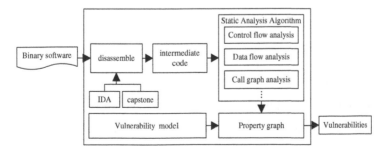

Fig. 1. Framework of detection method

based static analysis processes will be discussed in the third second part in this article. Vulnerability modeling and detection based on property graph will be discussed in the fourth part in this article. And the prototype system implementation and evaluation will be elaborated in the fifth part in this article.

3   Property Graph Representation for Binary Software

3.1   Intermediate Representation

Although we can analyze binary software directly on the assembly code, we usually do analysis after binary software is translated into intermediate representation. As far as we known, a large kinds of intermediate representation are widely used, the expression abilities of these languages are different according to their design purposes. For example, the vex language used in the Valgrind [4] binary instrument platform, the vine language used in the static analysis part of BitBlaze [2] platform, the BIL language used in BAP [3] binary platform, the LLVM IR used in the compiler framework, as well as REIL [5] and so on. These intermediate languages can be platform independent, can be converted from the assembly instructions through a direct instruction mapping, and can be used as the basis for the static analysis algorithm design. In this paper, we uses the REIL language as a start point and extends it to a universal intermediate representation language (UIRL) for all the analysis algorithms.

Just like REIL, each instruction of UIRL has no side effects which make it suitable for binary analysis. UIRL extends the REIL, the extension of the directive, including LSHL, RSHL, SEX, CALL, RET, PHI instructions, as shown in Table 1. Although original shift instruction can also express left and right shift operation, but it will bring inconvenience to the analysis and simplification, so the BSH compound semantic is separated to RSHL and LSHL instructions. The introduction of CALL and RET instructions is intended to simplify the internal process analysis. The introduction of PHI instruction is to facilitate the static single assignment analysis.

Table 1. Parts of intermediate representation instructions

Instruction	Operation Semantic	Description
LSHL	logical left shift	arithmetic instruction
RSHL	logical right shift	arithmetic instruction
SEX	signed expand	arithmetic instruction
CALL	function call	other instruction
RET	function return	other instruction
PHI	ϕ function function	other instruction

3.2   Static Analysis Based on Intermediate Representation

The static analysis algorithm is usually applied to draw some conclusion on the interested software behavior based on a set of constraints for specific application scenarios. But the Rice theorem [6] points out that for any non-trivial property of partial functions, there is no general and effective method to decide whether an algorithm computes a partial function with that property, as shown in Fig. 2. The value of the register variable eax after the 0×00400007 instruction is unknown. Therefore, the specific analysis can only be a certain degree of approximation of the real behavior of the software, there are usually three ways to deal with the specific:

1. **Conservative Analysis:** a conservative estimate of the possible value, then the range of eax values can be calibrated in the range of [0, 1], although the actual implementation results may only be 0. This approach appears to not miss a solution, and as far as possible to improve the accuracy of analysis.
2. **Accurate Analysis:** only gives a correct conclusion, then the calibration value range of eax is empty. Obviously this way assure to be accurate, but there will be a large number of omissions.
3. **Compromise Analysis:** make a trade-off between conservative estimation and accurate analysis based on the specific circumstances.

```
0x00400000 add [b32 esp, b32 0x4, b32 t1]
0x00400001 ldm [b32 t1, null, b32 t2]
0x00400002 and [b32 eax, b32 0x0, b32 t3]
0x00400003 str [b32 t3, b32 0x0, b32 eax]
0x00400004 bisz [b32 t2, null, b8 zf]
0x00400005 jcc [b8 zf, null, b32 0x00400007]
0x00400006 str [b32 0x1, b32 0x0, b32 eax]
0x00400007 nop [null, null, null]//eax=?
```

Fig. 2. Sample code for Rice theorem

The static analysis methods mainly used in this paper includes: control flow analysis, data flow analysis, function call analysis and so on.

The traditional control flow analysis is mainly used to collect the execution path of the program and to extract the control structure of the loops and branches, and make the relevant optimization analysis on the basis of those information. Through extensive analysis of the existing vulnerabilities, there is a certain correlation between the control flow structure and some attributes of the vulnerabilities, some of vulnerabilities can even be directly described by the control flow diagram.

Data flow analysis does not execute the target program, it just investigates the data transfer relationship directly. Usually interested information includes variable range, reachable variable definition, and available expressions and so on. In the compilation field the compiler performs the data flow analysis to remove dead code, allocate registers, optimize constant assignment and eliminate redundant code. Generally speaking, data flow analysis is to construct complete lattice according to specific analysis purpose and then program is transformed to monotone transfer function on the lattice to establish data flow equation. The program's properties can be inferred from the solution of those equations. For vulnerability analysis, data flow analysis considers the vulnerability attributes as the goal and these vulnerability attributes is obtained by solving the data flow equations, and then detects the potential vulnerabilities. In practice, data flow analysis and control flow analysis methods are usually used at the same time to improve the accuracy of the analysis.

Function call analysis is mainly used to collect dependencies between functions, and then inter-procedural control flow and data flow analysis are performed on the function call chains which meet some specific conditions to obtain the more accurate analysis results.

3.3 Property Graph Generation

Traditional static analysis methods generally only make use of the results from different static analysis algorithms in isolation and do not utilize the relationship between different algorithm's analysis results, which may miss a lot of global information to aid vulnerability detection. From this point of view, this paper will convert all the analysis results to a unified property graph. In the property graph nodes represent a concrete entity such as function, basic block, instruction and circulation. Edges represent the relationship between different entities such as function call, the domination relation, and the data flow relation and the properties of a node or edge are used to store additional information such as code address, instruction type, correlation sequence and so on.

As it can be observed that property graph is a kind of expression method with strong compatibility. As long as the analysis results can be expressed in the form of a normal graph or table, the result can be directly converted to the property graph. As a result, the process of property graph generation is actually a simple data transformation process, the nodes and edges in control flow diagram, data flow diagram, call flow graph and other diagrams are converted to nodes and edges with properties in the unified final property graph.

4 Vulnerability Model and Detect Based on Property Graph

Thorough understanding of the causes of specific types of vulnerabilities is the start point of vulnerability detection. Therefore, it is necessary to use accurate vulnerability model to guide the process of vulnerability detection. At present, due to binary software does not have a unified representation method, binary software vulnerability description basically are accomplished through natural language such as different vulnerability type descriptions in CVE. Although this can assist vulnerability classification but cannot meet the demand of automatic detection of vulnerabilities. In this paper, firstly, the binary is represented as a unified property graph, and then based on this property graph vulnerabilities are modeled and detected. By doing this the vulnerability characteristics can be described accurately and vulnerabilities can be detected directly using pattern matching and machine learning method. In this section, we introduce this new method for vulnerability modeling and detection.

4.1   Vulnerability Model

The vulnerabilities exist in software bring a critical threat to network space security. After analyzing typical vulnerabilities in CVE [7] and NVD [8] databases, we found out that the basic reason for software vulnerabilities is inadequate security check, such as not checking parameters of dangerous functions like strcpy or strcat carefully, and not verifying return value of a function correctly, which would lead to many memory-corruption vulnerabilities [9]. Among all these different kinds of vulnerabilities, there are some kinds of vulnerabilities which can be classified by some obvious characteristics, this paper will focus on a kind of return-value-unchecked vulnerability. The sample vulnerability code is shown in Fig. 3. For different parameters function sub_401000 returns the successfully allocated memory pointer or null pointer, and in the main function, it gets the return pointer by calling the sub_401000 function, and then assigns value to the return pointer. However, sub 401000 function return value may be a null pointer. Therefore, if the pointer is dereference directly before validated, it may produce null pointer dereference errors, and cause software crash.

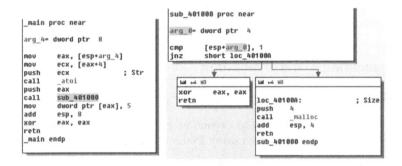

Fig. 3. Sample code for return-value-unchecked vulnerability

As we can see the return-value-unchecked vulnerability can cause many kinds of memory errors, the most common is null pointer dereference, and it is difficult to detect all types of those errors at binary level. Therefore, this paper focuses on those with obvious binary features. Normal pointer dereference must be used in memory, the feature of the assembly code is that the instruction contains memory operations, that is, there exists a data access pointer. After analyzing some typical vulnerabilities, the characteristics of the null pointer dereference caused by return-value-unchecked are summarized in nature language as follows:

1. Data access in the form of pointer, and the legitimacy of the pointer variable is not verified;
2. The pointer variable can take a variety of values, which may contain null value.

So in the property graph, this kind of vulnerability can be modeled as a query step:

1. Use the constant {ConstVar} as the start point, and search instruction node set {Instr} where {ConstVar} data flow to;

2. With whether the instruction access memory as the conditional, {Instr} is filtered out, and the memory address variable is extracted as {Var};
3. Search the direct dominant basic block set {BB} for basic blocks which instruction set {Instr} belong to, and extract the jump condition variable as {CVar};
4. Search variable in {Var} which has not data flow relation to {CVar} as {RVar}.

According to the characteristics of the vulnerability model, it can be inferred that there exists a null pointer dereference bug if the set {RVar} is not null.

4.2 Vulnerability Detection

Based on the property graph representation for binary software, the vulnerability detection process is actually a query process of the property graph. For example, to detect the presence of null pointer dereference bug, we only need to query the property graph according to the vulnerability modeling steps. If the query results is not empty, the null pointer dereference bug is detected. At the same time, the query can output vulnerabilities related variables, instructions, basic blocks and function, in order to analyze manually for further confirmation or to support dynamic analysis methods with further testing.

5   Implementation and Evaluation

5.1   Implementation

In order to construct a simple, easy to expand and fast algorithm design environment, the prototype system is implemented using Python dynamic programming language.

Disassembly Module: Disassembly is a very mature technology, this paper's disassembly module is based on IDA Pro [10] and capstone [11]. IDA Pro is the best decompilation tool, and the accuracy of recovered control flow structure is much higher than other tools. But in our implementation, the static analysis algorithm is based on the intermediate instructions, which need to decode the assembly instructions accurately, and then translate the instructions into the middle, and the IDA Pro cannot meet out design requirements. Therefore, we use IDA Pro to restore the control flow structure of the target software, the decoding of the instruction is completed by capstone engine.

Intermediate Representation Conversion Module: After the instruction decoding process, the UIRL translation is only a simple mapping process. First, for each assembly instruction a mapping function is designed, and the capstone engine decode binary to assembly instructions, and according to instruction type it is distributed to the mapping function to complete the instruction translation. Currently only 142 types of $\times$ 86 instruction are supported. The translation process of a function works as: first control flow graph is recovered, then sequentially traverse through the flow graph of basic blocks, and does the instruction translation, finally address assignment and code optimization is performed, etc. After basic conversion, in order to simplify the analysis algorithm, all code is converted into static single assignment form.

Static Program Analysis Module: This module includes many kinds of static analysis algorithms. The control flow analysis algorithm provides a consistent interface for any control flow graph, including transforming control flow graph into normal and directed graph, computing dominance relation, extracting the dominant tree, extracting the control boundary, extracting the edge, extracting the loop structure and so on. Data flow analysis is generally an interpretation process, it first completes the analysis of the basic block, and then complete the data flow between the basic blocks. As a result, we implemented a unified framework for data flow analysis, which noted as interpretation execution engine. The Interpretation engine is a symbolic execution module in nutshell, all of the UIRL instructions are interpreted as symbolic calculation process, the symbolic representation relationship between variables can be obtained at any time. Function call analysis algorithm extracts the function call relationship through the analysis of cross reference between the functions.

Property Graph Representation Module: In order to improve the efficiency of the vulnerability analysis process, the prototype system uses the graph database OrientDB [12] as the backend which is widely used to implement the storage and query operation for property graph.

Vulnerability Modeling and Detection Module: As described in section fourth, we use query steps in the graph database to model and detect vulnerabilities, so it is implemented as query scripts in the prototype system.

5.2   Evaluation

In order to evaluate the effectiveness of the prototype system, we select a vulnerable software and a real software to test the performance of the system. The major code of the vulnerable software is shown in Fig. 4. There is a call to a possible return to the null value function get_ptr in the code, but the caller does not have the correct verification for the return value. Real software selection is Win32k.sys, due to the return value is not full checked there exists an elevation of privilege vulnerability in the driver, the vulnerability ID is CVE-2014-4113.

```
int* get_ptr(int type){
    if (type == 1){    return NULL;
    }else{    return (int*)malloc(sizeof(int));
    }}
void handler_nullderef(const char **args){
    const char *arg = args[0];
    int *ptr;
    ptr = get_ptr(atoi(arg));
    *ptr = 5;
    return;
}
```

Fig. 4. Source code of vulnerable app

Table 2. Evaluation result

Target Soft	Version	File Size(KB)	Detected
VulnApp	1.1.0	208	Yes
win32 k.sys	6.1.7601.17514	2214	Yes

The evaluation results as shown in Table 2, the vulnerable software and the real driver software vulnerabilities are all accurately detected.

6 Summary

Based on the property graph representation of binary software, it can be more accurate and convenient to establish vulnerability models, and complex vulnerabilities can be detected by a simple property graph query operations. This paper proposed and realized a unified representation approach for binary software based on the property graph, which can represent all aspects of a target binary software in to a single graph. And our vulnerability detection method uses this representation as the foundation to model and detect vulnerabilities in binary software. And as a analysis platform, we cannot only detect vulnerabilities by queries, but also can combine machine learning algorithm of anomaly detection, clustering analysis algorithm to realize the intelligent vulnerability detection, which has been successfully applied for source code vulnerability detection [13]. Of course, vulnerability detection under product environment needs comprehensive utilization of static and dynamic detection methods, detection results of the method proposed in this paper can reduce the ranges where need to be tested by dynamic methods to improve the effectiveness of the dynamic methods.

References

1. Delaitre, A., Stivalet, B., Fong, E., et al.: Evaluating bug finders–test and measurement of static code analyzers. In: ACM 1st International Workshop on Complex Faults and Failures in Large Software Systems, pp. 14–20. IEEE (2015)
2. Song, D., Brumley, D., Yin, H., Caballero, J., Jager, I., Kang, M.G., Liang, Z., Newsome, J., Poosankam, P., Saxena, P.: BitBlaze: A new approach to computer security via binary analysis. In: Sekar, R., Pujari, Arun, K. (eds.) ICISS 2008. LNCS, vol. 5352, pp. 1–25. Springer, Heidelberg (2008). doi:10.1007/978-3-540-89862-7_1
3. Brumley, D., Jager, I., Avgerinos, T., Schwartz, Edward, J.: BAP: A binary analysis platform. In: Gopalakrishnan, G., Qadeer, S. (eds.) CAV 2011. LNCS, vol. 6806, pp. 463–469. Springer, Heidelberg (2011). doi:10.1007/978-3-642-22110-1_37
4. Nethercote, N., Seward, J.: Valgrind: a framework for heavyweight dynamic binary instrumentation. ACM SIGPLAN Not. **42**(6), 89–100 (2007)
5. Dullien, T., Porst, S.: REIL: A platform-independent intermediate representation of disassembled code for static code analysis. In: Proceeding of CanSecWest (2009)
6. Rice, H.G.: Classes of recursively enumerable sets and their decision problems. Trans. Am. Math. Soc. **74**(2), 358–366 (1953)
7. CVE. http://cve.mitre.org/

8. NVD. http://nvd.nist.gov/
9. Zhang, B., Wu, B., Feng, C., et al.: Statically detect invalid pointer dereference vulnerabilities in binary software. In: 2015 IEEE International Conference on Progress in Informatics and Computing (PIC), pp. 390–394. IEEE (2015)
10. IDA pro. https://www.hex-rays.com
11. Quynh, N.A.: Capstone: Next-gen disassembly framework. Black Hat USA (2014)
12. Tesoriero, C.: Getting Started with OrientDB. Packt Publishing Ltd., Birmingham (2013)
13. Yamaguchi, F.: Pattern-Based Vulnerability Discovery. Ph.D. thesis, Georg-August-University Göttingen (2015)

GA Based Optimization of Sign Regressor FLANN Model for Channel Equalization

Jagyaseni Sahoo[⊠], Laxmi Mishra, and Mihir Narayan Mohanty

ITER, Siksha 'O' Anusandhan University,
Jagamara, Bhubaneswar, Odisha, India
jagyanseni_sahoo@yahoo.in,
laxmimishra@soauniversity.ac.in,
mihir.n.mohanty@gmail.com

Abstract. The recent time has witnessed more reliable and higher data rate transmission standard wireless communication. The wireless channel is to be optimized accordingly. The ISI component at the output of equalizer is forced to zero on use of LTI system having appropriate transfer function. The deteriorating effect of inter symbol interference ia adequately compensated by the method of adaptive equalization. In this paper, the channel has been equalized using Sign Regressor Functional Link Artificial Neural Network model. QAM modulation technique is utilized in this piece of work. Further the weights of the model are optimized using Genetic Algorithm. The result of sign regressor adaptive algorithms have been compared and sign regressor FLANN shows better performance than other algorithm. Finally the optimized result of sign regressor FLANN model is exhibited for QAM in terms of error. Also, the eye pattern is shown f or the result as an evidence.

Keywords: Equalization · Adaptive equalization · Functional link neural network

1 Introduction

The reliability and quantity of service of modes in a wireless communication depends on the degree of minimization of channel distortion. These channel exhibits time varying characteristics and hence are represented by time dependent transfer function. The distortion caused by heterogeneous end users and the multi path users and the multipath interferences are the major bottlereck for high rate data transmission in wireless. LTF based channel equalizer provides a way to compensate the less incurred due to channel distortion [1]. However, it is efficient only in case of a linear distortion but does not provide a robust solution for the loss due to non-linear channel distortion. MSE based intention has been proposed in [2] which provides a better solution.

The design of channel equalizer based on neural network structure with adaptive algorithm have been found in literature [3–8], where the claim was MSE based criterion of channel equalizer performs better than the LMS based algorithm. The standard techniques like MMSE, DFE based algorithm were used by authors and also applied in

F. Xhafa et al. (eds.), *Recent Developments in Intelligent Systems and Interactive Applications*,
Advances in Intelligent Systems and Computing 541, DOI 10.1007/978-3-319-49568-2_18

MIMO based communication system [3, 4]. Similarly, neural network has been used for linear and non-linear channel equalization [5]. The modification of neural network model with different functions was the focus of researchers for the robustness. These models were named as FLANN model and is used in many cases [6–9].

In this work, the robust FLANN model is considered as the base model for equalization purpose. Further the optimization of the co-efficient using genetic algorithm provides better result. The following section explained the proposed model. For different digital modulation it has been tested. The efficient QAM technique is also applied.

The paper is organized as follows: Sect. 2 outlines the channel Equalization model and establishes the algorithms under consideration to determine the optimum filter coefficients with respect to a minimum mean square (MSE). Section 2.1 describes the Algorithm for Channel Equalization and the outlines of sign regressor FLANN Sect. 2.2 outlines the optimization using Genetic Algorithm. Section 3 discusses the simulation results equalization of sign regressor FLANN and the optimization and Sect. 4 concludes the work.

2   Channel Equalization Model

When the signal is passed through the channel, distortion is introduced in terms of amplitude and delay, which results with Inter Symbol Interference (ISI). ISI caused by multipath in band limited time dispersive channel distorts the transmitted data, causing bit errors at the receiver. ISI has been recognized as the major drawback in high speed data transmission over wireless channels. Hence, Equalizers combat ISI. An equalizer is implemented at the baseband or at IF in a receiver. And the basic receiving technique of any communication lies with noise signal performance [7].

The purpose of equalizer is to reduce ISI in the channel. Also the signal is as clear as possible. The basic block diagram of the communication system with the equalizer is shown in Fig. 1, where the signal for transmission is the noise represents, and the output represents), and the equalizer output is. The use of equalizer in the system can be against ISI. Hence it can be placed at the baseband level or at the intermediate frequency within the receiver. Keeping view of these facts initially the FLANN model is used and next to it further processing is proposed. In the following subsections these are explained.

2.1   FLANN Model

In this model the dimension of the input baseband signal is explained by introducing the linear functions. The neural network is modified with these additive functions. For improvement of the existing structure digital FIR filter is used within it. Due to this filter the performance of the equalizer has been enhanced and also reported in [10]. As a result, the steady state error has been reduced as well as the bit error rate was improved.

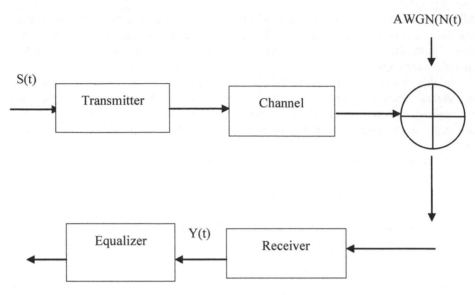

Fig. 1. Basic block diagram of channel equalization

Replacing the FIR filter with the filter bank, the FLANN model is changed. The model is used in this case is sign-regressor FLANN model with filter bank and is shown in Fig. 2. The weight of the model is represented as

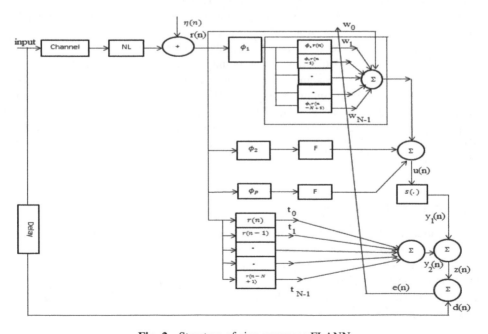

Fig. 2. Structure of sign regressor FLANN

$$w_{i+1} = w_i + \mu_1 \gamma(n)(e(n)y_1(n)) \tag{1}$$

Where

w_i = the initial weight

$e(n)$ = error

$\gamma(n)$ = change in adjacent amplitude and its value lies in between 0 to 1.

$\mu_1 \mu_2$ = learning rate parameters.

For 2P number of functions, the weight will be N(2P + 1).

The filter bank is used in the mentioned sign regressor FLANN model [10].

The output is represented as

$$z(n) = y_1(n) + y_2(n) \tag{2}$$

$$z(n) = w_0 r(n) + \sum_{i=1}^{N-1} w_t \phi_x[r(n-i)] + \sum_{i=0}^{N-1} t_i r(n-i) \tag{3}$$

Where $x = 1, 2, \ldots \ldots P$ and $w_0 = 1$

Now error produced as $e(n) = d(n) - z(n)$ and instantaneous error

Error can further be reduced by optimizing the weights which is being described in the below section.

Now weights are updated through sign Regressor LMS algorithm

$$w_{i+1} = w_i + \mu_1 \gamma(n) sign(e(n)y_1(n)) \tag{4}$$

$$t_{i+1} = t_i + \mu_2 sign(e(n)r(n)) \tag{5}$$

Where $\gamma(n)$ = change in adjacent amplitude and its value lies in between 0 to 1.

$\mu_1 \mu_2$ = learning rate parameters.

The output of the filter bank is useful for the FLANN model in error analysis.

2.2   Optimization Using GA

A genetic algorithm is most popular and robust technique for computation. It works like the genetic evolution process and survival for fittest procedure [11]. In this case the objective function is

$$f = \max(\text{err}) \tag{6}$$

$$\text{err} = W(n)(A(n) - D(n)) \tag{7}$$

The steps in the typical genetic algorithm is

1. Initialize the random weight of this filter bank.
2. Define the objective function.
3. Mutation.
4. Crossover.
5. Find the fitness value.
6. Repeat the steps from step 3 until the objective can be meet.

The objective weights are used in the model to reduce the error simultaneously. Based on the optimized model it has been computed and the result is shown in the following section.

3 Result and Discussion

The performance of filter bank based combination of FIR and FLANN equalizer is validated by using simulation studies. Initially the equalizer is trained with 5000 numbers of samples.

As the QAM is used in this case, the constellation diagram for symbol mapping in 16-QAM is shown in Fig. 3. It is visually excellent that proves the noise reduction.

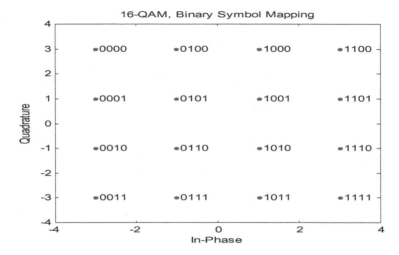

Fig. 3. Scatter plot of QAM modulation

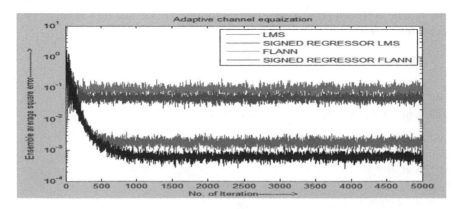

Fig. 4. MSE using LMS and FLANN in real environment

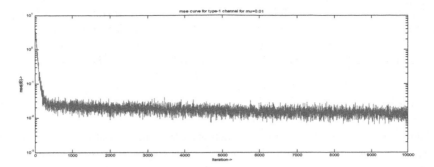

Fig. 5. Optimized MSE of sign regressor FLANN

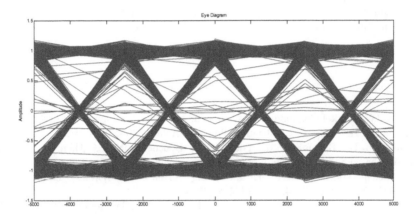

Fig. 6. Eye diagram of sign regressor FLANN

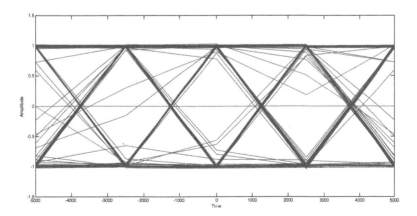

Fig. 7. Eye diagram of optimized FLANN

The comparison among different techniques in terms of MSE is shown in Fig. 4. Also the optimized MSE plot is shown in Fig. 5 that shows the efficiency.

As the ISI can be studied and observed from the eye diagram, the eye diagram are shown in Figs. 6 and 7. From Fig. 7, it is observed that the noise is reduced very well.

4 Conclusion

The sign regressor FLANN is used for reduction of ISI. The filter structure is trained by the sign-regressor LMS algorithm. In order to reduce the computational complexity of the FLANN structure, a filter-bank implementation has been used. The results in a reduction of computational needs by a factor of no of sine and cosine functions. The proposed equalization method is optimized. To achieve a good result the weights of the model has been optimized using GA. The optimized results shows their efficiency. Other optimization algorithm may be applied and the proposed technique can be compared in future work.

References

1. Proakis, J.G.: Digital Communications, 4th edn. McGraw-Hill, New York (2001)
2. Qureshi, S.U.H.: Adaptive equalization. Proc. IEEE **73**, 1349–1387 (1985)
3. Abdulkader, H., Benammar, B., Poulliat, C., Boucheret, M.-L., Thomas, N.: Neural networks-based turbo equalization of a satellite communication channel. In: 2014 IEEE 15th International Workshop on Signal Processing Advances in Wireless Communications (SPAWC), 22–25 June 2014, pp. 494–498 (2004)
4. Bandopadhaya, S., Mishra, L.P., Swain, D., Mohanty, M.N.: Design of DFE based MIMO communication system for mobile moving with high velocity. Int. J. Comput. Sci. Inf. Technol. **1**(5), 319–323 (2010)
5. Lyu, X., Feng, W., Shi, R., Pei, Y., Ge, N.: Artificial neural network-based nonlinear channel equalization: a soft-output perspective. In: 2015 22nd International Conference on Telecommunications (ICT), Sydney, NSW, pp. 243–248 (2015)
6. Carini, A., Sicuranza, G.L.: A new class of FLANN filters with application to nonlinear active noise control. In: 2012 Proceedings of the 20th European Signal Processing Conference (EUSIPCO), Bucharest, pp. 1950–1954 (2012)
7. Sahoo, S.K., Dash, S., Mohanty, M.N.: Adaptive channel equalization for nonlinear channel signed regressor FLANN. Int. J. Control Syst. Instrum. (IJCSI) **4**(2), 31 (2013)
8. Sahoo, S.K., Mohanty, M.N.: Effect of BER performance in RLS adaptive equalizer. Int. J. Adv. Comput. Res. **2**(6), 208–211 (2012)
9. Das, S., Sahoo, S.K., Mohanty, M.N.: Design of adaptive FLANN based model for non-linear channel equalization. In: Third International Conference on Trends in Information, Telecommunication and Computing, January 2010, pp 317–324 (2010)
10. Sahoo, S.K., Dash, A.: Design of adaptive channel equalizer using filter bank FIR sign-regressor FLANN. In: 2014 Annual IEEE India Conference (INDICON) (2014)
11. Thapaswini, P.P., Umadevi, S., Seerangasamy. V.: Design and optimization of digital FIR filter coefficients using Genetic algorithm. Int. J. Eng. Tech. Res. (IJETR), **3**(2) (2015)

Research of Chemical Fabric Style Prediction System Based on Integrated Neural Network

Jiang Chen and Hui Li[✉]

School of Information Engineering, Beijing Institute of Fashion Technology,
2 Yinghua Street, Chaoyang District, Beijing 100029, China
516603783@qq.com, liliuyanhui@sina.com

Abstract. In a common parlance the chemical fabric are described with the fabric structure parameters and the fabric styles and there are much subtle mathematical correlation between the two descriptions. The research on the correlation will be beneficial to the students of textile majors in the learning, fabric designers in the fabric design and fashion designers in the fabric choosing. The paper aims to do some research on the correlation in neural network technology, and to establish a chemical fabric style prediction system using MATLAB.

Keywords: Fabric structure parameters · Fabric styles · Integrated neural network · BP neural network · Fabric structural complexity factor

1 Introduction

The fabric structure parameters contain fabric composition, yarn count, fabric density, width and fabric weight, the fabric styles containing the main color, fabric texture, flower type, gloss type, drape coefficient, fabric smoothness, softness and transparency. The most structure parameters are provided by the fabric manufacturers, what's more, the scores of fabric styles, which range from 1 to 5, are graded by several experts of the Beijing Institute of Fashion Technology in a subjective evaluation way.

Among the parameters and fabric styles above, the width is often a constant, and the main colors, texture, flower type are descriptive information, no the grading information, so they are not used in the fabric style prediction.

2 Processing of the Samples Data

2.1 Preprocessing of Fabric Structure Parameters

The total 500 fabric samples of the study are derived from China International Trade fair for Apparel Fabrics and Accessories of "2009–2011 autumn and winter exhibition" and the "2010–2011 spring and summer exhibition". The fineness of warp and weft yarn need to be extracted from the yarn count in the fabric structure parameters, and the units of fabric structure parameters need unified, and convert to a general unit. The processing procedures are as follows:

F. Xhafa et al. (eds.), *Recent Developments in Intelligent Systems and Interactive Applications*,
Advances in Intelligent Systems and Computing 541, DOI 10.1007/978-3-319-49568-2_19

Fineness measurement has two unit systems, the fixed length and fixed weight systems. The larger the figure in fixed length unit system is, the thicker the yarn is, such as tex number (Tt) and denier (N_{den}); and the larger the figure in fixed weight unit system is, the thinner the yarn is, such as metric count (Nm) and English count (Ne). Tex is the general unit of fineness, and other units can be converted into the unit tex according to the formulas as follows:

$$Tt = \frac{N_{den}}{9}, \; Tt = \frac{1000}{N_m}, \; Tt = \frac{C}{N_e} \tag{1}$$

Among them, C is a parameter that is related to the moisture regains. And for chemical fabric C is 590.5.

The fabric density is the number of warp (weft) yarns along the weft (warp) yarn direction in the unit length. 10 cm-1, inch-1 and cm-1 are often used to measure the density of fabric, and 10 cm-1 is the general unit. For example, a fabric with the warp density 270 10 cm-1, means the number of warp yarns along the weft yarn direction in 10 cm length is 270. The conversion formulas are as follows:

$$1 \, inch^{-1} = 3.937 \, 10cm^{-1}, \; 1 \, oz = 28.375 \, g, \; 1 \, yd = 0.914 \, m \tag{2}$$

The weight of the fabric reflects the weight in unit area. g/m2, m/m, oz/m2 and oz/yd2 are often used to measure the weight of fabric. The weight conversion formulas are as follows [1]:

$$1 \, oz/m^2 = 28.35 \, g/m^2, \; 1 \, oz/yd^2 = 33.91 \, g/m^2, \; 1 \, m/m = 4.31 \, g/m^2 \tag{3}$$

2.2   Introduction of Fabric Structure Complexity Factor

The author proposes a definition of "fabric structure complexity factor" in the section, according to the potential correlation among the fabric structure parameters.

Suppose that G represents the weight of a woven fabric, and the unit is g/m^2; T_j and T_w represent the fineness of the warp and weft yarn, and the unit is tex; P_j and P_w represent the density of the warp and weft yarn, and the unit is 10 cm-1. Now deduce the relationship among them.

Suppose that a piece of fabric cloth, whose size is 1 m by 1 m, has the smooth surface and the uniform texture, and every warp and weft yarn in the fabric measures 1 m. And the total length of the warp and weft of the fabric are $10 \times P_j$ and $10 \times P_w$ respectively. The fabric weight can be calculated as:

$$G_c = 0.01 \times (T_j P_j + T_w P_w) \tag{4}$$

If taking the yarn twist, tightness and the fabric structure complexity in consider, generally speaking, the woven fabric weight should be larger than the calculated weight. The weight error is calculated as follows:

$$\eta = \frac{G - G_c}{G} \times 100\% \qquad (5)$$

14 pieces of chemical fabrics are selected randomly, and the weight errors are calculated. Then sort them according to the errors. As shown in the Table 1, the errors are all within ± 20 %. By the comparison of the fabric clothes, the author found that the weight errors can reflect the fabric structure complexity in a certain extent. As a result, we can take the weight errors as the fabric structure complexity factor.

Table 1. Weight errors of selected fabrics

Number.	Density of warp (/10 cm)	Density of weft (/10 cm)	Fineness of warp (tex)	Fineness of weft (tex)	Weight (g/m^2)	Calculated Weight (g/m^2)	Weight errors
......							......
T-11-6-0065	5.56	5.56	1259.84	700.79	135	109.01	19.25 %
T1-11-41-1207	11.67	16.67	700.79	440.94	188	155.29	17.40 %
T-14-12-0929	5.56	11.66	1023.62	417.32	124	105.53	14.90 %
T-11-21-0308	2.22	2.22	840	600	37	31.97	13.60 %
T-11-45-0723	8.33	8.33	680	480	110	96.62	12.16 %
T-11-78-1333	17.78	8.33	582.68	271.65	140	126.22	9.84 %
L-12-18-0571	38.86	41.64	213	225	189.19	176.46	6.73 %
T-11-17-0236	8.33	8.33	640	420	92	88.30	4.02 %
T1-12-24-1286	29.15	29.15	350.39	228.35	170	168.70	0.76 %
T1-11-18-0646	18.22	22.22	393.7	370.08	153	153.96	−0.63 %
T-5-1-0642	23.4	24	555	283	188.4	197.79	−4.98 %
T-11-86-1607	8.33	8.33	1023.62	433.07	110.51	121.39	−9.84 %
T-12-26-1519	8.33	49.47	1310	380	238	297.15	−24.85 %
T-14-12-0929	5.56	11.66	1023.62	417.32	124	105.53	14.90 %
......							......

2.3 Method for Dividing Fabric Subset

In this paper, 263 samples of chemical fiber fabrics are extracted from the total fabric samples and the fabric structure complexity factors are calculated. Then the chemical fiber fabric samples are divided into several subsets with 5 % increment of the fabric structure complexity factor. Next, the equal amounts of sub-neural networks are established for each chemical fabric subset. The processing of data and design of neural networks are proved effective to improve the accuracy of the prediction of fabric styles according to the final conclusion of the paper.

3 Comparison of Two Predictive Methods

3.1 Single BP Neural Network Prediction

In order to achieve the prediction from "fabric structure parameters" to "fabric style scores", the simplest way is to build a single BP neural network, and then the neural network is trained by all chemical fabric samples without division, to predict the fabric styles.

Network Training. The structural parameters of chemical fabrics are treated as the network input vectors which contain fabric composition, yarn count, fabric density, width, fabric weight. And the scores of fabric styles are treated as the network output vectors, which contain the scores of gloss type, drape coefficient, fabric smoothness, softness and transparency.

Data Preprocessing. In order to prevent over fitting in the training process and improve the generalization, the training data need to be normalized, in order that the data can be located within or near the interval $[-1, 1]$. At the same time, the input matrix is sparse matrix, and the principal component analysis can speed up the training speed.

57 chemical fabrics are selected to train and test on the single neural network. Figure 1 reflects the change of the mean square error during training, and Fig. 2 reflects the outputs of the test. We see that the training stops within only 12 iterations, and the minimum mean square error is 1.1434 at the 6th epoch, which is a little big.

In general, the composition of fabrics, the fiber yarn, the knitting method, and the dyeing and finishing of fabrics vary from each other in a very large extent. So the mathematical model of the feeling and visual style of the fabric is a serious nonlinear, strongly coupled and variable structure model, and the prediction accuracy of chemical fabrics is difficult to improve.

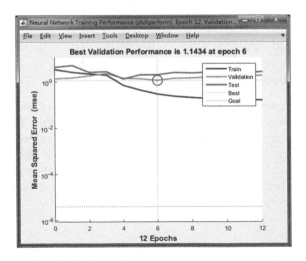

Fig. 1. Training diagram of single neural network

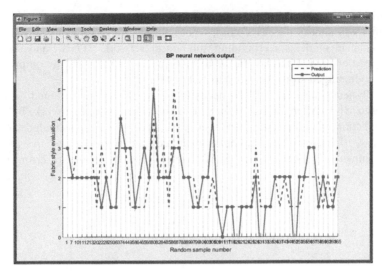

Fig. 2. Fitting error of single neural network

This experiment shows that it is hard to reach a nice predictive precision only using single neural network. So it is necessary to explore a new kind of neural network structure.

3.2 Integrated Neural Network Prediction

Because of the drawbacks mentioned above, the authors propose an integrated neural network model with a pre-classifier (Fig. 3). The input data will be pre-classified by integrated neural network classifier in order that every sample is presented to the corresponding sub-neural network, which is established and trained on the certain

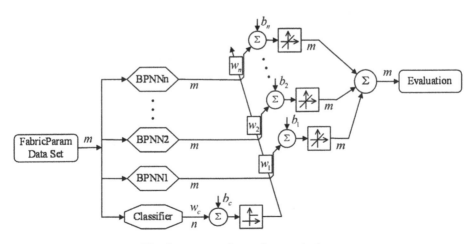

Fig. 3. Integrated neural network chart

subset, which is divided by different intervals of the fabric weight errors. The pre-classifier will control the perceptron, whose activate function is a step function, to open the channel of the corresponding sub-neural network BPNN$_i$, and the output result can be obtained. The classifier is a logic judgment unit, and its output are the single '1' code ranged from '0$\cdots$01' to '10$\cdots$0' [6].

Each sub-neural network BPNN$_i$ needs to be established and trained on the different subsets $S\{i\}$. After the establishment of integrated neural network, in order to test the prediction precision, the subset $S\{i|0<\eta<10\%\}$ (19 samples included) and the subset $S\{i|-10\%<\eta<0\}$ (18 samples included) are selected, and the prediction results as shown in the Figs. 4 and 5. We see that, the mean square errors fall into

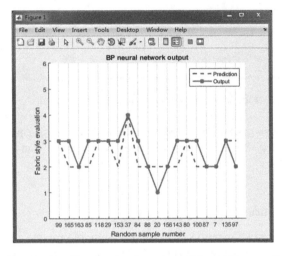

Fig. 4. Identification for the subset $S\{i|-10\%<\eta<0\}$

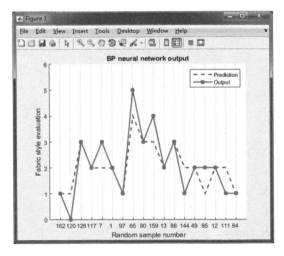

Fig. 5. Identification for the subset $S\{i|0<\eta<10\%\}$

0.4211 and 0.3889 respectively. It can conclude that the integrated neural network can achieve the chemical fabric prediction well and output the better results.

4 Conclusion

By comparison, we can draw a conclusion that integrated neural network predict the chemical fabric style better than a single BP neural network. The prediction has stronger adaptive identification ability and the results are more accurate.

Each fabrics composition, fiber yarn, structure and dyeing and finishing are different and the scores of the fabric styles also show a different distribution when experts evaluated the fabric style subjectively. So it is very difficult to find the mathematic model and uniform rules of the fabric samples. And the results reveal that the introduction of the fabric structure complexity factor is feasible and the integrated neural network can be better to achieve the prediction of the scores of the chemical fabrics. This is the innovation point of this paper.

Acknowledgements. The authors acknowledge the financial support of Beijing Key Laboratory of clothing materials research, development and evaluation under the grant number: 2011ZK-06. The authors also acknowledge the financial support of the project "The construction of the specialty of electronic information engineering under the Internet + background" under the grant number: NHFZ2016012.

References

1. Wang, G.: Fabric for Fashion Design, The 2nd version, pp. 40–52. China Textile & Apparel Press, Beijing (2010)
2. GB/T 22970-2010 textile fabrics coding-Parts of chemical fibre. National Institute of Standardization, Beijing, China (2014)
3. GB/T 31007.1-2014 Coding of textile fabrics-Part1: Cotton fabric. National Institute of Standardization, Beijing, China (2014)
4. GB/T 31007.2-2014 Coding of textile fabrics-Part2: Bast fiber fabric. National Institute of Standardization, Beijing, China (2014)
5. GB/T 31007.4-2014 Coding of textile fabrics-Part4: Wool fabric. National Institute of Standardization, Beijing, China (2014)
6. Haykin, S.: Neural Networks and Learning Machines, pp. 36–40. China Machine Press, Beijing (2014)

Internet and Cloud Computing

The Influencing Factor Analysis of Micro-blog User Adding Attend Object

Shuqin Li[1(✉)], Xiaohua Yuan[2], and Xiaobin Ling[1]

[1] College of Computer, Beijing Information Science
and Technology University, Beijing 100101, China
Lishuqin_de@126.com
[2] College of Information, Shanghai Oceanic University,
Shanghai 201306, China
xhyuan@shou.edu.cn

Abstract. This paper mainly analyzes the influence degree of both social information and interest characteristics, on micro-blog users' adding attention object, in which, as concern to social information, we mainly consider, basic factors, considering factors and influencing factors. And respectively, take Micro-blog user location, education background, work experience as a basic factor; take common concern between users and common friends as the considered factors; and the operations of forwarding and of @ as the influencing factors. Based on this, then intimate relationship and interest similarity are imported to characterize the social information and interest characteristics and to measure the degrees of intimation and interest similarity among micro-blog users, respectively. Then, based on each of the two relationships, recommendation methods are given for micro-blog user adding attention object. And at last, in the experiment, the results of the two recommendations are compared, which shows that during micro-blog user adding attention objects, the influence of social information is stronger than that of interested characteristics.

Keywords: Micro blog · Intimacy · Interest similarity · User recommendation

1 Introduction

Micro-blog has the characteristics of short, agility, and fastness, hence is adopted by most of the Internet users. On micro-blog platform, each user is both the information recipient, and can also the information promulgator, transfer. Data from Wikipedia shows that, taking SinaWeiboas as the example, the number of existing user is more than 500 millions, the daily active users arrives 46.2 millions, as the daily blog is more than 100millions [1]. Infront of such a huge user group and its micro-blog information, the micro-blog platform need to help user know, how to obtain their should attended object user and information, namely micro-bog platform need to make some reasonable recommendation to its user, and how to realize it has become one research focus.

At present, there are mainly two kinds of modified recommendations based on user relationship, one is based on user mode, the other is objected to the model of user relationship, and the difference of these two are as follows:

© Springer International Publishing AG 2017
F. Xhafa et al. (eds.), *Recent Developments in Intelligent Systems and Interactive Applications*,
Advances in Intelligent Systems and Computing 541, DOI 10.1007/978-3-319-49568-2_20

The former construct recommendation features mostly based on user basic information and user interest, and adjust the recommended strategy according to the change of user characteristics, and this recommendation mode take social network as the foundation [2]. For example, Guymicro-blog divide social network into two sub-networks, one is of similar users, and the other is of familiar users, and makes recommendation from the two sub networks respectively [3].

Ideas of the latter is that, first establish the corresponding relationship which among users and between the user and the information source respectively, then perform the community analysis, establish interest community or similar user group, and predict the behavior of individuals in population by the group behavior, thus make recommendation of individual user. Zhou et al. [4] use community detection model (based on label map) to simulate user interest model, through represent the model as a discrete topic distribution model, use the KL-divergence of this discrete topic distribution model to measure the similarity between users, and take the similarity as the recommendation index.

From aspects of social information and user interested characteristics, the author analyze the influent degree of social information and interest characteristics on users' adding attended object, and together with some comparing test, the author try to find out micro-blog platform recommending object to its user, which of the two factors is the more should be considered as one.

2 Relationship Between User Intimacy and Attended Objects

2.1 Influence Factors of Intimacy and the Representation

The inherent human need to have a sense of belongingness and are inclined to love others, and the intimate relationship is formed when these needs are satisfied [5]. This paper takes for that the intimate relationship among micro-blog users can be combined by basic factors, attended factors and influencing factors. We will review these three below.

2.1.1 Basic Factors

In micro-blog information; location, education background, work experience and etc. are taken as the basic factors, and assumed it can increase user's belonging senses. The main form of user location is about that user was born in which province and in which city. Using $L(u)$ to represent the location of user u then $L(u) \cap L(v)$ represent the location matching level between user u and user v. We set the levels of $L(u) \cap L(v)$ as that, if $L(u)$ and $L(v)$ are completely different, then match score is 0, else if only province are the same, then the score is 1, else the score is 2, for that both province and city in the location of the two uses' are the same.

In Sina Weibo, for both of education background and work experience, can include more than one record, for example, from which middle school, which university, and etc. We use $E(u)$ to represent the records set of education background, and $E(u) \cap E(v)$ to represent the matching level of $E(u)$ and $E(v)$. Similarly, we use $W(u)$ and $W(u) \cap W(v)$ to represent the work experience and the related matching level, respectively. Through counting the number of the same items in $\bar{L}(u) \cap L(v)$, $E(u) \cap E(v)$ and $W(u) \cap W(v)$, we can digitalize each of the basic information.

2.1.2 Considering Factors

The common attention and common friends among users can be used as considering factors, because more common attend represents more similar interest between user, and more common friends indicates more close relationship between users.

Using $C(u)$ as its attending set, and $F(u)$ as its friend set of user u, then for user u and user v, $C(u) \cap C(v)$ means their common attention, and $F(u) \cap F(v)$ means their common friend set.

2.1.3 Influent Factors

Interaction power is an important embodiment of close relationship [6]. In this paper, we take user's forwarding and @ operation as the influent factors. Because there are strong interaction between users in micro-blog, since that, if user will show interest in one micro-blog, he will forward or comment on it. Users can also use @ operation to remind other user to view their own message or comment too, thus, @ operation and forwarding operation become the mainly user connection.

Usually, there are four forms to construct a micro-blog, which are as following.

A. Text
B. Text@U_1.... @U_n.
C. Text//@V_nText$_n$...//V_0Text$_0$
D. Text@U_1.... @U_n //@V_nText$_n$...//V_0Text$_0$

Where A represents that, the target users only issue micro-blog, not perform other operation. B represents the target user issue micro-blog and has @ other users $U_1...U_n$. C represent that the micro-blog issued by v_0 has been forwarded and commented by each layer, and finally by the target user again. D represents at the foundation of C, the target user has further @ other users $U_1...U_n$.

For that we will only analyze the forwarding and @ operation of target user, thus, we only consider the three cases of B, C, and D. Besides this, for forwarding operation, we only consider the micro-blog source that direct forwarded by the target user, not consider those from other more upper sources. And for @ operation, we also set that, when count the @ operation number of one user, we only check the number of user @ others, not the number of @ that appear in the micro-blog content.

2.2 The Definition and Computation of Intimation

The intimation measures the intimation degree between users. In this paper, the authors use $R(u)$ to represent the micro-blog set forwarding by user u, and $r(u,v)$ represent the the micro-blog user v issued and user u forwarded. We use $M(u)$ represent the@ operation set performed by user u, and $m(u,v)$ represent the operation set that user u @ user v. Then the intimation degree between user u and user v can be calculated by formula 1.

$$i_{u,v} = \alpha \frac{|L(u) \cap L(v)|}{|L(u)|} + \beta \frac{|E(u) \cap E(v)|}{|E(u)|} + \gamma \frac{|W(u) \cap W(v)|}{|W(u)|} + \theta \frac{|F(u) \cap F(v)|}{|F(u)|} + \delta \frac{|C(u) \cap C(v)|}{|C(u)|} + \xi \frac{|r(u,v)|}{|R(u)|} + \varepsilon \frac{|m(u,v)|}{|M(u)|} \quad (1)$$

Where α, β, γ, δ, ε, ζ and θ are the weight, thus $\alpha + \beta + \gamma + \delta + \varepsilon + \zeta + \theta = 1$.

2.3 The Relationship Between User Intimacy and the Attended Object

When making a recommendation to one user based on intimacy, the recommendation score is taken as the intimate degree among users, and the results are ordered according to recommendation score descending, after that the recommendation set will be truncated to a given number, and be recommended to the user. Micro-blog recommendation algorithm based on intimate degree is that.

Algorithm 1. Intimation based user recommendation algorithm

Step1: Set the attended objects number $N_{tp} as N_{tp} = 0$, and that of the disinterested objects as $N_{fp} = 0$, to the user set U that to recommend, select and order the intimate relation degree descending of each user $N_u \in U$ from the intimate table in database, and from the selected result remove the records whose intimate degree is zero, thus obtain the close friends set $U_{un} = \{u_{u1}, u_{u2} \ldots u_{uu}\}$ of user u_u

Step2: Give the number x of required recommendation, when $x \leq n$, $U_{ux} = \{u_{ux}, u_{ux} \ldots u_{ux}\}$ is the obtained recommendation result, and jump to Step 5, otherwise, it means there is no enough user possible, thus all the user in U_n are remained and jump to Step 3

Step3: From formula 1, give the indirect intimate degree formula $Ui_{uw} = \tau i_{uv} + \omega i_{vw}$, Where τ and ω are the weight of $i_{u,v}$ and $i_{v,w}$ respectively, and $\tau + \omega = 1$

Step4: Calculation the indirect intimate degree between u_u and u_W ($u_w \in U_{vn'}$, $u_w \in U_{un}$), after remove the duplicate ones, we can obtain the indirect close friends set $UI_u = \{ui_{u1}, ui_{u2} \ldots ui_{un}\}$, reordering it descending, we can obtain $UI'_u = \{ui'_{u1}, ui'_{u2} \ldots ui'_{un}\}$

Step5: Take subset $UI_{u(x-n)} = \{ui_{u1}, ui_{u2} \ldots u_{u(x-n)}\}$ corresponding to $UI''_u = \{ui'_{u1}, ui'_{u2} \ldots ui'_{u,x-n}\}$. Finally we can obtain the result set $U_x = U_{x-n} \cup U_n$, which is taken as the recommended object

Step6: From the user relationship tables in database, obtain the number N of records that contain u_u, and take N as the number of total user attended. Check if (u_u, u_v) is contained in this table, if is, then attended number N_{tp} plus 1, otherwise, the disinterest number N_{fp} plus 1

3   The Relationship Between User's Interest Degree and Attended Object

In this paper, the user similarity mainly considers the similarity of blog content issued by the user, and define it as the interest similarity relationship and measure its degree.

3.1   Interest Similarity Calculation

In calculating the similarity between users, the first task is to break up the word of micro-blog, after that, to extract the feature words in micro-blog as user's interest, to construct the vector of feature words and finally to calculate the similarity between vectors, which we take as the interest similarity between users. We conclude this as Algorithm 2.

Algorithm 2. Interest similarity calculation algorithm

Step1:   Perform word segmentation, then, count the word of each micro-blog of user u_i, thus to obtain the segmentation set $C_{ui} = \{Cu_{i1}, Cu_{i2}...Cu_{in}\}$ of user u_i, after that, reorder C_{ui} according to the number of each word, thus obtain $C'_{ui} = \{Cu'_{i1}, Cu'_{i2}...Cu'_{in}\}$

Step2:   Using TF-IDF algorithm [7] to obtain user's feature word set, and take it as user's interest word set

Step3:   Transform the value vector of word into the vector probability $Pu_i = \{Pu_{i1}, Pu_{i2}...Pu_{in}\}$, and construct user interest word matrix as shown in formula 2

$$M = \begin{Bmatrix} C_1... & C_2... & C_i... & C_N \\ P_{u_{11}} & Pu_{12} & Pu_{1i} & Pu_{1N} \\ \vdots & \vdots & \vdots & \vdots \\ Pu_{M1} & Pu_{M2} & Pu_{mi} & Pu_{mn} \end{Bmatrix} \tag{2}$$

Where $C = \{C_1, C_2...C_N\}$ represents user's interest word set, and $Pu_j = \{Pu_{j1}, Pu_{j2}...Pu_{jn}\}$ means interest word vector of user u_j.

Step4:   Use the cosine vector similarity algorithm [8] to calculate the similarity degree between user interest vectors

3.2   Recommendation Based on Interest Similarity Relationship

In this recommendation, we firstly sort the interest similarity descending, then truncate the recommendation set according to the specified recommendation number, and then recommend the result to the user. The recommendation algorithm based on interest similarity relationship is as following.

Algorithm 3. Recommendation algorithm based on interest similarity relationship

Step1: Set the attended object number as $N'_{tp} = 0$ and disinterest object number as $N'_{fp} = 0$, give the user set u that to be recommended. From the similarity interest table in database, sort the records according to the interest similarity degree descending, then select the user set $U_{un} = \{U_{u1}, U_{u2}...U_{un}\}$, which are similar to user u_u. in U_{un}, the records of zero similarity degree are remained

Step2: Give the limit number of X by program, select the first X members from the recommendation set in Step 1, then obtain the recommendation result $U_{ux} = \{U_{u1}, U_{u2}...U_{ux}\}$

Step3: From the user relational table in database, calculate the number N of records that contain U_U, then check if (u_u, u_v) is contained in the table, if is, then N'_{tp} plus 1, otherwise, N'_{fp} plus 1

4 Experiment and Analyzing

4.1 Experimental Environment

This experiment is implemented using Java language and on eclipse platform. The experimental data are grabbed from SinaWeibo by a crawler written by the author of this article. Take the target user as the center, and adopting the mode of snowball, the crawler has grabbed 28604 users, 291065 attended relationship and 153277 micro-blog. In the experiment, we select 100 users as the recommended user, then to each user, perform the recommendation based on intimate relationship and interest similarity relationship respectively. Each type of experiment has been repeated 16 times, of which, at the first time, there are 5 object are recommended to each user, after that, each subsequent will increase of 5 objects.

4.2 Evalution Index of Result

In the experiment, we use feature value F1 (which taking accurate rate and recall rate as the variable), as the judgment index of recommended result [9]. As to the recommendation, the user to be predicted will adopt the corresponding behavior like that listed in Table 1.

Table 1. The 4 possible cases of user to be predicted.

Pay or not pay attention to	System recommend	System not recommend
Pay	True-Positive N_{tp}	False-Negative N_{in}
Not pay	False-Positive N_{ip}	True-Negative N_{tn}

As to user u, the accurate rate of recommended results can be calculated by formula 3.

$$P_u = \frac{N_{tp}}{N_{tp} + N_{fp}}$$ (3)

As to all users in the system, the accurate rate of recommended result can be calculated by formula 4.

$$P = \frac{1}{m}\sum P_u$$ (4)

Where M is the user number. As to user u, the recall rate of the recommendation result can be calculated by formula 5.

$$R_u = \frac{N_{tp}}{N_{tp} + N_{fp}}$$ (5)

To all the users, the recall rate of the recommendation result can be calculated by formula 6.

$$R = \frac{1}{m}\sum R_u$$ (6)

And the feature F_1 is defined as in formula 7.

$$F_1 = \frac{2 \times p \times R}{P + R}$$ (7)

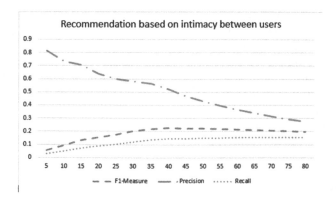

Fig. 1. Effect of user recommendation based on intimate relationship

4.3   Experimental Results

The result of user recommendation based on the intimation relationship is shown in Fig. 1.

And in experiment of user recommendation based on the interest similarity, we use Ansj [10], a segmentation software in JAVA. To each user, we only select the first 40 words occur with higher frequency as user feature words. We adopt algorithm TF-IDF to obtain user's feature word set as user's interest word set.

Considering the number difference of user micro-blog in the experiment, the vectors of word number are transformed into word probability vector, that is, the weight of

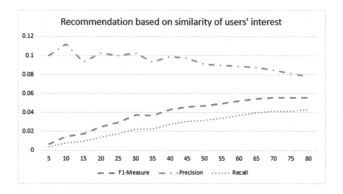

Fig. 2. Effect of user recommendation based on interest similarity relationship

each every, then using cosine vector similarity algorithm, we calculate the interest similarity relationship between two users. The recommendation results based on interest similarity relationship is shown in Fig. 2.

From the results of the two recommendations, we can see that the recommendation based on intimate relationship can obtain better effect. This shows that micro-blog users pay more attention to the users with whom they have intimate relationship with, not those they have similar interest with.

Although Micro-blog is a kind of virtual social network, however, on the micro-blog platform, in the choice of object that he/she wish to interact with, user will pay more attention to the social information, such as classmates, colleagues or countryman, not the common hobby, for example; basketball.

Therefore, when recommend attend object, micro-blog platform should consider more about user's social information, then user's interest information.

5   Conclusion

This paper mainly analyzes the influence degree of social information and interest characteristic of micro-blog users adding attend object, and conclude that, in micro-blog users adding attend object, the role of social information is stronger than that of the similar interests. The future research will consider synthetically the two aspects of social information and interest characteristics, hopeful to draw out some more reasonable recommending methods.

Acknowledgement. This work is supported by the Beijing municipal commission of education of science and technology general plan projects (71E1610970), supported by Opening Project of Beijing Key Laboratory of Internet Culture and Digital Dissemination Research (ICDD201507), supported by National Natural Science Foundation of China (61502039), and supported by Sensing & Computation Intelligence Joint Laboratory.

References

1. Wikipedia: SinaWeibo (2014). http://zh.wikipedia.org/wiki/SinaWeibo
2. Armentano, M., Godoy, D., Amandi, A.: Recommending information sources to information seekers in Twitter. In: Proceedings of the IJCAI: International Workshop on Social Web Mining, Barcelona, Spain (2011)
3. Guy, I., Zwerdling, N., Carmel, D., et al.: Personalized recommendation of social software items based on social relation. In: Proceedings of RecSys, pp. 53–60 (2009)
4. Zhou, T., Ma, H., Lyu, M., King, I.: UserRec: a user recommendation framework in social tagging systems. In: Proceeding of the 24th AAAI Conference on Artificial Intelligence, pp. 1486–1491 (2010)
5. Daniel, P.: The best of times, the worst of times: the place of close relationships in psychology and our daily lives. Can. Psychol. **48**(1), 7–18 (2007)
6. Guoxia, W., Heping, L.: Review of personalized recommendation system. Comput. Eng. Appl. **48**(7), 66–76 (2012)
7. Wikipedia: TF-IDF (2014). http://zh.wikipedia.org/wiki/TF-IDF
8. Wikipedia: Cosine Similarit (2014). http://zh.wikipedia.org/wiki/
9. Zhu, Y.X., Lu, L.Y.: Evaluation metrics for recommender systems. J. Univ. Electron. Sci. Technol. China **41**(2), 163–175 (2012)

History Path Reconstruction Analysis of Topic Diffusion on Microblog

Yichen Song[✉], Aiping Li, Jiuming Huang, Yong Quan,
and Lu Deng

College of Computer, National University of Defence Technology,
Changsha, China
491153214@qq.com

Abstract. With the growth of online social media, such as Twitter and Weibo, topic diffusion has concerned a lot attention. Based on the timing sequence and retweet relationship, the propagation history of one single Weibo have been rebuilt successfully in social network with various methods. However, the topic diffusion among users can hardly be reconstructed, as vast of different content Weibo need to be consider. In this paper, the authors propose a user topic diffusion history reconstruction method to reconstruct the propagation process of a topic. Based on the topic detection algorithm, different weibo can be clustered to a group as the data set. A score-based propagated model is proposed to obtain the information source node in a back propagation way with considering the influence of time, social relationship and text similarity.

Keywords: Social network · Topic information diffusion · Weibo · Propagated model

1 Introduction

With the development and wide spread of information technology and application, social network media is putting back the traditional paper print media progressively, becoming the warehouse of huge information resources. All kinds of social communication platform have a great development, such as Sina weibo, Tecent weibo, Facebook and Twitter. Compared with traditional media, social network enhances the spread effectiveness of information in time and breadth based on its freedom and openness. However, some fake information or bad information can't be got rid of and can cause panic among users. In conclusion, information propagation has influenced people's life in a remarkable way, and becoming a research hotspot in web field.

The research on diffusion processes in networks can be simply categorized into two groups: prospective analysis and retrospective analysis [1].

Prospective analysis research is mainly to minimize or maximize the network diffusion based on the propagation processes and network structural properties and algorithms. Many researchers have been down in this period yet, such as hot prediction and influence prediction and so on. There are several classic models: Linear Threshold Model [2] which based on the idea that one node cannot be activated unless its neighbor's influence sum exceeds its threshold, Independent Cascade Model [3, 4]

F. Xhafa et al. (eds.), *Recent Developments in Intelligent Systems and Interactive Applications*,
Advances in Intelligent Systems and Computing 541, DOI 10.1007/978-3-319-49568-2_21

which considers that each active node will affect its neighbors with a certain probability, while the influences are independent and decline over time, and Epidemic Model [5], i.e. SIS and SHIR, which can better describe the diffusion process through different infection mechanisms and individual user conditions with a relatively low accuracy.

Retrospective analysis is mainly about network inference such as identifying the source and constructing the diffusion process. Earlier works focus on identifying the source of diffusion [6–8] and on those relative technologies that have been developed based on comparative maturity, when more fundamental path reconstruction technology has been received attention lately. Most reconstructions are directly based on the retweet relationship, which is also named as the explicit propagation path of information, as the implicit path trends to be ignored that the accuracy hardly be verified. Safer [9] rebuilt the history with several snapshot at different time and proposed a submodularity based algorithm. Zhen Chen [1] transferred the problem into a Maximum A Posteriori (MAP) estimate problem and inferred the diffusion with a single snapshot that contains partial infection time. Han [10] gave a content based topic diffusion model in 2013, however, which is calculate the infect probability through the network structure rather than real contents which users speak out.

During this big data period, with great explosion of information and various characters of social network, the acceptance of topic information not only comes from those friends and followings, but also by other users who were activated before. Taking an example of weibo, users will get information from their friends after they login, and will also can skim through other blogs or hotspot topics that the website provide.

While the traditional information propagation research is focus on single or a few hot blog, in fact, it propagates more in a topic through users, containing huge and various blogs with multi-source. This paper use weibo as an example, analysing the reconstruction of topic diffusion path based on technologies like topic detection, score-based propagated mode and text similarity.

2 Problem Formulation

2.1 Basic Definition

Definition 1. Weibo Topic. A Weibo topic I is represented by several similar blogs which contains the same theme in a special time. $I = \{i_1, i_2, \ldots . i_n\}$, i_j represents a blog published by users related to topic I, and one user can give more relative blogs.

Definition 2. Topic Diffusion Path Diagram. A path diagram is a graph $G = (V,E)$, where $V = \{v_1, v_2, \ldots, v_n\}$ is a vertices set which represents users interested in the same topic, and $E \subseteq V * V$ is a edges set represents the information propagated direction through users, i.e. $e(v_i, v_j)$ represent the information passes from user vi to user vj.

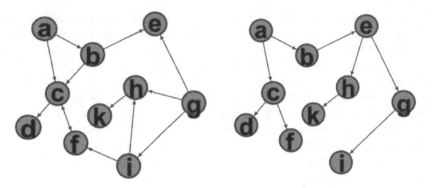

Fig. 1. (a) Weibo social network structure (b) A topic propagated path

The Fig. 1 describes, how does a topic propagate through network. As the social media like weibo has broken the restriction among territory and relationship, even when, there is no relation between two users, can spread information between each other. Like the users h and g, they can get information from user e, as shown in the Fig. 1(b) even though there are no connections among these three users as shown in the Fig. 1(a).

Definition 3. Information Coming Node. During the diffusion of a topic, we take first time that a user public a blog which is related to the certain topic as the time a user is infected. And the Information Coming Node (IC node) is the user node u from which the user v accept the information. Which means an edge $u \rightarrow v$ is exists and u is the IC node of v.

The topic in weibo is propagating through behaviors like reading, commenting, retweeting and thumbing. Among these behaviors, retweeting can be used to measure the diffusion path directly while the reading and comments can only influence users themselves that the information is hard to be obtained.

This research is based on a Chinese social website named sina weibo. We assume that the only information source of users is blogs in weibo that the other tools are ignored. In this paper, a topic detection method is first used to cluster the similar blogs into certain topic, and these topic clusters will be used to find the vertices in Topic Diffusion Path Diagram, which can also be said as being infected or accepted by the topic. A probabilistic score based diffusion model is proposed to find out the IC node and propagated path. The main idea of it, is to obtain explicit propagated path with retweet data and then find the implicit ones with the proposed model.

Algorithm 1. `Info_source search`

①   `V = time_sort(V);`

②   `For each v ∈ V`

③   `for each (u.time < v.time) & u ∈ V`

④   `calculate p (u → v);`

⑤   *endfor*

⑥   `u = max(p (u → v));`

⑦   `add(u,v) to E`

⑧   `endfor`

⑨   `return E`

2.2 Topic Detection

The information diffusion in weibo tends to be multi-sources, being composed by different blogs which contains the same topic. In order to study the diffusion path of a certain topic, the first thing needs to do is to cluster the weibo data to get dataset which include the users who has been infected by the topic and the blogs they published related with the topic. This step can be complete with different topic detection method [11, 12], such as LDA and so on. As such technology has been developed very mature and it is not our main research point that we will not develop the research at this period.

2.3 Probabilistic Score Based Diffusion Model

As mentioned before, the infected users are these who have publish blogs related to the topic, and they compose the user vertex set V. The blog information which represents first time that users interested into the topic can be found from the original dataset. Considering that all the blogs published before the time user v pays attention to the topic can deliver information to him, the influence function between u and v is like Eq. (1):

$$p(u \to v) = \begin{cases} 1, & \textit{if retweet exists} \\ \sum (p(t) + p(s) + p(r)), & \textit{else} \end{cases} \tag{1}$$

The propagated score is measured with time, text similarity and social structure.

Time Influence Score. Users are bare to millions information in weibo moment by moment, and these information can be delivered to users with a time sequence when they log-in. However, no users can skim all the new information published between his login so that some information will be missed. Through observation, users tend to read these blogs published by their friends first and after this will be the other blogs from index. Referring to the probabilistic reading model proposed by Zhu Xiang [13],

Gaussian distribution is chosen to be used to model the time influence score, as in Eq. (2):

$$p(t) = \begin{cases} e^{-\frac{(t-t_0)^2}{2\sigma^2(u,t)}}, & \text{if } t \geq t_0, \\ 0, & \text{else} \end{cases} \qquad (2)$$

Here, $p(t)$ is the probabilistic score of the blog user v wrote at time t is influenced by the blog user u wrote at time t_0. And $\sigma(u,t)$ is the time interval parameter which gives the most number of blogs user u can read when he login. It represents the nearest and farthest blogs that users can read and can calculate by claw the blog list of user at time t.

Text Similarity. The original blog of users tends to have a similar content with its information source, or the most influenced blog published by what we called as IC node. Based on the similarity of two blogs, the source of information can be backwards reasoning. Which can be understand as the higher similarity of the blog user v send at time t and the blog user u send at time t_0, the higher probability that the IC node of user v is user u.

Since the test contents of blogs have been filtrated at the first topic detection step, here we choose Jaccard Similarity to model the similarity score. In Eq. (3), the p(s) represents the similarity score of two blogs. S and T represent the word set of these two blogs, respectively. And the idea of this is divide the intersection set by the union set.

$$p(s) = \frac{S \bigcap T}{S \bigcup T} \qquad (3)$$

Social Relationship. Users tend to read and accept blogs published by their friends and people they used to pay attention to be measured by their retweet history. These behaviours makes the information passed from friends have higher influence than the other strangers' blogs they read through weibo network. So the probabilistic score is influenced by the social relationship which can be measure by the retweet history.

3   Experiment Results and Discussion

In order to improve the accuracy of the result, this paper takes time, social relationship and text similarity to build probabilistic score based diffusion model and find the IC node to build the topic diffusion diagram. This experiment restores the process of topic diffusion based on the data to settle the multisource problem.

3.1   Experiment Realization

The experiment use the blog data randomly clawed from weibo as experiment dataset. After basic data pre-processing work, the topic detection method is being used to cluster the blogs belonged to the same topic.

After topic detection, a certain topic and its blogs data is chosen as the topic data set, and then the info_source search algorithm is used into the topic data set to calculate the diffusion path and to formulate the history diffusion diagram.

3.2   Experimental Results and Analysis

The original dataset of the experiment is referring to the method used in Cindy [14], we get blogs about topic *I*. In order to test the performance of our proposed probabilistic score based diffusion model, a test data set is extract from the original dataset. The blogs which contain the retweet parameter is chosen to test the accuracy. We compare the IC node with the retweet user, and get the accuracy of 45 %, which means the accuracy of our methods in judge the implicit information source of weibo users is about 45 %.

Table 1. Result Count of propagated path

No source	Retweet	Following	Index
7 %	7 %	6 %	80 %

Table 1 presents the result of info_source search and the finding way of user's IC node. During the experiments, in order to simplify the complexity, the time interval in 2.3.1 is set to a fix 24 h instead of claw data for every user, which is also the reason of why there are these source nodes. As there are about 7 % retweet data, the accuracy can be improved to 49 %.

Fig. 2. Graph of a topic propagated path

Figure 2 describes the part of the diffusion diagram, different colours are used to represent the different ways getting information, by directly retweeting, from follow-ings, or from randomly surfing the weibo website. And they are represented by green, yellow and pink, respectively. The green are the nodes whose information flow is for sure that the accuracy is 100 %, the yellow are the information received from their followings has been infected before, the pink represents these users who can't get information from their social net so that we take the strangers in weibo as their IC node. But these data changes with different topics.

As shown in Fig. 2, the diffusion diagram can give clear condition of how information propagated through social website. At the same time, as time and text similarity factors are considered, the users hold the same idea will be clustered together, which can benefit the analysis and management of public opinion. The topic diffusion diagram can help in macroscopic information analysis, public opinion guidance and the prediction of the other diffusion of the similar topic.

4   Conclusion

This paper accept the idea of classic SI model to assume that the user will not recover once when infected and the information source is mainly among weibo. Combining topic detection and the proposed probabilistic score diffusion model to study the history path reconstruction analysis of topic diffusion on microblog.

The proposed probabilistic score diffusion model consider different interferences user may get when accept information with time, social relationship and text similarity to obtain the most possible IC node. With the IC node, the information diffusion path of a topic in social network can be obtained and rebuilt the real diffusion process.

However, the experiment is relatively simple, and there exists several problems, like how to improve the effect with the social network character, how to increase the accuracy of time influence, and considering the heat of blogs into the model. These are the basic initial point of our future research.

Acknowledgements. This work was supported by the National Natural Science Foundation of China (No. 61472433) and the National Basic Research Program of China ("973" Program, No. 2013CB329604).

References

1. Chen, Z., Tongy, H., Ying, L.: Full diffusion history reconstruction in networks. In: Big Data (2015)
2. Granovetter, M., Sonng, R.: Threshold models of interpersonal effects in consumer demand. J. Econ. Behav. Organiz. **7**(1), 83–99 (1986)
3. Goldenberg, J., Libai, B., Muller, E.: Talk of the network: a complex systems look at the underlying process of word-of-mouth. Mark. Lett. **12**(3), 211–223 (2001)
4. Goldenberg, J., Libai, B., Muller, E.: Using complex system analysis to advance marketing theory development: modeling heterogeneity effects on new product growth through stochastic cellular automata. Acad. Mark. Sci. Rev. **9**(3), 1–18 (2001)
5. Xiong, F., Liu, Y., Zhang, Z.-j., Zhu, J., Zhang, Y.: An information diffusion model based on retweeting mechanism for online social media. Phys. Lett. A **376**(30), 2103–2108 (2012)
6. Fioriti, V., Chinnici, M.: Predicting the sources of an outbreak with a spectral technique (2012). arXiv preprint arXiv:1211.2333

7. Lokhov, A.Y., MéZard, M., Ohta, H., et al.: Inferring the origin of an epidemic with dynamic message-passing algorithm (2013). arXiv preprint arXiv:1303.5315

8. Comin, C.H., Costa, L.D.F.: Identifying the starting point of a spreading process in complex networks. Phys. Rev. E **84**(5), 56105 (2011)

9. Sefer, E., Kingsford, C.: Diffusion archeology for diffusion progression history reconstruction. In: IEEE 14th International Conference on Data Mining (ICDM), Shenzhen, China, December 2014, pp. 530–539 (2014)

10. Zhongming, H., Hui, Z., Meng, Z.: A hot topic propagation model based on topic contents. CAAI Trans. Intell. Syst. **8**(3), 233–239 (2013)

11. Zhao, W., Jiang, J., Weng, J., He, J., Lim, E.-P., Yan, H., Li, X.: Comparing twitter and traditional media using topic models. In: Proceedings of the 33rd European conference on Advances in Information Retrieval, pp. 338–349 (2011)

12. Wentang, T., Zhenwen, W., Fengjin, Y.: A partial comparative cross collection LDA model. J. Comput. Res. Dev. **50**(9), 1943–1953 (2013)

13. Xiang, Z., Yan, J., Yuanping, N.: Event propagation analysis on microblog. J. Comput. Res. Dev. **52**(2), 437–444 (2015)

14. Lin, C.X., Mei, Q., Han, J., Jiang, Y., Danilevsky, M.: The joint inference of topic diffusion and evolution in social communities. In: ICDM 2011 (2011)

A Novel Session Identification Scheme with Tabbed Browsing

Xie Yonghong[1,2(✉)], Wan Yifei[1,2], and Zhang Dezheng[1,2]

[1] School of Computer and Communication Engineering,
University of Science and Technology Beijing, Beijing 100083, China
xieyh@ustb.edu.cn
[2] Beijing Key Laboratory of Knowledge Engineering for Materials Science,
Beijing, China

Abstract. As network-oriented pattern identification has become more popular in practical applications, tabbed browsing analysis as an effective method in service computing, deserves more attention. The session identification plays an important role to achieve tabbed browsing. However, the fixed threshold session identification scheme in the field of the session may achieve unsatisfactory performance due to the inappropriate division for sessions in some cases. To avoid such limitation, a novel session identification scheme is proposed in this article. Considering the different operation practices in tabbed browsing, our projected scheme designs an effective scheme through the use of different segmentation thresholds for different users. We present the detailed design algorithm to explain, how effective optimization can be achieved in this novel scheme. Moreover, we test the effectiveness of our scheme on two large-scale data sets from the computing environments in NASA and USTB to demonstrate its optimization performance.

Keywords: Session identification · Fixed threshold · Tabbed browsing

1 Introduction

Session identification, whose purpose is to divide the user's access records into some single sessions, is a part of data preprocessing in the field of users' behavior analysis [1]. The Session identification directly affects the subsequent clustering, behavior analysis and various operation results. The accurate session identification results, in addition to reflect the user's behavior characteristic [2], also can provide a good data base for the user clustering, which makes the clustering results more practical. However, fixed threshold session identification scheme in the field of the session can't adapt to the changes of the current users' behavior due to the inappropriate division for sessions in some cases. For example, the sessions which should be identified as the same session may be divided into different sessions, may be identified as the same session by fixed threshold session identification scheme.

Many session identification schemes have been proposed so far, such as Dynamic timeout-based a session identification algorithm [3], Performance enhancement in session identification [4], An improved referrer-based session identification algorithm

F. Xhafa et al. (eds.), *Recent Developments in Intelligent Systems and Interactive Applications*,
Advances in Intelligent Systems and Computing 541, DOI 10.1007/978-3-319-49568-2_22

using MapReduce [5], Traffic session identification based on statistical language model [6], Improved method for session identification in web log preprocessing [7], Session identification based on linked referrers and web log indexing [8]. The methods mentioned above, however, cannot eradicate the drawbacks of fixed threshold session identification scheme and achieve inappropriate division accuracy.

Considering the different operation practices in users' tabbed browsing, the paper proposes a novel session identification scheme through the use of different segmentation thresholds for different users and compare the fixed threshold session identification scheme with the novel session identification scheme in detail. We test the effectiveness of our scheme on two large-scale data sets from the computing environments in NASA and USTB to demonstrate its optimization performance. It is proved that the novel session identification scheme achieve appropriate division and raise the accuracy of the session identification by experiments.

2 Related Works

In this section, the characteristics of traditional browsing and tabbed browsing are described in detail. Furthermore, we have analyzed statistical data of questionaires which are necessary for the understanding of current users' browsing habits.

2.1 Traditional Browsing and Tabbed Browsing

In the early stage of the internet, users tend to browse the web page from the beginning to the end when they visit a website, and then open next web page they are interested to browse and repeat the process, which is the traditional browsing process. However, as the technology of the tabbed browsing has become more popular in practical applications, tabbed browsing changes users' browsing habits gradually. Users tend to open multiple web pages they are interested in and then browse the web page they opened one by one. The comparison of two browsing modes is shown in Fig. 1.

2.2 Survey of Users' Browsing Modes

We had conducted a questionnaire about user's browsing habits on Questionnaire Star (a professional online survey, evaluation, voting platform). As of October 2014, more than 530 copies of valid questionnaires covering 33 different areas from home and abroad were recollected. Statistical diagrams of survey results are shown in Fig. 2.

From the statistical results, respondents in the age of 15 to 30 years accounted for 90 % in the total number, which is consistent with the main age distribution of current Internet users. From the view of browsing mode, on the one hand, nearly 80 % of respondents prefer tabbed browsing. On the other hand, the current mainstream browsers generally used the form of tab from statistical data of browsers which were completed by Baidu from January to October in 2014 [9], which indicate that tabbed browsing has become more popular than the traditional browsing. Furthermore, users are more inclined to search for the target information directly when they visit websites,

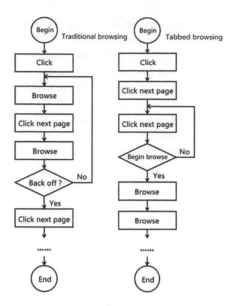

Fig. 1. The contrast of users' browsing

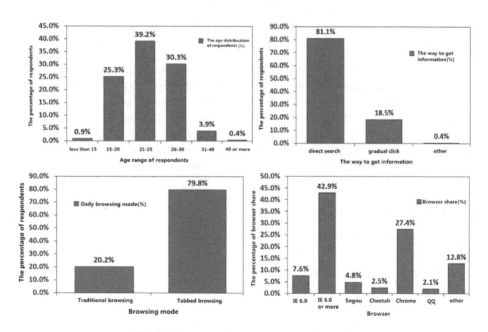

Fig. 2. The contrast of users' browsing mode

which indicates that users' choice of the entrance to websites is no longer limited by traditional home page and hierarchical classification structure. As a result, each level page may be the starting point for users to access.

2.3   Tabbed Browsing and Session Identification

Tabbed browsing as an increasingly popular pattern has changed the selection criteria of time threshold in the session identification. Current researchers tend to take browsing time of web page and session duration into account when they research session identification. But the tab-based browsing makes discrete distribution of the browsing time of web page centralize to the end of all clicks.

3   Fixed Threshold Session Identification Scheme

Time threshold was set artificially in fixed threshold session identification scheme. If the time difference (browsing time of web page) between the next web page and current web page in the user access sequence exceeds the time threshold, then the next web page is regarded as the starting point of a new user's session. Browsing time of web page is not recorded directly in the web log, but the access time of web page is recorded in the web log. Therefore, browsing time of web page is usually calculated according to the difference of access time of web page between the current web page and the next web page, namely $T_i = Tm_{i+1} - Tm_i$, where T_i is the browsing time of users' web page i, and Tm_{i+1} and Tm_i represents the access time of the next web page $i + 1$ and current web page i respectively. However, the browsing time of web page which is calculated by the fixed threshold session identification scheme may be inconsistent with the actual browsing time of web page due to a variety of reasons, for instance, users may leave for a while after they click on the web page, or browser crashes. Thus, we set time threshold to determine the target web page during the users' session.

$$T_{max} = \frac{\sum_{i=1}^{n} T_i}{n} \tag{1}$$

Where T_i represents browsing time of web page ranking the top 10 % according to the reverse order of browsing time. When $T_i > T_{max}$, let $T_i = T_{max}$. In addition, the browsing time of the last web page can't be calculated according to the method described above since there is no subsequent web pages when users access the last web page.

$$T_{last} = \frac{\sum_{i}^{|s|-1} T_i}{|s| - 1} \tag{2}$$

Where $|s|$ is the length of the access sequence.
Therefore, all the value of T_i can be drawn as

$$T_i = \begin{cases} Tm_i - Tm_i, 0 < i < |s| \\ T_{last}, i = |s| \\ T_{max}, T_i > T_{max} \end{cases} \tag{3}$$

4 Novel Session Identification Scheme

For the users' clicking sequence $P = \{p_1, p_2, p_3, \ldots, p_n\}$, when they visit websites, there is

$$\Delta CT_i = CT_{i+1} - CT_i, i = 1, 2, \ldots, n - 1 \qquad (4)$$

In the Eq. (4), n is the length of the sequence, and ΔCT_i represents the time when users click on the page p_i, and ΔCT_i represents the clicking time interval. T_i as browsing time of web page in the fixed threshold session identification scheme is a special case of ΔCT_i. The paper argues that each user's clicking habits are different and there should not set a uniform threshold to split users' sessions.

The paper presents an algorithm to calculate the session identification threshold: here clicking interval set $I = \{\Delta CT_1, \Delta CT_2, \Delta CT_3, \ldots, \Delta CT_{n-1}\}$. First of all, we rank the elements of according to the ascending order, and get a ordered vector $I' = \{\Delta CT'_1, \Delta CT'_2, \Delta CT'_3, \ldots, \Delta CT'_{n-1}\}$.

Algorithm1. the algorithm of calculating session identification threshold ΔCT_{THR} is given as follows:

Input: Clicking interval set: $I = \{\Delta CT_1, \Delta CT_2, \ \Delta CT_3, \ldots, \Delta CT_{n-1}\}$

Output: The value of session identification threshold:.

begin

    I'=sort I according to the ascending order;

    $\Delta CT_{THR} = 0$;

    $i=1$;

    while($i \leqslant$ n-1 and $\Delta CT_1 < 30$)

        $\Delta CT_{THR} = \Delta CT_i$;

        $i = i_+ 1$;

    end while

    if($\Delta CT_i \geqslant 30$)

        $\Delta CT_{THR} = (\Delta CT_{THR} + \Delta CT_i)/2$;

    end if

end begin

After the improvements, the session identification scheme will generate a user's clicking interval threshold ΔCT_{THR} according to the user's access characteristics. The threshold ΔCT_{THR} divide the access sequence which is more than 30 min into two sessions by referring to the characteristics of the fixed threshold session threshold (30 min). Meanwhile, dividing the access sequence on the basis of users' own access habits could be compatible with the fixed threshold session identification scheme and the current users' habits. After obtaining the threshold ΔCT_{THR}, we compare each element of I with the threshold ΔCT_{THR} by traversing the clicking interval set $I = \{\Delta CT_1, \Delta CT_2, \Delta CT_3, \ldots, \Delta CT_{n-1}\}$, if $\Delta CT_i > \Delta CT_{THR}$, then we consider that ΔCT_{THR} is the split point of the previous session and the next session.

Algorithm2. the algorithm of the novel session identification scheme is given as follows:

Input: the user's clicking sequence $P = \{p_1, p_2, p_3, \ldots, p_n\}$

Output: the split point ΔCT_i of the previous session and the next session and the split sequence H_j;

```
begin
    compute ΔCT_THR according to Algorithm1;
    j=0;
    for(i=1:n-1)
        if( ΔCT_i > ΔCT_THR )
            j=j+1;
            H_j={p_{i+1}} ;
        else
            H_j=H_j U {p_{i+1}} ;
        end if
    end for
end begin
```

5 Experiment and Evaluation

5.1 Date Preprocessing

Experimental environment include Windows 8.1, Mysql and Java. The experimental data set include the NASA (National Aeronautics and Space Administration) data set (1891714 records of official website) and USTB (University of Science and Technology Beijing) data set (4947573 records undergraduate teaching website).

Before the experiment, data pre-processing must be carried out to obtain "pure" data. The task of data pre-processing is to delete the data that has nothing to do with the algorithm of novel session identification scheme from the data set.

5.2 Metrics

Heuristic evaluation method has been implemented to evaluate experimental results of session identification, and we evaluate experimental performance by calculating the absolute coincidence rate and the relative coincidence rate between the identified session and the real session. With the increasing rate of both absolute coincidence rate and relative coincidence rate, we can get the better the performance of algorithm of session identification. The two coincidence rate and its computational method are presented as follows:

(1) Relative coincidence rate of session refers to relative coincidence degree between the session $H_d = \{p_1, p_2, p_3, \ldots, p_n\}$ which is identified by the session identification scheme and real session $H_r = \{p'_1, p'_{22}, p'_3, \ldots, p'_m\}$. The relative coincidence of session can be denoted as Eq. (5):

$$coin(H_d, H_r) = \frac{|H_d \cap H_r|}{m} \tag{5}$$

Obviously, the range of the coincidence rate of session is between 0 and 1. If coin $(H_d, H_r) = 1$, then the session was fully identified. In addition, it is noteworthy that $|H_d \cap H_r|$ of the Eq. (5) isn't the set of the same web page in H_d and H_r but the set of the same ordered web page in H_d and H_r. After all, we can't ignore the sequence of session when we concentrate on session identification.

Let S_d^H represent the identified session set, S_r^H represent the real session set, then $coin(H_d, S_r^H)$ represents the relative coincidence between session and session set and $coin(S_d^H, S_r^H)$ represents the relative coincidence between session sets. The computational methods are Eqs. (6) and (7) respectively.

$$coin(H_d, H_r) = \max\{coin(H_d, H_r)|H_r \in S_r^H\} \tag{6}$$

$$coin(S_d^H, S_r^H) = \frac{coin(H_d, S_r^H)}{n} \tag{7}$$

will be used to evaluate the effectiveness of session identification scheme.

(2) Absolute coincidence of two sessions refers that the relative coincidence of session between $H_d = \{p_1, p_2, p_3, \ldots, p_n\}$ and $H_r = \{p'_1, p'_2, p'_3, \ldots, p'_m\}$ is equal to 1, and the first and the last web page of H_d are the same as those of H_r. That satisfies:

$$(coin(H_d, H_r) = 1) \wedge (p_n = p'_m) \tag{8}$$

Absolute coincidence rate between two session sets refers to the ratio between the number of sessions which satisfies Eq. (8) and the number of sessions which belong to real session set S_r^H.

5.3 Performance Evaluation

We test the effectiveness of our scheme on two large-scale data sets from the computing environments in NASA and USTB to demonstrate its optimization performance. The number of sessions which were identified by the fixed threshold session identification scheme and the novel session identification scheme are shown as Table 1.

Table 1. The number of sessions

Data set	The number of real sessions	The number of sessions identified by fixed threshold scheme	The number of sessions identified by novel scheme
NASA	8361	8472	8417
USTB	2265	2254	2263

Figure 3(a) shows the number of session identified by fixed threshold scheme and novel scheme and real session on NASA data set (the early data set), and Fig. 3(b) shows that on USTB data set (the recent data set).

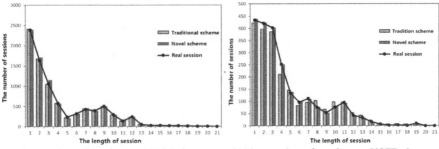

(a) The number of session on NASA data set   (b)The number of session on USTB data set

Fig. 3. The number of session on two large-scale data sets

From the results, the distribution of the session length appears to be different obviously due to the difference of website properties which was represented by NASA data set and USTB data set. Furthermore, it has little differences that NASA data set was identified by novel session identification scheme and fixed threshold session identification scheme, but the differences in recent data set (USTB data set) is more obvious. The length of session identified by fixed threshold session identification scheme is longer than that identified by novel session identification scheme in USTB data set, because users usually stay for a short time when they visit the website, but the fixed threshold make the sessions which should have been divided counted in a session so that the length of session increases and the number of session decreases.

From the Table 2, the coincidence rate of novel session identification scheme is close to that of fixed threshold session identification scheme on NASA data set. However, on the USTB data set, absolute coincidence rate and relative coincidence rate

Table 2. Session identification coincidence rate assessment results

Data set	Fixed threshold scheme		Novel scheme	
	Relative coincidence rate	Absolute coincidence rate	Relative coincidence rate	Absolute coincidence rate
NASA	0.932	0.870	0.933	0.872
USTB	0.941	0.865	0.972	0.906

of novel scheme is relatively larger than that of fixed threshold scheme. Through experiments mentioned above, the novel session identification scheme we proposed can better adapt to the recent and the early data sets, and the performance of novel session identification scheme on the recent data set is better than that on early data sets.

6   Conclusion

Nowadays, the fixed threshold session identification in the field of the session achieve unsatisfactory performance due to the improper division for sessions in some cases, and tabbed browsing analysis as an effective method in service computing deserves more attention. So we take the clicking habits of users into account and proposed a novel session identification scheme based on tabbed browsing in this article. The novel session identification scheme was compared with the fixed threshold session identification scheme detailedly. Furthermore, we test the effectiveness of our scheme on two large-scale data sets from the computing environments on NASA and USTB to demonstrate its optimization performance.

Acknowledgement. This work is funded by the National Key Technologies R&D Program of China under Grants 2013BAI13B06 and 2015BAK38B01, and the Fundamental Research Funds for the Central Universities under Grant 06500025.

References

1. Lu, L., Yang, Y.: Research on data preprocessing in web log mining web. Comput. Eng. **26** (4), 66–67 (2000)
2. Wang, X., Kang, L., Song, M., Wang, W.: Interactive behavior characteristic mining based on percolation theory for information spreading. J. Comput. Inf. Syst. **11**(1), 133–140 (2015)
3. Xinhua, H., Qiong, W.: Dynamic timeout-based a session identification algorithm. In: International Conference on Electric Information and Control Engineering, ICEICE 2011 - Proceedings, pp. 346–349 (2011)
4. Mary, S.P., Baburaj, E.: Performance enhancement in session identification. In: International Conference on Control, Instrumentation, Communication and Computational Technologies, ICCICCT 2014, pp. 837–840 (2011)
5. Huang, P., Chen, D., Le, J.: An improved referrer-based session identification algorithm using MapReduce. In: Proceedings - International Conference on Natural Computation, pp. 1072–1076 (2013)

6. Lou, X., Liu, Y., Yu, X.: Traffic session identification based on statistical language model. In: Motoda, H., Wu, Z., Cao, L., Zaiane, O., Yao, M., Wang, W. (eds.) ADMA 2013. LNCS (LNAI), vol. 8347, pp. 264–275. Springer, Heidelberg (2013). doi:10.1007/978-3-642-53917-6_24
7. Yuan-kang, F., Xue-gang, H., Qi-tao, X.: Improved method for session identification in web log preprocessing. J. Comput. Eng. 35(7), 49–51 (2009)
8. Baidu statistical Flow Research Institute: PC platform browser market share data. http://tongji.baidu.com/data/browser/
9. Shieh, J.-C.: From website log to findability. Electron. Libr. 30(5), 707–720 (2012)

Research of Optimizing the System Partition in Android System

Hong Zhu, Quanxin Zhang$^{(\boxtimes)}$, Lu Liu, Khaled Aourra,
and Yu'an Tan

Beijing Laboratory of Network and Information Security,
School of Computer Science, Beijing Institute of Technology,
Beijing 100081, People's Republic of China
{zhuhong0204,victortan}@yeah.net, zhangqx@bit.edu.cn,
skye66@126.com, khaledaourra@yahoo.fr

Abstract. With Google's Android system applied in the field of intelligence, the boot speed is one of the significant factors that affect the user experience. Traditional boot optimization in Android system cannot meet the requirements of starting the special service in the double system. Based on meticulous study of the System partition in Android system, we have designed a simplified system partition which is suitable for the Android dual-system on condition that it don't affect the use of customers.

Keywords: Android system · Dual-boot · System partition · Optimization method · Simplify

1 Introduction

Android is the open source mobile operating system based on Linux platform. The Android system is based on the Linux kernel, but it's not a standard Linux. The Android operating system with its influential, stability, wide compatibility and so on, promptly wins many people's favor. But its architecture is very large. During the stage of booting, it initialize a lot of function modules, which are actually not necessary for some platforms [1].

The Android system is based on Linux open-kernel operating system, the platform is composed by the operating system, middleware, user interface and application software [2]. It adopted the structure of the software stack which is mainly divided into three layers: the operating system, middleware, application software. In the term of function, it has four main functional layers from bottom up [3, 4]. There are Linux kernel, Android Runtime and Libraries, Application Framework, Application which are shown in Fig. 1.

After the Linux kernel started completely, the first user-level process init read and parse the script files, init.rc and init.xxx.rc, as shown in Fig. 2 [5]. Creating some key processes through the fork system, lots of Android Native Service which is defined in init.rc file will be created first, such as Console, Service Manager, Vold, Zygote, Media Server and so on [6]. The base of Android system is the Zygote process and its child process-the Android system service [7, 8].

© Springer International Publishing AG 2017
F. Xhafa et al. (eds.), *Recent Developments in Intelligent Systems and Interactive Applications*,
Advances in Intelligent Systems and Computing 541, DOI 10.1007/978-3-319-49568-2_23

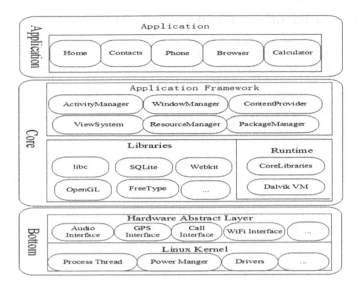

Fig. 1. The Android system structure

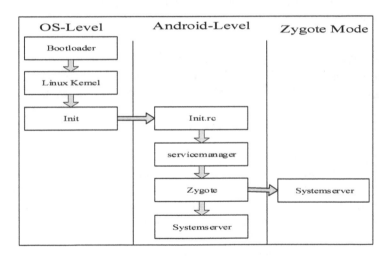

Fig. 2. The Android system booting sequence

The purpose of the system partition to simplify is that of the big system partition. When we need to dual-boot the phone and under the condition of without affecting the Android system to run, so we can transplant the simplified system partition to some place of the original operating system.

2   The System Partition in Android System

The Android system usually includes boot partition, kernel partition, system partition, data partition, cache partition and sdcard partition. And the boot partition enables the phone to boot, as the name suggests. It includes the bootloader and the kernel. Without this partition, the phone will not able to boot. The data partition contains the user's data – this is where the contacts, messages, settings and apps that you have installed. The cache partition is used for system upgrade and storing Over-the-Air upgrade package [9, 10].

The system partition mainly contains the entire operating system, other than the kernel and the bootloader. This includes the Android user interface as well as all the system applications that come pre-installed on the device. Wiping this partition will remove Android from the device without rendering it unbootable, and you will still be able to put the phone into recovery or bootloader mode to install a new ROM.

2.1   Simplify the System Partition

The mobile phone models in the experiment is Samsung Galaxy S6 G9200, and the version of the Android system is 5.0.2. We can use *df* command to view the system partition as shown in Fig. 3.

```
root@zerofltechn:/system # df /system
Filesystem    Size   Used     Free   Blksize
/system       2.9G   2.6G   292.4M      4.0K
```

Fig. 3. System partition usage

From Fig. 3, we can identify that the Android system allocates space of 2.9G for system partition, and the space of 2.6G is used. The role of system partition is same as the Disk C in Personal Computer, which is used for storing the system files. The partition basically includes the whole Android Operating System except kernel and ramdisk, which includes Android user interface and all pre-installed system applications.

The system/app directory deposits the system program, the system source program and mobile phone manufacturer's custom software. The system/priv-app directory deposits system key application such as launcher, system user interface, setting provider and so on. The system/lib directory deposits library files in the Android package program. The system/framework directory deposits the framework which is used for boot Android system, such as some .jar files. The system/etc. directory mainly deposits Android system configuration files, such as APN access point and many other important system configuration files. The system/build.prop file is property file. In the Android system .prop files are very important, that record settings and change of the system. The system/fonts deposits system font, we can download others *ttf* format to replace the font in the font to modify the system fonts.

From Fig. 4 we can calculate all files size in the system directory. For example, priv-app contains 29 percentage which the total size is about 2.6G so that the size of priv-app is 766 M. So the key of simplifying the system partition is the priv-app file, app file, framework file, lib file and lib64 file. Through a long period of exploration and attempt, the system partition in Android system is reduced to about 790 M from 2.6G as shown in Fig. 5. "du–sh/system" is one Linux command, it is often used of check the size of specific directory.

By calculating file size from Fig. 4 and Table 1, we can obtain that the size of system/priv-app is reduced to 90 M from 766 M, which is reduced by 88 %. The size of system/app is reduced to 12 M from 613 M, which is reduced by 98 %. The size of system/lib is reduced to about 99 M from 200 M, which is reduced by almost 50 %. The size of system/fonts is reduced to about 12 M from 47 M, which is reduced by 74 %.

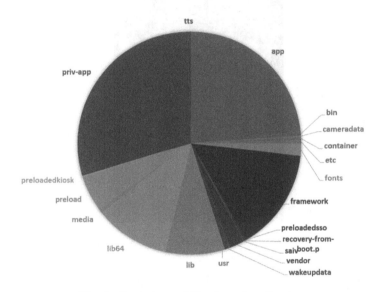

Fig. 4. Percentage of files in system directory

```
root@zerofltechn:/ # du -sh /system
790M    /system    _
```

Fig. 5. System partition size after simplify

3 Experiments and Analyses of Simplifying System Partition

After simplifying the system partition in Android system, the phone desktop is same as Fig. 6. There are only Setting, Camera, Photo, Video and Record program in the mobile phone. All Other programs in the phone is deleted. We can't use the simplified

Table 1. Files of the system partition after simplifying

File name	Size	File name	Size	File name	Size
app/	12 M	build.prop	8 K	etc./	13 M
framework	351 M	bin/	15 M	cameradata/	4.2 M
fonts/	12 M	lib/	99 M	lib64/	116 M
lkm_sec_info	4.0 K	media/	8 K	priv-app/	90 M
saiv/	2.0 M	usr/	19 M	vendor/	57 M
xbin/	924 K				

Fig. 6. The phone desktop after simplified

phone to calling up, sending messages and surfing Internet and so on. So the role of the phone is taking photos and recording in some special condition.

Although the space used by the current system partition is reduced to about 800 M, but the overall size of the system partition has not changed, still is 2.9G. So the next task is to change the whole system partition size, and check integrity of system partition when booting the phone.

At this time we need two Linux command: e2fsck and resize2fs. The command e2fsck is used to check validity of image file. The command resize2fs is the tool for resizing the ext4 file system. For example, we can use the command - "resize2fs system.img 800 M" to change the size of system.img file.

We can extract image file from the system partition, can put it into data partition of another mobile phone. We usually use the phone without simplified system under normal circumstances. As soon as coming across some special condition, we can switch from un-simplified to the Android simplified system and the time of starting up the phone also become shorten.

4   Conclusion

This paper has simplified an Android system in Samsung mobile phone based on already exist Android system. Experiments shows that, it is feasible. More kinds of optimization methods of startup Android system should be research overall so that the time of starting up the phone become more and more short.

Acknowledgements. This research was supported by the National Natural Science Foundation of China (No. 61370063, No. 61272511, No. 91438117, No. 91538202) and the Scientific Research Foundation for the Returned Overseas Chinese Scholars, State Education Ministry.

References

1. Cai, C.: Research and Optimization of Booting and Application Loading Performance Based on the Android system. Southeast University (2013)
2. Yang, F.: Android Internals: System. China Machine Press, Beijing (2010)
3. Android Developers. Android Architecture (2011). http://developer.android.com/guide/basics/whais-android
4. Zeng, J., Shao, Y.: Study of android's system architecture and application development. Microcomput. Inf. **27**(9), 1–3 (2011)
5. Singh, G., Binpin, K., Ahawan, R.: Optimizing the boot time of Android on embedded system. In: 2011 IEEE 15th International Symposium on Consumer Electronics (ISCE), Singapore, pp. 503–508 (2011)
6. Wang, Q.: A brief analysis on the framework and its Kerncl of Android embedded system. Comput. Dev. Appl. **24**(4), 59–61 (2011)
7. Han, C., Liang, Q.: Android System-Level Depth Development: Transplant and Debug. Electronic Industry Press, Beijing (2011)
8. Wu, Q., Zhao, C., Guo, Y.: Android Security Mechanism and Application Practices. China Machine Press, Beijing (2013)
9. Guo, G.: The Optimization Study of Android Boot Process. Lanzhou University, Lanzhou (2012)
10. Raja, H.Q.: Android Partitions Explained: boot, system, recovery, data, cache&misc (2011). http://www.addictivetips.com/mobile/android-partitions-explained-boot-system-recovery-data-cache-misc

The Research of User Satisfaction Model in Hybrid Cloud Environment

Li Hao, Huang Shaowei[⊠], and Zhang Jianhong

Software College, Yunnan University, Kunming, China
c3349354@gmail.com

Abstract. It's a major problem to evaluate properly the merits of the hybrid cloud computing service, the quality of service (QoS) reflect the performance of cloud computing directly, but which can exactly reflect the user's satisfaction is the quality of experience (QoE), the customer satisfaction evaluation model based on QoE hybrid cloud computing can be properly assessed. This paper will introduce cloud bank model architecture, and put forward a kind of evaluation model based on QoE that can be used for customer satisfaction evaluation in the hybrid cloud.

Keywords: Hybrid cloud · User satisfaction · Quality of experience · Cloud bank

1 Introduction

In essence, a hybrid cloud computing is to build new architecture in multiple cloud computing service providers. Because hybrid cloud usually involves with different domain of technologies and suppliers, and face the business process requirements that customers keep updating, this means that there is the key to solve this problem by a platform with complex management ability. And the problem that how to establish a complete hybrid cloud index evaluation system is need to solve.

2 The Brief Introduction to the Cloud Bank

Cloud bank is a model for managing computing resources, it applies the bank model of economics to the trading of hybrid cloud computing resources, establish the market mechanism of cloud architecture from the viewpoint of macroeconomics, and through the macroeconomic regulation to control the relationship between aggregate supply and demand.

As shown in Fig. 1 Cloud Bank Model mainly includes five modules: physical resource pool, the pool of SLA, resist risk, scheduling and pricing policy.

The operation in each module of model is realized by the QoS and based on SLA. The providers of resource is stored in a SLA pool, in the form of SLA document. The SLA pool is equivalent to a virtual description of the physical resource pool, risk prevention and cloud bank resource scheduling that is implemented by calling the SLA document. The upper pricing strategy specifies the most reasonable price for the

© Springer International Publishing AG 2017
F. Xhafa et al. (eds.), *Recent Developments in Intelligent Systems and Interactive Applications*,
Advances in Intelligent Systems and Computing 541, DOI 10.1007/978-3-319-49568-2_24

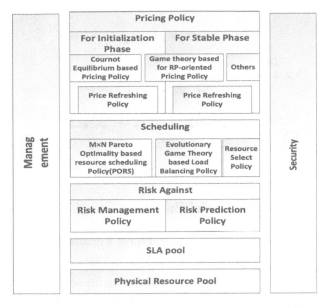

Fig. 1. The architecture diagram of cloud bank model

resource based on market supply and demand situation in order to achieve the maximum interests of the participants.

For the SLA related QoS metrics in cloud services, in the process of SLA, the service provider and service consumer may negotiate the content of the agreement. But compared with QoS, QoE parameters can reflect the user's service experience more.

3   The Index System Building of the QoE of the Cloud Bank

Because of the cloud bank service quality has a great dependence on the network performance, so the network performance indicators that includes throughput and transfer delay, connectivity, etc. [1]. It should be received enough attention.

QoS evaluates business based on the technical performance of cloud computing. QoE stresses how to analyze business based on the point of view of customer service, and regard the business as a whole and to consider all aspects of the impact on business performance. In this sense, QoS is a subset of QoE, as QoS has to ensure or enhance QoE and be applied in the network of technical indicators. Therefore, to get the desired user experience quality is the key to ensure the profitability of the operator, so this is also an important issue needed to be further studied in cloud bank.

As shown in Fig. 2, the QoE levels and parameters and QoS parameters can be expressed as a three layer structure model, the top layer is the QoE indicators layer, the middle layer is QoE parameter layer, which describes the related information of user experience, can obtain the content of user's perception by user survey, and acquire the user satisfaction by the comprehensive analysis; the third layer is the QoE parameter which reflects the business performance, generally refers to the end-to-end quality of service (QoS) [2]. It can be seen that the QoE parameters can directly affect the QoS parameters.

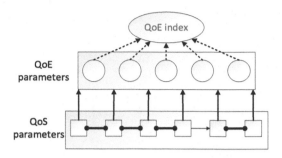

Fig. 2. QoE/QoS three layer structure model

4 The QoE Evaluation Model Based on Cloud Bank

4.1 The Brief Introduction of Linear Structural Equation Model

The user satisfaction model ACSI is in fact a linear structure model, model including the latent variables, the manifest variable, the direct influence, and the load. Hidden variable is the hypothetical variable that cannot be observed, the manifest variable is the variable that can be observed [3]. In ACSI, hidden variables include customer expectation, quality perception, value perception, customer satisfaction, customer complaint and customer loyalty. The external factors that affect their and can be measured are the significant variables.

Exogenous variable is the variable which is not affected by other variables, endogenous variable is defined as a variable which is determined by one or more variables. In ACSI, in addition to customer expectation are exogenous variables, other variables are endogenous variables.

The linear structural equation model is a set of equations that reflect the relationship between implicit and explicit variables. The purpose is to deduce the hidden variable by the measurement of the manifest variable, and to verify the model. The linear structural equation model can be expressed by mathematical expressions [4].

4.2 The QoE Evaluation Model

As shown in Fig. 3, in the latent variable, due to the introduction of the QoS parameters, perceived quality should be divided into service quality perception and technology quality perception, they respectively describe the degree of excellence of user's judgment on the service and system technical, in addition to the original indicators of ACSI model structure, it's required to add a variable called "technical index", used to represent the reliability, response time, security and usability of cloud services [5].

In this model, the user expectations and the technical specifications are exogenous variables, and the rest are endogenous variables.

And the manifest variables in the model are (Table 1):

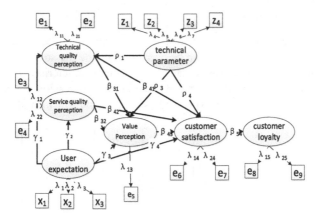

Fig. 3. The QoE evaluation model in cloud bank

Table 1. The manifest variables

The manifest variables	Content
x_1 (customer expectation)	The cloud services network speed/the maximum bandwidth provided by the cloud bank
x_2 (customer expectation)	The number of resources that used by user such as the computing, storage, network, and monitoring/the total number of resources provided by the operator such as cloud computing, storage, network, and monitoring
x_3 (customer expectation)	The number of user using the cloud bank resources/the total number of users
e_1 (technical quality perception)	Successful times of using resources/the total number of using resources
e_2 (technical quality perception)	Delay time/expected completion time
e_3 (service quality perception)	Cloud bank service satisfaction rate
e_4 (service quality perception)	The number of terminating the cloud bank business/the number of cloud bank resources used
e_5 (value perception)	The cost of the user to use the cloud bank resources/all of the user's cloud bank charges
e_6 (customer satisfaction)	The time for the user to use this cloud bank resource/the total time for the user to use cloud bank resources
e_7 (customer satisfaction)	The number of tasks completed in the expected time/the total number of tasks submitted
e_8 (customer loyalty)	If it is, set the value to 1; otherwise 0
e_9 (customer loyalty)	If it is, set the value of 1; whereas 0
z_1 (technical parameters)	The number of successful execution of the cloud service/the total execution times
z_2 (technical parameters)	The percentage of remote access time
z_3 (technical parameters)	If the security level of the cloud bank services is SSL, its value is 1, if the level is WSS, then the value is 2, and the rest are 0
z_4 (technical parameters)	The average value of multiple response time

4.3   The Structure Model

Structural equation can be used to measure the relationship between the measure indexes and the latent variables, and can be used to calculate the structural equation and measurement equation simultaneously. Because of this characteristic of the structural equation model, it can get more accurate result than the regression analysis [6].

The above QoE model can be expressed as a mathematical formula:

$$
\begin{bmatrix} \eta_1 \\ \eta_2 \\ \eta_3 \\ \eta_4 \\ \eta_5 \end{bmatrix} = \begin{bmatrix} 0 & 0 & 0 & 0 & 0 \\ 0 & 0 & 0 & 0 & 0 \\ \beta_{31} & \beta_{32} & 0 & 0 & 0 \\ \beta_{41} & \beta_{42} & \beta_{43} & 0 & 0 \\ 0 & 0 & 0 & \beta_{54} & 0 \end{bmatrix} \times \begin{bmatrix} \eta_1 \\ \eta_2 \\ \eta_3 \\ \eta_4 \\ \eta_5 \end{bmatrix} + \begin{bmatrix} \gamma_1 \\ \gamma_2 \\ \gamma_3 \\ \gamma_4 \\ 0 \end{bmatrix} \times \xi + \begin{bmatrix} \rho_1 \\ 0 \\ \rho_3 \\ \rho_4 \\ 0 \end{bmatrix} \times \omega + \begin{bmatrix} \zeta_1 \\ \zeta_2 \\ \zeta_3 \\ \zeta_4 \\ \zeta_5 \end{bmatrix}
$$

(1)

It can be abbreviated as:

$$
\eta = B\eta + \Gamma\xi + P\omega + \zeta
$$

(2)

$\eta_1, \eta_2, \eta_3, \eta_4, \eta_5$ respectively represent technical quality perception, service quality perception, value perception, customer satisfaction, and customer loyalty. ξ represents the customer expectation, ω represents the technical parameter, and ζ represents the disturbances term.

4.4   The Measurement Model

$$
\begin{bmatrix} x_1 \\ x_2 \\ x_3 \end{bmatrix} = \begin{bmatrix} \lambda_1 \\ \lambda_2 \\ \lambda_3 \end{bmatrix} \times \xi + \begin{bmatrix} \delta_1 \\ \delta_2 \\ \delta_3 \end{bmatrix}
$$

(3)

$$
\begin{bmatrix} z_1 \\ z_2 \\ z_3 \\ z_4 \end{bmatrix} = \begin{bmatrix} \lambda_4 \\ \lambda_5 \\ \lambda_6 \\ \lambda_7 \end{bmatrix} \times \omega + \begin{bmatrix} \varphi_1 \\ \varphi_2 \\ \varphi_3 \\ \varphi_4 \end{bmatrix}
$$

(4)

$$
\begin{bmatrix} e_1 \\ e_2 \\ e_3 \\ e_4 \\ e_5 \\ e_6 \\ e_7 \\ e_8 \\ e_9 \end{bmatrix} = \begin{bmatrix} \lambda_{11} & 0 & 0 & 0 & 0 \\ \lambda_{21} & 0 & 0 & 0 & 0 \\ 0 & \lambda_{12} & 0 & 0 & 0 \\ 0 & \lambda_{22} & 0 & 0 & 0 \\ 0 & 0 & \lambda_{13} & 0 & 0 \\ 0 & 0 & 0 & \lambda_{14} & 0 \\ 0 & 0 & 0 & \lambda_{24} & 0 \\ 0 & 0 & 0 & 0 & \lambda_{15} \\ 0 & 0 & 0 & 0 & \lambda_{25} \end{bmatrix} \times \begin{bmatrix} \eta_1 \\ \eta_2 \\ \eta_3 \\ \eta_4 \\ \eta_5 \end{bmatrix} + \begin{bmatrix} \varepsilon_1 \\ \varepsilon_2 \\ \varepsilon_3 \\ \varepsilon_4 \\ \varepsilon_5 \\ \varepsilon_6 \\ \varepsilon_7 \\ \varepsilon_8 \\ \varepsilon_9 \end{bmatrix}
$$

(5)

5   Experiment

The QoE measurement and the QoS measurement is the significant basis for the realization of QoS control (admission control, resource allocation, traffic engineering, etc.), SLA ensure and QoS billing [7].

We will use cloudsim to simulate and built the experimental environment in this study. It provides the function of the virtual engine, which is proficient to realize the virtual service on the node of the data center. These virtualized services is multiple, cooperative and independent. CloudSim simulator using hierarchical structure, consists of Sim Java, Grid Sim, Cloud Sim and User code.

User code layer is the expansion of the user's own research on the platform. When the platform is regenerated, you can call in the classes, methods, member variables, etc., that have been written by themselves in simulation program. In this paper, the experimental testing is in the user code layer which modifies the module and inserts the measurement code and then run the simulation program which can get the measured data.

Using SAS software, to input measurement model (1), (3), (4) and (5) with LINEQS statement, and then import 16 measurements of cloudsim simulation experiment obtained by the sample value of the manifest variables. SAS will output 5 endogenous variables, the weight of the interaction between the two exogenous variables and the load of the corresponding parameters of the variable.

Enter the data obtained by the CloudSim, you can get the QoE model as shown below (Fig. 4).

At the same time, the evaluation parameters of the model can be obtained:

According to the value of the index in Table 2, we can draw the conclusion that the QoE model can fit well.

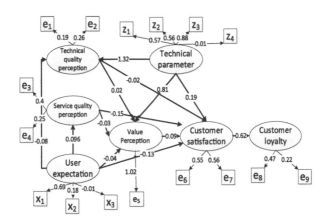

Fig. 4. The influence factors and the loads of lattice variables in QoE model

Table 2. The evaluation parameters of QoE model

Index	SRMR	BCFI	RMSEA	AGFI
Value	0.0231	0.9584	0.0318	0.9814

6 The QoE Estimation of Cloud Bank System

According to the above results, we can calculate the QoE value of each cloud bank user (Table 3).

Table 3. The example data of cloud computing

Variable	x_1	x_2	x_3	e_1	e_2	e_3	e_4	e_5	e_6	e_7	e_8	e_9	z_1	z_2	z_3	z_4
User A	0.41	0.05	0.4	0.65	0.43	0.24	0.45	0.69	0.43	0.27	1	0	0.56	0.59	2	3
User B	0.39	0.35	0.49	0.5	0.26	0.7	0.42	0.65	0.27	0.4	1	0	0.21	0.43	0	8

The process of calculating the exogenous variables of user A is as follows:

$$\zeta = x1 * 0.69 + x2 * 0.18 + x3 * -0.01$$
$$= 0.41 * 0.69 + 0.05 * 0.18 + 0.4 * -0.01 = 0.29$$

In the same way, the value of five endogenous variables and one exogenous variable can be got:

$$\eta_1 = 0.23, \ \eta_2 = 0.2, \ \eta_3 = 0.71, \ \eta_4 = 0.38, \ \eta_5 = 0.47, \ \omega = 2.38$$

$$QoE = \left(\sum_{i=1}^{n} P_i w_i / \sum_{i=1}^{n} w_i \right)$$

Let $n = 7, w_i = 1$, using QoE (Ca) to express the QoE value of A, can get:

$$QoE(Ca) = (0.29 + 0.23 + 0.2 + 0.7 + 0.38 + 0.47 + 2.38)/7 = 0.66$$

Similarly, you can get the QoE value of B:

$$QoE(Cb) = (0.3 + 0.16 + 0.39 + 0.67 + 0.37 + 0.47 + 0.28)/7 = 0.38$$

7 Conclusion

Based on rough comparison of measured data, we can observe that the user B has a higher user experience than the user A, but the calculated QoE value can make it more intuitive and quick. In addition, cloud service providers also can conduct a further analysis of the lower QoE users to help to identify the reasons for the reduction of user experience. For example in the QoE value calculation, the technical parameters of the user B value is less than the user A, and according to the technical parameters of manifest variables measured data we can see user A use the technical level of cloud services is much higher than user B. From this we can know that the cloud service providers can improve the security and performance of cloud services to improve the user experience, in order to attract more users.

So we can observe that the QoE model has a good reference value for the service providers of cloud bank.

References

1. Paxson, V.: Towards a framework for defining Internet performance metrics. In: Proceedings of INET, vol. 96, pp. 50–56 (1996)
2. Du, Y., Zhou, W., Chen, B., et al.: A QoE based evaluation of service quality on WAP in wireless network. In: 5th International Conference on Wireless Communications, Networking and Mobile Computing, WiCom'09, pp. 1–4. IEEE (2009)
3. Yong, L.: Study on the customer satisfaction index model and its evaluation method. China Mining University, Beijing (2008). In Chinese
4. Hou, J.T., Wen, Z., Cheng, Z.J.: Structural equation model and its application. Science and Education Press (2004)
5. Yunting, L.U., Zhenjun, L.I.: AQoE Evaluation model of cloud computing based on the user satisfaction model. Computer & Digital Engineering (2014)
6. Wenjing, W.: QoE evaluation and optimization of mobile cloud computing. Beijing University of Posts and Telecommunications (2013). In Chinese
7. Lun, Z.F.M.S.Y.: Mobile telecommunication network QoE measurement and quantification method. J. Electron. Measur. Instrum. **24**(3), 230–236 (2010)

Research into Information Security Strategy Practices for Commercial Banks in Taiwan

ShiannMing Wu[(⊠)] and Dongqiang Guo

College of Business Administration, National Huaqiao University, Fujian, China
wuming.hqu@gmail.com

Abstract. With the explosion of the internet, computers and networks have become an indispensable part of life, and due to the rise of E-commerce, enterprises can increase productivity and competitiveness via internet. All things related to the internet have produced new and inspiring opportunities in data access, commercial operations, and productivity. The recent big data, cloud computing, mobile commerce, internet, Blockchain, FinTech, and other network applications, as well as P2P transactions, all force electronic transaction patterns to undergo drastic change. The formulation of an information security strategy is of great importance to an organization's success and continued existence.

Keywords: E-commerce · Information security · Blockchain · Fintech · P2P transaction

1 Introduction

In recent years, major countries have vigorously promoted new information security policies to respond to the threat of information security, for instance, the information security issue has been improved to the level of national security to emphasize nationally related personal information and the protection of privacy rights; security issues related to cloud computing are gradually valued. Regarding the international standards for information security, ISO27001, ISO27002, ISO15408, ISO13335, COBIT, ITIL, PCIDSS, etc., are widely applied. ISO27001 is the most commonly used international standard to promote domestic information security management systems; the spirits of ISO/IEC 27001: 2013 included:Conform to ISO Annex SL for simplify integration with other risk management systems, refer to ISO 31000 unknown risk management requirements, sets risk monitoring and measurement indicators. The information related to security standards should be dominated by FISMA, HIPAA, COSO, SOX, and other relevant acts. NIST SP 800 s document, as well as other technical criteria are followed by commercial banks in Taiwan.

2 Literature Review

According to Lin and Chang (2009, pp. 10–3), confidentiality, integrity, and availability are thought to be the three elements of information security, called the three principles of information security, and referred to as CIA. Abidance of the main

F. Xhafa et al. (eds.), *Recent Developments in Intelligent Systems and Interactive Applications*,
Advances in Intelligent Systems and Computing 541, DOI 10.1007/978-3-319-49568-2_25

principles can master the main points of information security. The universal goal of any computer security policy is to protect the CIA of data stored in information system (Wu 2015); to put it more concretely, (1) confidentiality (2) integrity. (3) availability. According to the ISO 27001 standard, a set of all-around information security management framework is provided (Jayawickrama 2006), which is widely used by countries and enterprises around the world. At present, ISO 27001 has become the common language of information security management (Humphreys 2008).

The responsibilities between the three defensive lines in IT risk management are unclear; the problems are, as follows: placing particular stress on the independent ex post IT risk review and audit, belittling advance risk warnings and control difficulties of processes embedded in an event, as well as the real-time warning treatment, insufficiency in the construction of IT management and automation, lower manual or semi-automatic risk review efficiency and business value, shortage of total involvement, etc. (Ching 2013, p. 12). The purpose of this research is to discuss the informatization security level of the banking industry in Taiwan, the opinions regarding information risk, including the possible threat of information risk of the current and future banking industry in Taiwan, and the relative protective strategies and measures adopted by this industry, in order to understand the appropriateness in the formulation of information security strategies for the banking industry in Taiwan (Fig. 1).

3 Implementation Procedures

The implementation of an information security plan has huge challenges, and must consider many aspects, including encryption, application security, disaster recovery, and other compliance issues. Enterprises must abide by many regulatory requirements, such as, HIPAA, PCI DSS, and Sarbanes-Oxley. The security standards most commonly quoted by the industry are, as follows: (1) COBIT. (2) ISO 27001 series provides a wide scope of information security frameworks to be applied to all types and sizes of enterprises. (3) NIST SP 800 series, collected large quantities of information security standards and best practice documents. The American government uses NIST SP 800-53 to follow the 200 requirements stipulated in FIPS. Enterprises can use this channel program to comply with various kinds of regulations, such as HIPAA, Sarbanes Oxley, PCI DSS, and GLBA (Fig. 2).

Commercial banks in Taiwan are required to implement strict information security in all domestic industries, as transaction security and client's personal data are valued by competent authorities, and it is necessary to conform to the laws and regulations. Therefore, regarding the issue of risk security guarantees; it is necessary to consider the features of organizations, importance of the business, degree of impact and damage, endurable reply time (acceptable down time), reply cost, and other factors, to formulate and implement the emergency response measures of various businesses. Yang (2001) put forward a diagram on the security control relationship of an information system, where control items include: (1) division of functions and responsibilities in the information department; (2) system development, program modification, and control; (3) controlled system documents; (4) controlled data access and processing;

184   S. Wu and D. Guo

(5) controlled data Input and Output; (6) controlled files and equipment; (7) controlled System hardware and system software; (8) controlled recovery plan.

With the progress of automation and growth of business volume, banks gradually pay more attention to information security in practice, as well as during the course of the standardization of laws and regulations. In the infrastructure of implementing information security, project blocks related to information security are summarized in a diagram, and implemented one by one.

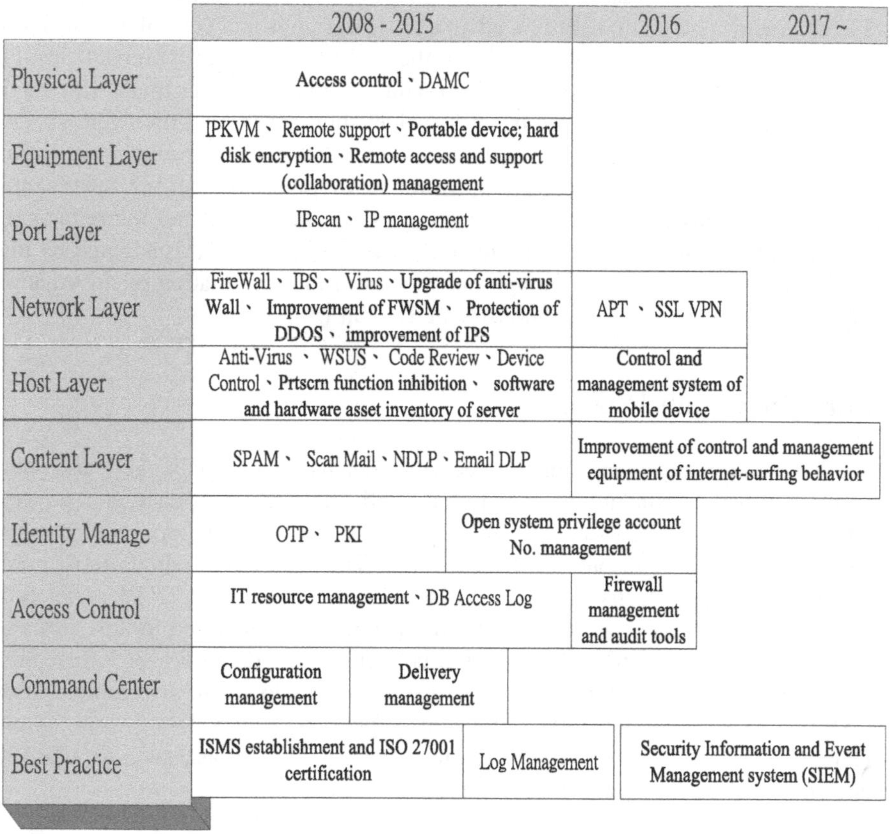

	2008 - 2015	2016	2017 ~
Physical Layer	Access control、DAMC		
Equipment Layer	IPKVM、 Remote support、Portable device; hard disk encryption、Remote access and support (collaboration) management		
Port Layer	IPscan、 IP management		
Network Layer	FireWall、 IPS、 Virus、Upgrade of anti-virus Wall、 Improvement of FWSM、 Protection of DDOS、 improvement of IPS	APT、 SSL VPN	
Host Layer	Anti-Virus、 WSUS、 Code Review、Device Control、 Prtscrn function inhibition、 software and hardware asset inventory of server	Control and management system of mobile device	
Content Layer	SPAM、 Scan Mail、NDLP、Email DLP	Improvement of control and management equipment of internet-surfing behavior	
Identity Manage	OTP、 PKI	Open system privilege account No. management	
Access Control	IT resource management、DB Access Log	Firewall management and audit tools	
Command Center	Configuration management	Delivery management	
Best Practice	ISMS establishment and ISO 27001 certification	Log Management	Security Information and Event Management system (SIEM)

Fig. 1. Informatiaon security architecture blueprint

Risk sources place emphasis on the need for the formulation of an information security strategy. Regarding the formulation of an information security strategy, many scholars have put forward to assist organizations to establish a feasible information security policy in a systematic manner. Generally speaking, the main factors influencing an organization's information security strategy come from four parts of the organization's information environment, including the information resources of the organization's internal environment, the organization's features, the information technology opportunities of external environments, and the organization's industrial status.

Regarding risk awareness and the application of technology standards, Rosen (2004) thought that, while information technology effectively reduces the risks of the traditional business of commercial banks, information technology in itself brings new risks for commercial banks. Information System Audit and Control Association (COBIT) states the implementation guidelines and audit standards, as based on the concept of IT governance, and is oriented to IT governance in the course of IT construction. COBIT5 plays the role of a connecting link to ensure the consistency of enterprise operations and IT strategy.

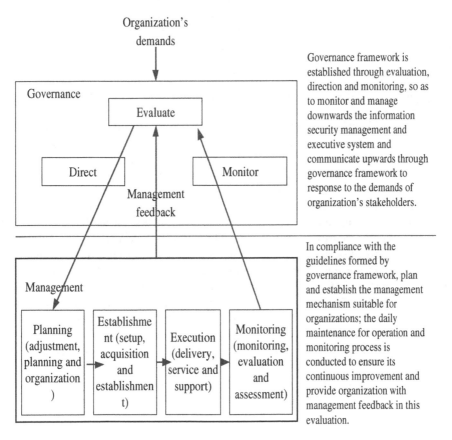

Organization's demands

Governance

Evaluate

Direct

Monitor

Management feedback

Governance framework is established through evaluation, direction and monitoring, so as to monitor and manage downwards the information security management and executive system and communicate upwards through governance framework to response to the demands of organization's stakeholders.

Management

Planning (adjustment, planning and organization)

Establishment (setup, acquisition and establishment)

Execution (delivery, service and support)

Monitoring (monitoring, evaluation and assessment)

In compliance with the guidelines formed by governance framework, plan and establish the management mechanism suitable for organizations; the daily maintenance for operation and monitoring process is conducted to ensure its continuous improvement and provide organization with management feedback in this evaluation.

Fig. 2. Key fields for information security governance and management Source: ISACACOBIT5

Risk refers to the uncertainty of issues influenced by internal and external factors faced by organizations of all types and sizes during the process of achieving its goals, including its degree of change and the possibility of accident occurrence. Based on the main principles of ISO regarding management system standards and design, the structure of ISO31000 also adopts the PDCA principle for continuous improvement, including the two interactive cycles of Risk Management and Managing Risk. According to the guideline principles, a risk management framework is designed to carry out risk management (Huang 2010).

4   Conclusion

Obstacles still exist in the execution of information security. In accordance with investigations conducted by various domestic industries, the Institute for Information Industry considers incalculable ROI, insufficient expenditures, and shortage of professionals as the three difficulties for the implementation of information security. Based on incalculable ROI, insufficient expenditure, shortage of professionals, enterprises are advised to consider adopting the model of outsourcing information security, as outsourcing can be used to reduce the establishment costs of information security, and to solve the problem of the shortage of talent. In personnel training, enhancement of the concept of information security is the most important, followed by the introduction of technology.

As an expert in information security, Huang (2015) suggested that enterprises must master 3 information security management principles: (1) data monitoring and big data analysis, which are used to identify any abnormal behaviors hidden in normal flow, in order to determine, in advance, the points that may suffer from hacker attacks, and to repair them for the prevention of risks; (2) continuous education and advocacy of information security is conducted among staffs; while a good information security prevention mechanism cannot avoid intentional or unintentional leakages by internal staffs; information security personnel should continue to learn and intensify their cognition of information security attacks, and master the latest trends, in order that they can know how to prevent security breaches; (3) increase the visibility of information security incidents; invisibility of information security risks do not represent inexistence. Enterprises can steadfastly implement the scope criteria, as stipulated by laws and regulations, to achieve the goal of information security in spite of not introducing international standards to assess information security risks.

This research only summarizes identified relevant research discussions and practices. Due to the limited length of this paper, there are many cross-domain projects or technical issues which are not discussed, thus, future scholars are expected to share more research achievements in order to rapidly develop information applications for a safer digital society.

References

Research, Development and Evaluation Commission: Executive Yuan, Risk Management Operation Manual Ver 3.0
http://www.rdec.gov.tw/DO/DownloadControllerNDO.asp, 31 December 2013
Global Technology Audit Guideline (GTAG): The Institute of Internal Auditors-Chinese Taiwan, June 2011

Li, P.C.: Discussion on IT Control Framework COBIT. Taiwan Stock Exchange Corporation, September 2009

Hong, K.S., Chao, R.Y.: Discussion on the theory of information security management. Comments Inf. Manage. **12**(6), 17–47 (2003)

Liu, Y.L.: Formation mechanism and comprehensive evaluation of synergistic effect for virtual enterprises knowledge. Commer. Econ. **9**, 99–101 (2009)

Hang, G.M.: Analysis on high-level development strategy for bank information security. Agri. Dev. Financ., **1** (2006)

Huang, K.P.: Discussion on requirement and application of ISO31000 risk management, Bimonthly on Development of Sustainable Industry, no. 53, December 2010

Yu, S.W.: Risk management idea and framework for advanced Countries. National Central University (2007)

Chen, C.M., Yeh, K.C.: The Late International Standards on Information Security Management–Brief Discussion on the Emphasis of New Version of ISO 27001, Deloitte & Touche Communication, December 2013

The Global Risks Report: World Economic Forum, January 2015

Cryptanalysis of an Efficient and Secure Smart Card Based Password Authentication Scheme

Chi-Wei Liu[1], Cheng-Yi Tsai[1], and Min-Shiang Hwang[1,2(✉)]

[1] Department of Computer Science and Information Engineering, Asia University, 500, Lioufeng Road, Wufeng, Taichung 41354, Taiwan, R.O.C.
mshwang@asia.edu.tw
[2] Department of Medical Research, China Medical University Hospital, China Medical University, No. 91, Hsueh-Shih Road, Taichung 40402, Taiwan, R.O.C.

Abstract. The user authentication scheme has been widely applied to verify the users' legality. In order to enhance the security, the smart card has widely used in an authentication scheme. Recently, Liu et al. shown that some weaknesses exist in Li et al.'s scheme. An efficient and secure user authentication scheme with a smart card presented by them is more efficient and secure than other schemes. However, the security issues of their scheme proposed by them also exist, so we will demonstrate that their scheme is vulnerable to the replaying attack.

Keywords: Password · Smart card · User authentication

1 Introduction

The user authentication scheme has been widely applied to verify the users' legality. Many password-based user authentication schemes have been proposed to verify the remote users' identification [1–16]. However, the password is easy to be exposed by guessing attack. In order to enhance the security, the smart card has widely used in an authentication scheme [18–30].

Recently, a robust smart-card-based remote user password authentication scheme [5] was proposed by Chen et al. However, Li et al. pointed out some weaknesses (i.e., forward secrecy and wrong password login problem) in Chen et al.'s scheme [14]. Li et al. also proposed an enhanced smart card based user authentication scheme [14]. However, Liu et al. shown that Li et al.'s scheme was unable to against the man-in-the-middle and insider attacks [17]. An efficient and secure user authentication scheme with a smart card proposed by them is more efficient and secure than other schemes. However, the security issues of their scheme proposed by them also exist, so we will exhibit that, their scheme is vulnerable to the replaying attack.

The rest of this paper is organized as follows. In Sect. 2, we briefly review Liu et al.'s user authentication scheme. In Sect. 3, we analyze and show that some security weaknesses in Liu et al.' user authentication scheme. Finally, we present our conclusions in Sect. 4.

© Springer International Publishing AG 2017
F. Xhafa et al. (eds.), *Recent Developments in Intelligent Systems and Interactive Applications*,
Advances in Intelligent Systems and Computing 541, DOI 10.1007/978-3-319-49568-2_26

2   Review of Liu-Chang-Chang Scheme

In this section, Liu et al.'s user authentication scheme (Liu-Chang-Chang Scheme) with a smart card [17] has been briefly reviewed. Liu-Chang-Chang's user authentication scheme has three participants: a user (U for short), a smart card (C for short), and a server (S for short). The scheme is composed of four phases such as registration, login phase, authentication phase, and e password change phase. The notations used in this paper are listed in Table 1.

Table 1. The notations used in this paper

Notations	Meaning
U_i	The user i
ID_i	The identity of the user i
PW_i	The password of the user i
S	The providing service server
X	The server's master secret key
Ti & T_s	The timestamp of the user I and server, respectively
Sk	The shared session key
h(.)	A collision-free one-way hash function
$\oplus$	An XOR operation
$\|$	The message concatenation operation

The Registration Phase: In this phase, the server S makes a smart card for a new user (U_i). The smart card contains four parameters, $\{B_i, C_i, h(.), r\}$, where $B_i = A_i \oplus h$ $(r \| PW_i)$; $A_i = h(ID_i \oplus x) \| h(x)$; $C_i = h(A_i \| ID_i \| h(r \| PW_i))$; h(.) denotes a collision-free one-way hash function; r denotes a random number; ID_i and PW_i are user's identity and password, respectively. The registration phase is executed as follows.

The Login Phase: In this phase, a user (U_i) wants to login the server via public Internet. The login phase is executed as follows.

(1)   The user Ui sends the login request parameters, IDi and PWi to the smart card.
(2)   The smart card computes A'i and C'i as follows: A'I = Bi $\oplus$ h(r $\|$ PWi); C'I = h (A'I $\|$ IDi $\|$ h(r $\|$ PWi)). Next, the smart card checks whether C'I is equal to Ci. If C'I is equal to Ci, the smart card continues to execute Step 3, otherwise, the smart card terminates this login request.
(3)   The smart card computes Di and Ei as follows: Di = h(IDi $\oplus$ α); Ei = A'I $\oplus$ $\oplus$ Tc, where Tc denotes the current timestamp of the smart card and α denotes a random number.
(4)   The smart card sends IDi, Di, Ei and Ti to the server S.

The Authentication Phase: Upon receiving the message, {IDi, Di, Ei, Tc}, from User (Ui), the server S executes this authentication phase as follows.

(1) The server checks IDi format and the timestamp Tc whether or not in valid time. If both conditions are not hold, the server S rejects the login request.
(2) The server computes Ai, α', and Di' as in Fig. 1. Next, the server checks D'I whether equals to Di. If the equation is not hold, the server S rejects the login request.
(3) The server randomly selects β and computes Fi and Gi as in Fig. 1. Next, the server S sends {Fi, Gi, Ti} vis public channel to user Ui.
(4) The user Ui the timestamp Ts whether or not in valid time. If this condition is not hold, the user terminates this session.
(5) The user computes β'and F'I. Next, the user checks F'I whether equals to Fi. If this condition is true, the user Ui confirms the server S is legit.
(6) The server S and the user Ui compute the session key sk = h(α ∥ β ∥ h (Ai ⊕ IDi)).

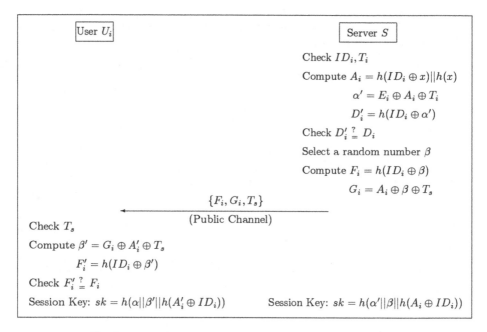

Fig. 1. The authentication phase of Liu-Chang-Chang's scheme

3   Cryptanalysis of Liu-Chang-Chang Scheme

In this section, it is demonstrated that the user authentication scheme proposed by Liu-Chang-Chang's [17] cannot resist the replaying attack when the hacker intercepts {IDi, Di, Ei, Ti} between smart card and server S and {F, G, Ts} between user Ui and server S. The first replaying attack is listed as follows.

Step 1.   When the smart card sent the message, {IDi, Di, Ei, Ti}, to the server S in the login phase, the hacker intercepts {IDi, Di, Ei, Ti} between smart card and server S via public channel.

Step 2.   The hacker computes a new E'I as follows:

$$E'I = Ei \oplus Ti \oplus Th$$
$$= (A'I \oplus \alpha \oplus Ti) \oplus Ti \oplus Th$$
$$= A'I \oplus \alpha \oplus Th$$

Here, Th denotes the timestamp of Hacker's device. Next, the hacker sends the forged message {IDi, Di, E'i, Th} to replace the intercepted {IDi, Di, Ei, Ti}.

Step 3.   The server S will check successfully the equation in Steps (1) and (2) in the authentication phase. Thus, the server will be deceived by the hacker (Fig. 2).

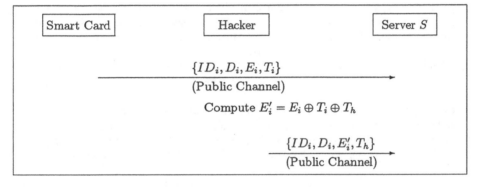

Fig. 2. The replaying attack when the hacker intercepts {IDi, Di, Ei, Ti}

The second replaying attack is similar to the first replaying attack. The attack listed as follows.

Step 1.   When the server S sent the message, {Fi, Gi, Ts}, to the user Ui in the authentication phase, the hacker intercepts it between server S and user Ui via public channel.

Step 2.   The hacker computes a new G'i as follows:

$$G'i = Gi \oplus Ts \oplus Th$$
$$= (Ai \oplus \beta \oplus Ts) \oplus Ts \oplus Th$$
$$= Ai \oplus \beta \oplus Th$$

The hacker sends the forged message {Fi, G'i, Th} to replace the intercepted {Fi, Gi, Ts}.

Step 3.   The user Ui will check successfully the equation in Steps (4) and (5) in the authentication phase. Thus, the user Ui will be deceived by the hacker.

4   Conclusion

We have demonstrated that the user authentication scheme proposed Liu-Chang-Chang [17] have a weakness. Their scheme cannot resist the replaying attack when the hacker intercepts {IDi, Di, Ei, Ti} between smart card and server S and {F, G, Ts} between user Ui and server S.

Acknowledgments. This study was supported by the National Science Council of Taiwan under grant MOST 104-2221-E-468 -004 and MOST 103-2221-E-468 -026.

References

1. Ahmed, A., Younes, A., Abdellah, A., Sadqi, Y.: Strong zero-knowledge authentication based on virtual passwords. Int. J. Netw. Secur. **18**(4), 601–616 (2016)
2. Amin, R.: Cryptanalysis and efficient dynamic ID based remote user authentication scheme in multi-server environment using smart card. Int. J. Netw. Secur. **18**(1), 172–181 (2016)
3. Anwar, N., Riadi, I., Luthfi, A.: Forensic SIM card cloning using authentication algorithm. Int. J. Electron. Inf. Eng. **4**(2), 71–81 (2016)
4. Chang, C.-C., Hsueh, W.-Y., Cheng, T.-F.: An advanced anonymous and biometrics-based multi-server authentication scheme using smart cards. Int. J. Netw. Secur. **18**(6), 1010–1021 (2016)
5. Chen, B.L., Kuo, W.C., Wuu, L.C.: Robust smart-card-based remote user password authentication scheme. Int. J. Commun. Syst. (in press). http://dx.doi.org/10.1002/dac.2368
6. Feng, T.H., Ling, C.H., Hwang, M.S.: Cryptanalysis of Tan's improvement on a password authentication scheme for multi-server environments. Int. J. Netw. Secur. **16**, 318–321 (2014)
7. He, D., Zhao, W., Wu, S.: Security analysis of a dynamic ID-based authentication scheme for multi-server environment using smart cards. Int. J. Netw. Secur. **15**, 282–292 (2013)
8. Huang, H.F., Chang, H.W., Yu, P.K.: Enhancement of timestamp-based user authentication scheme with smart card. Int. J. Netw. Secur. **16**, 463–467 (2014)
9. Hwang, M.S., Chong, S.K., Chen, T.Y.: Dos-resistant ID-based password authentication scheme using smart cards. J. Syst. Softw. **83**, 163–172 (2000)
10. Hwang, M.S., Li, L.H.: A new remote user authentication scheme using smart cards. IEEE Trans. Consum. Electron. **46**, 28–30 (2000)
11. Li, C.T., Hwang, M.S.: An online biometrics-based secret sharing scheme for multiparty cryptosystem using smart cards. Int. J. Innovative Comput. Inf. Control **6**, 2181–2188 (2010)
12. Li, C.T., Hwang, M.S.: An efficient biometrics-based remote user authentication scheme using smart cards. J. Netw. Comput. Appl. **33**, 1–5 (2010)
13. Li, L.H., Lin, I.C., Hwang, M.S.: A remote password authentication scheme for multi-server architecture using neural networks. IEEE Trans. Neural Netw. **12**, 1498–1504 (2001)
14. Li, X., Niu, J., Khan, M.K., Liao, J.: An enhanced smart card based remote user password authentication scheme. J. Netw. Comput. Appl. (in press). http://dx.doi.org/10.1016/j.jnca.2013.02.034
15. Lin, I.C., Hwang, M.S., Li, L.H.: A new remote user authentication scheme for multi-server architecture. Future Gener. Comput. Syst. **19**, 13–22 (2003)
16. Ling, J., Zhao, G.: An improved anonymous password authentication scheme using nonce and bilinear pairings. Int. J. Netw. Secur. **17**(6), 787–794 (2015)

17. Liu, Y., Chang, C.-C., Chang, S.-C.: An efficient and secure smart card based password authentication scheme. Int. J. Netw. Secur. **19**(1), 1–10 (2017)
18. Lu, Y., Yang, X., Wu, X.: A secure anonymous authentication scheme for wireless communications using smart cards. Int. J. Netw. Secur. **17**(3), 237–245 (2015)
19. Osei, E.O., Hayfron-Acquah, J.B.: Cloud computing login authentication redesign. Int. J. Electron. Inf. Eng. **1**(1), 1–8 (2014)
20. Prakash, A.: A biometric approach for continuous user authentication by fusing hard and soft traits. Int. J. Netw. Secur. **16**, 65–70 (2014)
21. Shen, J.J., Lin, C.W., Hwang, M.S.: Security enhancement for the timestamp-based password authentication scheme using smart cards. Comput. Secur. **22**, 591–595 (2003)
22. Shen, J.J., Lin, C.W., Hwang, M.S.: A modified remote user authentication scheme using smart cards. IEEE Trans. Consum. Electron. **49**, 414–416 (2003)
23. Stanek, M.: Weaknesses of password authentication scheme based on geometric hashing. Int. J. Netw. Secur. **18**(4), 798–801 (2016)
24. Tang, H., Liu, X., Jiang, L.: A robust and efficient timestamp-based remote user authentication scheme with smart card lost attack resistance. Int. J. Netw. Secur. **15**, 446–454 (2013)
25. Wang, Y., Peng, X.: Cryptanalysis of two efficient password-based authentication schemes using smart cards. Int. J. Netw. Secur. **17**(6), 728–735 (2015)
26. Wei, J., Liu, W., Hu, X.: Secure and efficient smart card based remote user password authentication scheme. Int. J. Netw. Secur. **18**(4), 782–792 (2016)
27. Wijayanto, H., Hwang, M.-S.: Improvement on timestamp-based user authentication scheme with smart card lost attack resistance. Int. J. Netw. Secur. **17**(2), 160–164 (2015)
28. Yang, C.C., Chang, T.Y., Hwang, M.-S.: The security of the improvement on the methods for protecting password transmission. Informatica **14**, 551–558 (2003)
29. Zhu, H., Zhang, Y., Zhang, Y.: A provably password authenticated key exchange scheme based on chaotic maps in different realm. Int. J. Netw. Secur. **18**(4), 688–698 (2016)
30. Zhuang, X., Chang, C.C., Wang, Z.H., Zhu, Y.: A simple password authentication scheme based on geometric hashing function. Int. J. Netw. Secur. **16**, 271–277 (2014)

An Improved Password Authentication Scheme for Smart Card

Cheng-Yi Tsai[1], Chiu-Shu Pan[1], and Min-Shiang Hwang[1,2(✉)]

[1] Department of Computer Science and Information Engineering, Asia University, 500, Lioufeng Road, Wufeng, Taichung 41354, Taiwan, R.O.C.
mshwang@asia.edu.tw
[2] Department of Medical Research, China Medical University Hospital, China Medical University, No. 91, Hsueh-Shih Road, Taichung 40402, Taiwan, R.O.C.

Abstract. In recent times, Wei et al. proposed a secure smart card based on remote user password authentication scheme. Their scheme is more secure than other schemes. In this article, we will prove their scheme is vulnerable to password guessing attack, privileged insider attack, and denial of service attack. Furthermore, we will propose an improved scheme to eliminate the security vulnerability.

Keywords: Password · Smart card · Timestamp · User authentication

1 Introduction

One of the most commonly used solutions to protect in distributed network environments from illegal access is user authentication scheme [1, 2]. There are many password-based authentication schemes proposed to verify user's identification [3–9]. However, the password is easily exposed by guessing attacks. In order to enhance the security, the smart card has widely used in an authentication scheme [10–17].

Recently, Chen et al. proposed a smart-card-based user authentication scheme [18]. However, Li et al. pointed out some weaknesses in Chen et al.'s scheme [19]. Li et al. also proposed an enhanced smart card based on a user authentication scheme [19] to resist the above flaws existing in Chen et al.'s scheme. However, Wei et al. showed that Li et al.'s scheme is powerless against the off-line password guessing attack [20]. They also proposed an efficient and secure smart card based remote user password authentication scheme. Their scheme is more efficient and secure than other schemes. However, we find the security of their scheme also exist. In this article, we will prove their scheme is vulnerable to password guessing attack, privileged insider attack, and denial of service attacks.

2 Review of Wei-Liu-Hu Scheme

In this section, we briefly review Wei et al.'s smart card based remote user password authentication scheme (Wei-Liu-Hu Scheme) with smart cards [20]. There are three participants in Wei-Liu-Hu's user password authentication scheme: a user (U for short),

© Springer International Publishing AG 2017
F. Xhafa et al. (eds.), *Recent Developments in Intelligent Systems and Interactive Applications*,
Advances in Intelligent Systems and Computing 541, DOI 10.1007/978-3-319-49568-2_27

a smart card (SC for short), and a server (SV for short). The scheme consists of four phases, namely the initialization and registration phase, the login phase, the authentication phase, and the password change phase.

The initialization and registration phase: In this phase, the server SV makes a smart card SC for a new user (U_i). The initial-ization and registration phase are executed as follows.

(1) The user U_i chooses his/her identity ID_i and password PW_i and sends ID_i and PW_i to the server SV.
(2) SV computes $A_i = h(ID_i \oplus x) + h(ID_i \| PW_i)$, where $h(.)$ denotes a secure hash function; x denotes a master secret key; p, q, x, and $h(.)$ are selected by SV.
(3) SV sends the smart card SC to U_i. The smart card contains: $\{A_i, p, q, h(.)\}$.

The login phase: In this phase, a user (U_i) wants to login into the server SV to obtain some service; the user first attaches his/her smart card to a device reader and inputs his/her identity ID_i and password PW_i. The login phase is executed as follows:

(1) The user U_i sends the login request parameters, his/her identity ID_i and password PW_i to the smart card SC.
(2) SC computes B_i, D_i, F_i, M_i as follows: $B_i = A_i - h(ID_i \| PW_i) = h(x \| ID_i)$; $D_i = h(ID_i)^a \bmod p$; $F_i = D_i + B_i$; $M_i = h(ID_i \| F_i \| T_1)$, where T_1 denotes the current timestamp of SC, and a denotes a random number selected by SC.
(3) SC sends $\{ID_i, F_i, M_i, T_1\}$ to the server SV.

The authentication phase: Upon receiving the login request message $\{ID_i, F_i, M_i, T_1\}$ from User (U_i), the server SV executes this authentication phase as follows:

(1) The server SV checks whether ID_i format and the timestamp T_1 are correct or not. If both of conditions hold, SV continuously authenticates the following steps.
(2) SV checks whether $M_i' = h(ID_i \| F_i \| T_1)$ is equal to M_i or not. If it does not hold, the server SV rejects the login request. Otherwise, SV computes V_i and M_s as follows: $V_i = h(ID_i)^b \bmod p$ and $M_s = h(ID_i \| D_i' \| V_i \| Z_i \| T_s)$ where b is a random number; $D_i' = F_i - h(x \| ID_i) = h(ID_i)^a \bmod p$; $Z_i = (D_i)^b \bmod p$; T_s denotes the current time of SV. Next, SV sends $\{ID_i, V_i, M_s, T_s\}$ to user U_i.
(3) The user Ui checks whether ID_i and T_s are correct or not. If this condition hold, U_i computes $Z_i = (V_i)^a \bmod p$ and checks whether $M_s' = h(ID_i \| D_i \| V_i \| Z_i' \| T_s)$ is equal to M_s or not. If it does not hold, U_i terminates this session. Otherwise, U_i computes $R_i = h(ID_i \| F_i \| V_i \| Z_i \| T_2)$ and sends $\{ID_i, R_i, T_2\}$ to SV.
(4) SV checks whether ID_i and T_2 are correct or not. If this condition hold, the user computes and checks whether $R_i' = h(ID_i \| F_i \| V_i \| Z_i \| T_2)$ is equal to R_i or not. If it does not hold, the user terminates this session. Otherwise, SV and U_i compute the session key $sk = h(ID_i \| D_i \| V_i \| Z_i)$.

3   Cryptanalysis of Wei-Liu-Hu Scheme

In this section, we will analyze Wei-Liu-Hu's user authentication scheme [20]. Wei et al. claimed that their scheme is resistant to offline password guessing attacks, replay attacks, impersonation attacks, parallel attacks, perfect forward secrecy, and known-key security. Next, we show that Wei-Liu-Hu's user password authentication scheme is vulnerable to denial of service attacks, privileged insider attacks, and password guessing attacks.

Denial of service attack: In Wei-Liu-Hu's scheme, there are three parameters needed to be checked for resisting the denial of service attacks.

(1) Both of the server and user check whether the user's identity ID_i is in correct format in Steps 1 & 3 of the authentication phase.
(2) The server checks whether the timestamp T_1 and T_2 whether are in valid time in Steps 1 and 4 of the authentication phase. The user checks whether the timestamp T_s is in valid time in Step 3 of the authentication phase.
(3) The server checks whether the parameters M_i and R_i are equal to M_i' and R_i' or not in Steps 2 and 4 of the authentication phase. The main function of M_i and R_i is to avoid the transmission messages $\{ID_i, F_i, M_i, T_1\}$ and $\{ID_i, R_i, T_2\}$ in Step 3 of the login phase and Step 3 of the authentication phase forged or replayed. Similarly, the user checks whether the parameter M_s is equal to M_s' or not in Step 3 of the authentication phase. The main function of M_s is to avoid the transmission message $\{ID_i, V_i, M_s, T_s\}$ in Step 2 of the authentication phase forged or replayed.

Next, we show that Wei-Liu-Hu's scheme is vulnerable to the denial of service attack. The adversary may send the modified login request message $\{ID_i, F_i', M_i', T_1'\}$ to a server, where $M_i' = h(ID_i \parallel F_i' \parallel T_1')$; T_1' is the current timestamp; F_i' is any number selected by the adversary. In this case, the modified message $\{ID_i, F_i', M_i', T_1'\}$ will be checked successfully by the server in Step 2 of the authentication phase. Thus, the server will continually execute the other steps of the authentication phase. The adversary can create or update the false information for login. The denial of service attack might result from the more computation load the server performs.

Privileged insider attack: In a real world, many users have registered in many servers for accessing different services and applications. For convenience of remembering these passwords, it is a common practice that many users use the same passwords in these servers. Thus, if the system manager or privileged insider of a server knows the password of the user, he/she may try to impersonate the user by using the password. In the initialization and registration phase of Wei-Liu-Hu's scheme, the user sends his/her identity ID_i and password PW_i to the server SV directly. Thus, the privileged insider could get the user's password. Therefore, Wei-Liu-Hu's user password authentication scheme is vulnerable to the privileged insider attack.

Password guessing attack: Suppose an adversary has stolen the user's smart card. The adversary may guess the user identity ID_i's password PW_i, and then observes the communication between the smart card and the server. If the password is not correct, the user will terminate this session in Step 3 of the authentication phase. The adversary may guess the user's password again and repeats to observe the communication between the smart card and the server. When the password is guessed, the smart card will perform all steps in the authentication phase. In other words, the adversary may guess the user's password and then observe whether the Step 4 of the authentication phase is performed or not. Therefore, Wei-Liu-Hu's user password authentication scheme is vulnerable to the password guessing attack.

4   An Improvement of Wei-Liu-Hu's Scheme

In this section, we will modify Wei-Liu-Hu's user password authentication scheme to resist the denial of service, privileged insider, and password guessing attacks.

4.1   Initialization and Registration Phase of the Improved Wei-Liu-Hu's Scheme

In this phase, the server SV makes a smart card SC for a new user (U_i). The smart card SC contains four parameters, $\{A_i, p, q, h(.), r, W_i\}$, where $A_i = h(ID_i \oplus x) + h(ID_i \parallel (h(PW_i) \oplus r))$; $W_i = h(PW_i \parallel r)$; $h(.)$ denotes a secure hash function ($h(.)$: $\{0, 1\}^* \rightarrow Z_p^*$); p and q are two large prime numbers such that $p = 2q + 1$; x denotes a master secret key ($x \in Z_q^*$); r is a random number; ID_i and PW_i are user's identity and password, respectively. p, q, x, and $h(.)$ are selected by the server SV. ID_i, PW_i, and r are selected by the user U_i. The main difference between Wei-Liu-Hu's scheme and the improved scheme is that the password is not sent to the server directly. The user sends $\{ID_i$ and $(h(PW_i) \oplus r)\}$ to the server. The server does not know the random number r and then is hard to guess the user's password. Therefore, the improved Wei-Liu-Hu's scheme is resistant to the privileged insider attack.

4.2   The Login Phase of the Improved Wei-Liu-Hu's Scheme

In this phase, a user (U_i) wants to login into the server SV for obtaining some services; the user first attaches his/her smart card to a device reader and inputs his/her identity ID_i and password PW_i. The login phase is executed as follows:

(1)   The user U_i sends the login request parameters, his/her identity ID_i and password PW_i to the smart card SC. SC computes $Wi' = h(PW_i \parallel r)$ and checks whether W_i' is equal to W_i. If it holds, SC executes the next steps. If the user fails to verify ID_i and PW_i for 3 times, the user will lock the smart card SC.
(2)   SC computes B_i, D_i, F_i, M_i as follows: $B_i = A_i - h(ID_i \parallel h(PW_i) \parallel r) = h(x \parallel ID_i)$; $Di = h(ID_i)^a \bmod p$; $F_i = D_i + B_i$; $M_i = h(ID_i \parallel F_i \parallel T_1) \oplus B_i$, where T_1 denotes the current timestamp of SC and a denotes a random number.
(3)   The smart card sends $\{ID_i, F_i, M_i, T_1\}$ to the server SV.

The main difference in the login phase between Wei-Liu-Hu's scheme and the improved scheme is that the smart card checks the correct password. The adversary only has three times to guess the user's password in Step 2 of the login phase.

Therefore, the improved Wei-Liu-Hu's scheme is resistant to the password guessing attack.

4.3   The Authentication Phase of the Improved Wei-Liu-Hu's Scheme

Upon receiving the authentication request message $\{ID_i, F_i, M_i, T_1\}$ from User (U_i), the server SV executes this authentication phase as follows:

(1)   The server checks whether ID_i format and the timestamp T_1 are in valid time or not. If both of conditions hold, the server SV continuously authenticates the following steps:

(2)   The server checks whether $M_i' = h(ID_i \parallel F_i \parallel T_1) \oplus h(x \parallel ID_i)$ is equal to M_i or not. If it does not hold, the server SV rejects the login request.

The other steps are the same as the authentication phase of Wei-Liu-Hu's scheme. The main difference in the authentication phase between Wei-Liu-Hu's scheme and the improved scheme is that $M_i = h(ID_i \parallel F_i \parallel T_1) \oplus h(x \parallel ID_i)$ instead of $h(ID_i \parallel F_i \parallel T_1)$. The adversary does not know $h(x \parallel ID_i)$. Therefore, the improved Wei-Liu-Hu's scheme is resistant to the denial of service attack.

5   Conclusion

In this article, we have shown that there are some weaknesses in Wei-Liu-Hu's user authentication scheme [20]. Their scheme cannot withstand the denial of service, privileged insider and password guessing attacks. We also propose an improvement of Wei-Liu-Hu's user password authentication scheme to resist the above weaknesses.

References

1. Prakash, A.: A biometric approach for continuous user authentication by fusing hard and soft traits. Int. J. Netw. Secur. **16**, 65–70 (2014)

2. Osei, E.O., Hayfron-Acquah, J.B.: Cloud computing login authentication redesign. Int. J. Electron. Inf. Eng. **1**(1), 1–8 (2014)

3. Ahmed, A., Younes, A., Abdellah, A., Sadqi, Y.: Strong zero-knowledge authentication based on virtual passwords. Int. J. Netw. Secur. **18**(4), 601–616 (2016)

4. Feng, T.H., Ling, C.H., Hwang, M.S.: Cryptanalysis of Tan's improvement on a password authentication scheme for multi-server environments. Int. J. Netw. Secur. **16**, 318–321 (2014)

5. Li, L.H., Lin, I.C., Hwang, M.S.: A remote password authentication scheme for multi-server architecture using neural networks. IEEE Trans. Neural Networks **12**, 1498–1504 (2001)

6. Ling, J., Zhao, G.: An improved anonymous password authentication scheme using nonce and bilinear pairings. Int. J. Netw. Secur. **17**(6), 787–794 (2015)

7. Stanek, M.: Weaknesses of password authentication scheme based on geometric hashing. Int. J. Netw. Secur. **18**(4), 798–801 (2016)

8. Zhu, H., Zhang, Y., Zhang, Y.: A provably password authenticated key exchange scheme based on chaotic maps in different realm. Int. J. Netw. Secur. **18**(4), 688–698 (2016)
9. Zhuang, X., Chang, C.C., Wang, Z.H., Zhu, Y.: A simple password authentication scheme based on geometric hashing function. Int. J. Netw. Secur. **16**, 271–277 (2014)
10. Amin, R.: Cryptanalysis and efficient dynamic ID based remote user authentication scheme in multi-server environment using smart card. Int. J. Netw. Secur. **18**(1), 172–181 (2016)
11. Anwar, N., Riadi, I., Luthfi, A.: Forensic SIM card cloning using authentication algorithm. Int. J. Electron. Inf. Eng. **4**(2), 71–81 (2016)
12. Chang, C.-C., Hsueh, W.-Y., Cheng, T.-F.: An advanced anonymous and biometrics-based multi-server authentication scheme using smart cards. Int. J. Netw. Secur. **18**(6), 1010–1021 (2016)
13. He, D., Zhao, W., Wu, S.: Security analysis of a dynamic ID-based authentication scheme for multi-server environment using smart cards. Int. J. Netw. Secur. **15**, 282–292 (2013)
14. Huang, H.F., Chang, H.W., Yu, P.K.: Enhancement of timestamp-based user authentication scheme with smart card. Int. J. Netw. Secur. **16**, 463–467 (2014)
15. Yanrong, L., Yang, X., Xiaobo, W.: A secure anonymous authentication scheme for wireless communications using smart cards. Int. J. Netw. Secur. **17**(3), 237–245 (2015)
16. Wang, Y., Peng, X.: Cryptanalysis of two efficient password-based authentication schemes using smart cards. Int. J. Netw. Secur. **17**(6), 728–735 (2015)
17. Wijayanto, H., Hwang, M.-S.: Improvement on timestamp-based user authentication scheme with smart card lost attack resistance. Int. J. Netw. Secur. **17**(2), 160–164 (2015)
18. Chen, B.L., Kuo, W.C., Wuu, L.C.: Robust smart-card-based remote user password authentication scheme. Int. J. Commun. Syst. **27**(2), 377–389 (2014)
19. Li, X., Niu, J., Khan, M.K., Liao, J.: An enhanced smart card based remote user password authentication scheme. J. Netw. Comput. Appl. **36**(5), 1365–1371 (2013)
20. Wei, J., Liu, W., Xuexian, H.: Secure and efficient smart card based remote user password authentication scheme. Int. J. Netw. Secur. **18**(4), 782–792 (2016)

Design and Construction of Graphical Cloud Computing Platform

Zhang Lanyu[(✉)]

Mathematics School and Institute of Jilin University, Changchun, China
lanyu@jlu.edu.cn

Abstract. With the flourishing enhancement of cloud computing technology, the clouded graphical computation has been realized. It became a new mode of cloud application to improve resource utilization rate and reduce users' cost by taking advantage of the GPU in the cloud for graphical computation. This paper discusses the software design of the graphical cloud computing platform and the method of platform construction, describes the advantages and disadvantages of GPU virtual mode by taking the graphical cloud platform as an example, and explores the utilization rate of cloud platform.

Keywords: Cloud computing · Virtual GPU · Graphical computation

1 Introduction

At present, with the development of 3D production and rendering software, the production of 3D animations and games has became popular, instead of being only monopolized by several large companies. However, with the improvement of requirements on films and games by audiences and players, the requirements of hardware producing 3D videos and games are improved gradually, which causes the increase of proportion taken by costs which are put in to the hardware of animations and games among the total production costs, so that many personal animation producers with insufficient fund have to invest the limited fund for hardware, i.e. graphic workstation. The high cost of hardware causes vast resource waste in animation and game industries. On the other hand, with the flourishing development of cloud computing, people naturally consider if graphic computation can be conducted in the cloud to reduce the cost of graphical computation and improve the resource utilization rate. In order to establish large-scale graphical computation workstation on the "cloud" and set up multiple virtual machines which can realize graphical computation, users of computation service is required to log in those virtual machines remotely and to carry out graphical computation through them, which is commonly so-called graphical cloud computing. It virtualizes single graphical workstations to plural virtual graphical workstations with weak performance through virtualization software. When a user needs to use the workstation, he logins the virtual graphical workstation through software remotely and carries out graphical computation with the computing resources established on the other end of network. Those remote graphical workstations are known as "cloud".

F. Xhafa et al. (eds.), *Recent Developments in Intelligent Systems and Interactive Applications*,
Advances in Intelligent Systems and Computing 541, DOI 10.1007/978-3-319-49568-2_28

Based on above reasons and integration with cloud computing technology, virtual GPU technology, and VPN technology, the graphical cloud computing platform, which is designed and applied in this paper, realizes graphical cloud computing by setting up virtual machines on the graphical workstation, providing better graph and image processing technology for virtual machines through the GPU technology, protecting the "cloud" through VPN, and visiting virtual machines via VPN by using remote software. In order to reduce users' cost of graphical computation, increase resource utilization rate and achieve the optimum results of "graphical cloud computing".

2 Key Technologies

2.1 GPU Virtualization

Generally speaking, GPU virtualization enables many users to share the same GPU hardware resource remotely by optimizing instruction set mechanism, namely, virtualization based on GPU hardware. GPU virtualization can be realized only with the coordination of both software and hardware:

Hardware: Support of underlying GPU hardware architecture is demanded with the capacity of using a GPU separately according to different demands and operating different applications on a GPU.

Software: Support of driver software is demanded to provide interaction between applications and underlying hardware instructions, including division and allocation of hardware environment, operation management and data extraction, etc.

Virtualization of GPU resources can be realized only on the basis of virtualization of both hardware and software of GPU.

With the improvement of production process, the demand for GPU virtualization has been generated, since excess of GPU resources has occurred to personal users; however, professional GPU is too expensive to bear for some small enterprise users. GPU virtualization can mollify that contradiction to a certain degree. The typical applications of GPU virtualization include: Video coding and transcoding.

Each virtual GPU can complete a stream of video coding. After virtualization mechanism, each GPU can process several streams of video coding, which largely improves working efficiency on production and playing of video.

Remote rendering: The support of large-scale graphical processing software and high-end graphical workstation platform is required for professional CG and art production with high product cost. However, GPU virtualization mechanism can be applied to provide virtual operating environment for medium and small producers, which can realize product production with low cost. [1]

2.1.1 Soft 3D and vSGA vDGA

The diversified GPU virtualization methods include Soft 3D, virtual sharing graphics accelerator (vSGA), and virtual dedicated graphics accelerator (vDGA). With different working principles, they are pertinent to different situations.

Soft 3D allows the virtual machine (VM) in the host to simulate GPU with software and physics card is not require, which cannot provide high processing performance, but a free alternative solution for applications which only need ordinary 3D graphical processing capacity.

vSGA: For employees using lightweight 2D and 3D application programs, vSGA allows client to use physical GPU which is installed on the server. It can divide GPU into many different virtual adapters, and allocate image memory for each VM. Since vSGA uses the same display driver program as Soft 3D, it can easily realize the transition from software acceleration to hardware acceleration. For example, if the GPU memory of server has ran out, other VMs will automatically select to use Soft 3D, in order to prevent fault.

vDGA: Users who demand their graphical intensive applications can possess the performance on individual workstations can use vDGA. It allows the client to visit GPU directly, and use GPU directly during virtual desktop communication. However, different from vSGA, vDGA does not share GPU. The processing capacity of GPU can only satisfy a virtual desktop.

2.2 Virtual Private Network (VPN) Technology

VPN is used to establish dedicated network on public network for encrypted communication; and it is widely applied to enterprise network. VPN gateway realizes remote visit through encryption of data package and transition of the target address of data package. With several classification methods, VPN is mostly classified according to protocols. Besides, it can be realized through a lot of methods such as server, hardware, and software. [3]

3   Platform Architecture

3.1   Network Architecture

Graphical cloud computing platform consists of 3 parts including: cloud server, network connecting the client and the network and VPN server for safety guarantee, and client software or thin client.

Users visit the VMs established in server after passing the certification of VPN server by connecting VPN server through thin client or PC software, as shown in Fig. 1.

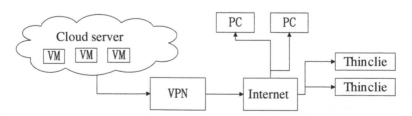

Fig. 1. Network architecture of cloud computing

3.2   Cloud Platform Architecture

It establishes multiple virtual GPUs by the respective virtualization of all GPUs in workstation through GPU virtualization manager in graphical workstation. Besides, GPU virtualization is not equal to graphics card virtualization. Sometimes, a graphical card will carry many GPUs; and during the virtualization process, it is GPUs equipped for the graphics card but not the graphics card is virtualized. For example, if a graphics card contains 4 GPUs and each of them can be virtualized as 3 GPUs, each graphical workstation can provide remote graphical computation for 12 users at most, as shown in Fig. 2.

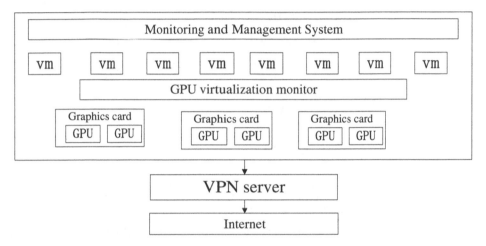

Fig. 2. Graphical cloud computing platform architecture

4   Software Structure

Cloud computing platform needs to login VM establish in graphical workstation through software by applying for user name and password through manager (this username will be used when a VM is logged in through VPN), and selecting VM performance required by the user, including: GPU, CPU, memory and storage space; then starting the software, connecting the front-end VPN server through network, and inputting the username and password applied before; after passing the VPN certification, logging in the server; and then the software will start the VM established on the server by the user directly through the username and password input before, and display the interface of VM on the user's PC desktop, as shown in Fig. 3. [4]

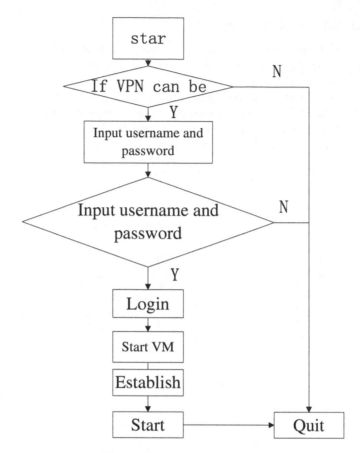

Fig. 3. PC software architecture

5 Conclusion

This paper introduces the fundamental conception of graphical cloud computing, analyzes its required technologies, researches and obtains the network and architecture of graphical cloud computing platform through GPU virtualization technology, analyzes its technological principles and operation flow and establishes a complete graphical cloud computing system through innovative software system, which lays a solid foundation for improving computing resources utilization rate and establishing an efficient and high-end graphical computing system.

References

1. Beau, W.: Develoing IP Multicast Networks. Cisco Press, Indianapolis (2000)
2. Block, D.G.: Speed VOD With P2P (2003)
3. Hu, J.: P2P technology based video on demand technology. China Sci. Technol. Inf. **4**, 58–61 (2007)
4. Sun, M., Gu, M., Gao, J.: Domestic and foreign research status online video resource development programs. Res. Libr. Sci. **7**, 96–101 (2012)

Research on Real-Time Flow Abnormal Traffic Detection System Based on DDoS Attack

Xin Yue[✉], Xiuliang Mo, Chundong Wang, and Xin Yao

Tianjin University of Technology, Tianjin, China
1521481125@qq.com

Abstract. This paper describes DDoS attacks in SYN-Flood, ICMP-Flood, NTP-Flood etc. Putting forward a comprehensive and effective traffic anomaly detection algorithm, combining the exponentially weighted moving average algorithm (EWMA), building an abnormal traffic monitoring system. Tianjin Education metropolitan area network is considered as data source to the experiment. To assess the capability of the detection system, we undertake the experiments, and the result shows that the system can detect abnormal flow effectively, and monitor network dynamically.

Keywords: Abnormal netflow monitoring · DDos attack · Flow monitoring

1 Introduction

DDoS (Distributed Denial of Service) attacks in large data environment as the representative of the large traffic attacks are gradually getting more attention [1]. In the three key elements of the security information ("confidential" and "integrity" and "availability") and DDoS targeting is "availability", with the client/server model, an attacker can built a cluster victim hosts composed by thousands of Puppet Master initiated by malicious attacks in a short period of time.

Anuradha, S.G. ect has created a morphological change detection system for real time traffic analysis [2]. Zhongshan University Haina Tang et al. realized a deployed large-scale network real-time traffic analysis system [3] and analysed the characteristics of user behavior by means of active data streams, IP packet size, the types of applications and other statistics [4]. In the massive data processing, Kafka (the representative of message oriented middleware) and HBase (the representative of the non-relational distributed database) are more appropriate choice. Rick Hofstede in the year of 2013 put forward the article of a real-time intrusion detection system based on NetFlow and IPFIX [5]. The main drawback of this architecture system model is that optimization of special hardware support must be used to security analysis module in order to achieve the high-speed processing network link. Apparently, this program is not suitable for large-scale deployment and not have a horizontal expansion capability. In view of the above problems, this paper takes the corresponding measures, as follows:

1. The system uses Kafka and HBase as the basis selection. In addition, the platform has a good scalability. Only increasing the number of physical machines, the upper layer application can calmly deal with the input data with different orders and magnitudes.

F. Xhafa et al. (eds.), *Recent Developments in Intelligent Systems and Interactive Applications*, Advances in Intelligent Systems and Computing 541, DOI 10.1007/978-3-319-49568-2_29

2. Based on the system architecture, a design scheme of network security of big data computing platform is putting forward, which is based on the open source of big data technology. The traffic monitoring system can realize the detection and identification of multiple types of DDoS attacks from load index of real-time network, the abnormal fluctuation of traffic flow, and the discrimination of attack types.

2 Construction of Abnormal Traffic Monitoring Platform

Logical view of the platform, data warehouse [6] module as data flow core of the system (acquisition, transmission and storage), the role of the system is not for the data storage system and innovation to break the tight coupling between the data producer and consumer. Another core component is the resource management, it owned by each of the physical computer memory and CPU is abstracted into computing resources can be used for the upper layer of the caller, the full integration of first come first serve (FCFS), priority algorithm, fair algorithm, different scheduling strategies [7] in response to the cluster users' resource request. Computing layer undertakes all data processing. At the same time, this paper will incorporate off-line and real-time computing framework, computational framework and distributed machine learning [8] system into the platform architecture. Data processing is the most important part of the overall system structure as shown in Fig. 1. The data processing procedure is shown in Fig. 2:

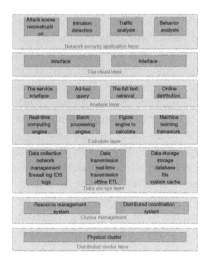

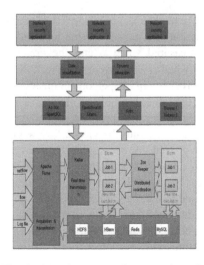

Fig. 1. Overall   architecture   diagram platform.

Fig. 2. The data processing procedure chat

3   Abnormal Traffic Monitoring Algorithm

3.1   Exponentially Weighted Moving Average

Exponentially Weighted Moving Average algorithm (EWMA) [9] is a commonly used sequence data processing algorithm, the algorithm exponentially in accordance with numerical weights, higher weights near the current point in data, away from the current data with smaller weights, the core idea of EWMA algorithm is to calculate the estimation of current value before estimation of value and the current moment. The following formula can be obtain by applying EWMA to abnormal traffic detection:

$$\bar{x}_t = \alpha \times x_t + (1 - \alpha) \times \bar{x}_{t-1} \tag{1}$$

$$e_t = x_t - \bar{x}_t \tag{2}$$

$\bar{x}$ respects the weighted average of t moments, x_t represent the measured value of the t time, $\alpha = 2/(N+1)$ and α less than 1. In order to calculate the weighted average value of t in the current time, the data of the past 2 h are brought into account, this paper makes the weight factor of the parameter N = 120. Formula 2 is mainly used to calculate the t moments of the estimated error. After getting the weighted average and the estimated error, we can calculate the upper bound of the real-time load index in the current time:

$$T_{upper,t} = \overline{x}_t + \max(c \cdot \sigma_{e,t}, M_{\min}) \tag{3}$$

The parameter $\sigma_{e,t}$ represents the standard deviation of all the time values before the introduction of a constant value to avoid the value to be too small. By comparing the actual measured value x_t and the upper limit value of its variation $T_{upper,t}$, we can identify whether the current t network flow is abnormal jump. As can be seen from the above process, this paper only considers the limit of the load index $T_{upper,t}$ and not discuss the lower limit of the index. As a typical traffic network attacks, during DDoS attacks time, that will produce instantaneous high flow. A network connection request through a router into the victim host segments, which makes the router create a large number of NetFlow records in the short term, causing sharp fluctuations in the record creation num, PPS and other key indicators. So we only need to calculate the index changes of upper bound can be achieved for large scale DDoS attack detection. Compared to other complex traffic anomaly detection algorithm, this network real-time load index for object detection algorithm has better real-time computing attribute, which makes the large-scale network environment to achieve real-time detection of DDoS traffic network attacks.

3.2   Implementation of Traffic Anomaly Location Algorithm

Exponentially weighted moving average algorithm can help the system to find abnormal changes in network load index, but network anomaly is not enough, taking the premise of effective measures is a specific type of clear network attack. So on the basis of the above, in this paper, the realization of the attack recognition module based on predefined rules. The module contains a number of detection rules, which can be achieved on the identification of seven common types of DDoS attacks (Connection Flood, Stream Flood, SYN Flood, ACK Flood, UDP Flood, ICMP Flood, NTP Flood).

SYN attacker in the attacking process deliberately not complete the three-way handshake, the victim host waiting in the three times handshake state (the previous allocation of resources will continue to be occupied), according to the TCP protocol until timeout after the victim host to release resources. Test rules are shown in Fig. 3.

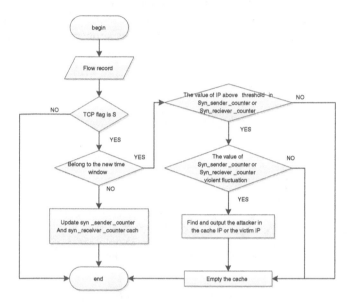

Fig. 3. The recognition rule of SYN-flood attack

4 Experiment

4.1 System Throughput

The testing process of the system of the distributed cluster environment consists of four computers, the Hadoop distributed file system (HDFS) is composed by four nodes. The Tianjin Education metropolitan area network in 2015 December 13 00:00:00 to 2015 December 13 23:59:59 NetFlow data were completed storage down, the data set as an input source, pressure testing of NetFlow real-time processing module. This test data used in this paper consists of 52835120 NetFlow, covering Tianjin Education metropolitan area network the C class segment.

In this paper, we start up with 4 Worker nodes in the Storm cluster, and run a Worker instance on each computer. Spout and Bolt on the Executor (thread). Testing process, processing of the 24 h Netflow data last for a total of 29 min and 55 s (from 7:39:30 December 16, 2015 to 8:09:25 December 16, 2015). Initialization phase

(Introduction) and end stage (8:09:00), the average throughput of real-time traffic monitoring system is per minute 1650067 a NetFlow records, the fastest processing speed (1782258 records), and the slowest (1405952 records) stage of occurred separately in the nine time window and the 21st time window. By viewing the record storm system logs, the author found that, the host is responsible for data transmission of netty client process in 7:57. As a result, the upstream Kafka-Spout sent to the part of the instance of the host computer worker data loss, and throughput of the system affected. Supervisors who in charge of monitor the host status detected the event immediately attempt to restart the netty client. Kafka-Spout using ACK timeout monitoring mechanism to perceive the data lost after the data replay, in this way, the Supervisor and Spout of the common collaboration system eventually recovered the normal operation. It can be seen from the graph that Storm has stronger fault tolerance, which can achieve error recovery in a relatively short time. The system throughput curve in minutes is shown in Fig. 4.

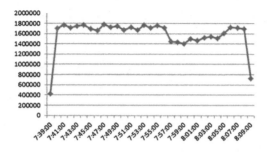

Fig. 4. Minute system throughput

4.2   Validity Verification of DDoS Attack Detection Module

This paper evaluates the detection capability of the system based on the NSFOCUS anti denial of service System (NSFOCUS Anti-DDoS). The attack events recorded by ADS NSFOCUS mainly include the time stamp, the attack type, the source IP, the destination IP and other information.

Since the attack ADS log mainly contains Connection-Flood, ICMP-Flood, Stream-Flood, SYN-Flood, UDP-Flood. So the analysis work only focus on the five types of attacks. After the compression of the NSFOCUS ADS log from December 24, 2015 to December 29, 2015, the number of attacks on various types of DDoS is shown in Table 1.

As it can be seen that, ICMP flood attacks and SYN flood attacks are the largest among all the DDoS attacks against the Tianjin University of Technology in the two kinds of attacks, reaching 80 %. The other three attacks are relatively few and will not be violent.

DDoS detection system processes the Netflow data and the system's detection results and validation data sets were comprised and analyzed. The detection accuracy is shown in Table 2.

Table 1. Statistic on the number of attacks from December 24th to December 29th

Data	Connection-Flood	ICMP-Flood	Stream-Flood	SYN-Flood	UDP-Flood
12.24	2	14	3	11	3
12.25	1	26	3	13	1
12.26	3	9	1	18	4
12.27	2	2	1	16	1
12.28	1	5	3	16	3
12.29	2	12	1	14	4

Table 2. Statistic of DDOS attacks

Data	Connection-Flood	ICMP-Flood	Stream-Flood	SYN-Flood	UDP-Flood
12.24	100 %	92.9 %	66.7 %	54.5 %	100 %
12.25	100 %	73.1 %	66.7 %	53.8 %	100 %
12.26	100 %	77.8 %	100 %	72.2 %	75 %
12.27	100 %	100 %	100 %	56.3 %	100 %
12.28	100 %	80 %	100 %	56.3 %	100 %
12.29	100 %	66.7 %	100 %	71.4 %	100 %

5 Summarize

In this paper, a DDoS attack detection algorithm is proposed for Netflow data, which is composed of a number of weighted moving average (EWMA) algorithm. EWMA algorithm core idea is the previous moment estimation value and the current observation value to calculate the estimated value. The paper implements a real-time traffic monitoring system for Tianjin education metropolitan area network. The experimental results effectively detect abnormal flow, and monitoring network dynamically. But there are still some problems remain to be solved: according to dynamic interactive visualization, multiple data association analysis should be further researched.

References

1. Li, Y.: Security network technology based on the principle of information security control. World of Biotechnology (2016)
2. Anuradha, S.G., Karibasappa, K., Eswar Reddy, B.: Morphological change detection system for real time traffic analysis. In: CGVIS 2015, pp. 237–242 (2015)
3. Qiuhua, X.: Research on the key technology of DDoS attack defense – DDoS attack detection. Shanghai Jiao Tong University (2007)
4. Haina, T., Xiaola, L., Hui, L., et al.: High performance & real-time traffic analysis in large scale networks. In: International Conference on Educational and Information Technology. Chongqing, China, pp. 457–460 (2010)
5. Hofstede, R., Bartosy, V., Sperotto, A., et al.: Towards real-time intrusion detection for NetFlow and IPFIX, Zurich, Switzerland, pp. 227–234 (2013)

6. Seah, B., Selan, E.: Design and implementation of data warehouse with data model using survey-based services data, Luton, England, pp. 58–64 (2014)
7. Yinglong, X.: Study on statistical machine learning methods for distributed heterogeneous data. Tsinghua University (2006)
8. Zongqin, F.: Research on resource scheduling algorithm in cloud environment. Beijing Jiaotong University (2014)
9. Wang, J.: Properties of EWMA controllers with gain adaptation. IEEE Trans. Semicond. Manuf. **23**(2), 159–167 (2010)

An Improved Virtualization Resource Migration Strategy and Its Application in Data Center

WuXue Jiang[1], YinZhen Zhong[2(✉)], WenLiang Cao[1], and ShuFei Li[1]

[1] Department of Computer Engineering,
Dongguan Polytechnic, Dongguan 523808, China
jiangwx@dgpt.edu.cn, 253925564@qq.com,
478403926@qq.com
[2] Department of Finance, Dongguan Polytechnic, Dongguan 523808, China
zhongyz@dgpt.edu.cn

Abstract. In order to improve the efficiency of virtualization of data center and to reduce the cost of computation migration, this paper studies all kinds of virtualization solutions and migration strategies to make a comprehensive comparison and propose a kind of p-order autoregressive prediction model. This model can plan the three important indicators as a whole, namely: CPU, memory and network bandwidth, so as to make comprehensive optimization for virtual system computation migration, which can design a virtualization plan for data center, thus this plan can be effective for data center storage system's upgrading, which also can improve the utilization rate of CPU, memory as well as network bandwidth token the data center of transportation department of Guizhou province as an example, the feasibility and effectiveness of the proposed scheme can be proved.

Keywords: Virtualization · Data center · Server migration · Data recovery · Resource scheduling strategy

1 Introduction

Guizhou province is a severe geological disaster area, which mostly affected by i earthquake. In order to ensure the efficient scheduling of the transportation of province highway and waterway in the event of natural disasters, the transportation department of Guizhou province (hereinafter shorted for "the department of province"), the construction of the emergency command system is constantly advancing, with the application of many industries, while the information security system is going to put into practical operation, the number of data center management application system of the department of the province is increasing, with the increasing application each will purchase one or more servers as the routine method, which will make the number of servers to increase rapidly with the construction of service system, which will eventually lead to a sharp rise in the information infrastructure construction as well as in the cost of operation, all of which will have higher requirement on the reasonable allocation and effective application of provincial data center's information infrastructure as well as the resource of the server.

© Springer International Publishing AG 2017
F. Xhafa et al. (eds.), *Recent Developments in Intelligent Systems and Interactive Applications*,
Advances in Intelligent Systems and Computing 541, DOI 10.1007/978-3-319-49568-2_30

At present, the emergency command center construction of the provincial department has been completed, the schedule of moving the provincial data center to the emergency command center building is imminent, how to complete the movement with many servers without interrupting the service application from the provincial data center to the command center smoothly, how to intensify the integration of the existing server resources and improve the utilization efficiency of the whole server system, so as to save the cost of running the system during the migration process, at the same time, building up the computer center for the data center with low-carbon energy so as to become a green data center, moreover, based on the above basis, planning and constructing IT infrastructure construction with good adaptability, which should be easier to be expand, safe and stable has become a pressing danger for the provincial data center.

2 Analysis on Present Situation

There are many problems that exist in the provincial data center. At present, the number of all kinds of application servers that are in the charge of the provincial information center has reached to 76. Among them, the number of CPU whose cores are more than 32 is 2, the number whose cores are more than 16 is 10; the number whose size of memory is above 16G is 18, while the number of the memory size that is more than 8G is relatively less; the size of disk is basically wandering among 800G, all of which can indict that huge difference in the allocation of the existing center server. There are more lagging equipment that are soon to be obsoleted.

Since the scope of transportation information system continues to enhance and the information service is gradually advanced, the existing equipment of provincial data center is bound to withstand the tremendous pressure, the quantity of the servers that need management and guarantee will be more and more, which will bring series of problems to the normal operation and maintenance of the whole transportation information system: (1) The utilization rate of the existing resource is low, the phenomenon of wasting exists, besides the energy consumption is large; (2) The number of servers has increased sharply, resulting in the increase of management cost, refrigeration cost as well as energy consumption cost; (3) The stability of the system is poor, downtime may occur at any time.

3 Technology Strategy for Virtualization and Data Migration

3.1 The Main Researching Contents Are as Follows

(1) Analyzing the existing problems as well as the current construction situation of host computer hardware resources and application system of provincial data center.
(2) Making comparative analysis on the current mainstream technology of virtual commercial products, offering main performance indicators, the functional requirements, the application scope, as well as its advantages and disadvantages.

(3) Making research and analysis on virtualization technology so as to integrate the general idea of storage resources of host computer in the provincial data center, the principle of implementation, the structure of system as well as the content of construction.

(4) Combined with the specific operation of the provincial data center host storage hardware resources and business system, proposing the resource virtualization integration of host computer storage resources and the concrete execution mode of dynamic migration, construction steps and working content, so as to put forward the practical and feasible construction (contrast) scheme in different stages.

(5) Making analysis on the benefit after the implementation of the scheme.

3.2   The Backup and Recovery of Sever

VMware Consolidated Backup can provide a simple virtual machine backup solution, which can be used together with the third-party backup agent software provided by the separate backup proxy server (not the server of the running ESX Sever) installed an agent inside the virtual machine. It can manage the virtual machine centrally through the proxy server backup, by providing support for Fiber Channel SAN, eliminating the need to back up the communication on the network, which can be allow to recover the individual files and directories through providing the file-level integration and incremental backups, so as to allow recover, the entire virtual machine image in the event of a catastrophic failure by providing image-level backup, so as to provide a highly flexible backup and restore functionality. When the backup is performed, Consolidated Backup can make virtual machine be stored in the SAN storage, then mounting the virtual machine snapshot from SAN to the local directory on the backup proxy server (namely, backup for the virtual machine's file-level of Microsoft Windows), or exporting the snapshot from the virtual machines to the backup proxy server (namely, backup for virtual machine image), finally, the backup software can remove the virtual machine snapshot from the backup proxy server of virtual machine and exit the snapshot mode. Besides, the changes can be submit to the backup disk or tape when the computer is in the snapshot mode.

3.3   The Migration of Server

vSphere vMotion can realize the migration of the running virtual machines from one physical server to another physical server in the case of zero downtime and continuously available service, which also can guarantee the integrity of the transaction.

With the aid of the real-time migration function of vSphere, it can move the whole running virtual machine from one physical server to another without stopping machine. And the virtual machine can keep its network identity and connection, so as to ensure the realization of the seamless migration process. Through high-speed network transmission, the memory of activity and the precise execution state of the virtual machine can be transmitted, so that the virtual machine can run from the running source vSphere host computer to the target vSphere host computer. vSphere vMotion can realize the migration of the running virtual machines from one physical server to another physical server in the case of zero downtime and continuously available service, which also can guarantee the integrity of the transaction. And its application migration process can be shown in Fig. 1. Besides, both BT-MPA and SE-MMA algorithm can improve the efficiency of deployment and migration effect.

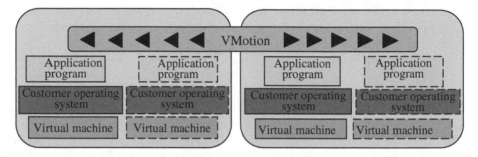

Fig. 1. The principal of migration of server based on vSphere vMotion

3.4 The Autoregressive Prediction Mode $AR(p)$ of P-Order

$AR(p)$ is one of the most common algorithms based on time series prediction algorithms, which is based on the observed value of time series to predict the parameters shown as follows:

$$\hat{y}_t = \varphi_1 y_{t-1} + \varphi_2 y_{t-2} + \cdots + \varphi_p y_{t-p} + \varepsilon$$

Among them, $\hat{y}_t$ is the current predictive value, which is the random variable as its own past observation value $y_{t-1}, y_{t-2}, \ldots$ in the same sequence at different time, p is the order of the model, ε is the noise.

Reconstruction of the mean value of the sample sequence $X(j) = \frac{1}{n} \sum_{i=1}^{n} X(i)$.

The memory takes 5S as the cycle period to take sampling, the cycle of adopting of CPU is 1 s.

Setting v_c, v_m, v_w as the integration of server of CPU, memory and network bandwidth, selecting v as the target node pool. $v = v_c \cap v_m \cap v_w$.

Combined with weight λ_i, calculating the comprehensive weight W_j

$$W_j = \lambda_1 w_1 + \lambda_2 w_2 + \lambda_3 w_3$$

Then the forwarding probability of each node is: $p_j = \dfrac{W_j}{\sum\limits_{n=1}^{n} W_j}$.

4 Experiment Simulation

On the server cluster with 15 physical host computers, the virtual machine can carry out multiple experiments by using the non-load balancing strategy, the migration scheduling strategy based on the threshold equilibrium strategy and the prediction strategy. Having statistics on the host CPU and memory utilization rate, so as to compare the cluster load balancing effect under three situations. CPU utilization rate and memory utilization rate is shown in Figs. 2 and 3.

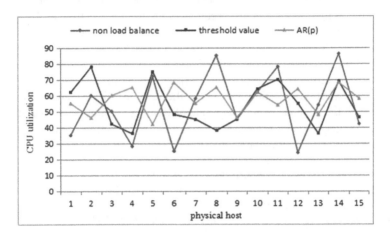

Fig. 2. Comparison chart of CPU utilization rate

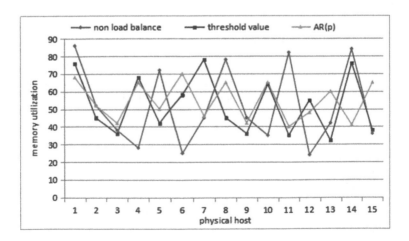

Fig. 3. Comparison chart of memory utilization rate

From the chart we can analyse, the resource utilization rate of each host computer is not balanced under the non-load balancing strategy, the utilization rate of some host computer is higher, while the utilization rate of others is very low, which can result in the low quality of service, with serious waste of resources. The effect of the strategy based on the threshold equilibrium is slightly better, among them, the effect of the load balancing strategy based on $AR(p)$ prediction is the best, which can improve the service quality of the system and make full use of system's resources.

5 Summary

In this paper, the authors analyzes and research on present situation of the construction of the existing server hardware resources as well as the existing problems of the provincial data center, making comparison and analysis on the current mainstream industry virtual technology products, making research and analysis on using virtualization technology integration department of data center, with the general idea of storage resources of host computer, the principle of implementation, the structure of system as well as the content of construction, proposing the resource virtualization integration of host storage and the concrete execution mode of dynamic migration, construction steps and working content, so as to provide the practical construction scheme of the stage, which has great practical significance and theoretical instruction value.

Acknowledgement. This work is financially supported by the Zheng-Xiao-Hang-Qi project of Dongguan Polytechnic (ZHENG201607, ZXHQ201505), and by the project of science and technology for social development in Dongguan City in 2013 (No. 2013108101045), and by the key teaching reform project of Dongguan Polytechnic (No. JGZD1639).

References

1. Xiaochuan, S., Hongyan, C., Bin, X., Chen, J., Liu, Y.: Multidimensional game theory based resource allocation scheme for network virtualization. J. Comput. Inf. Syst. **9**(9), 3431–3441 (2013)
2. Li, L., Jianya, C., Cui, H., Huang, T., Liu, Y.: Resource scheduling virtualization in service-oriented future internet architecture. J. Chin. Univ. Posts Telecommun. **22**(4), 92–100 (2015)
3. Absalom, E., Frincu Marc, E., Junaidu, S.B.: Characterization of grid computing resources using measurement-based evaluation. Multi-agent Grid Syst. **12**(1), 13–34 (2016)
4. Xue, J.W., Hui, H., Fei, Y.P., Feng, L.J.: A cloud computing architectural model based on virtual machine pool. Adv. Appl. Sci. Ind. Technol. **798**, 668–671 (2013)
5. Enzo, B., Danilo, A., Nicola, C.: Minimum-energy bandwidth management for QoS live migration of virtual machines. Comput. Netw. **93**, 1–22 (2015)
6. Alain, T., De Palma, N., Ibrahim, S., Daniel, H.: Software consolidation as an efficient energy and cost saving solution. Future Gener. Comput. Syst. **58**, 1–12 (2016)
7. Sherif, A., Bechir, H., Mohsen, G., Taieb, Z.: Efficient virtual network embedding with backtrack avoidance for dynamic wireless networks. IEEE Trans. Wireless Commun. **15**(4), 2669–2683 (2016)

Application of the Big Data Technology for Massive Data of the Whole Life Cycle of EMU

Han Liang, Liu Feng[⊠], and Zhang Chun

Department of Computing Science,
Beijing Jiaotong University, Beijing 10044, China
fliu@bjtu.edu.cn

Abstract. In view of the EMU products have the huge amounts of data from different manufacturing information system. In the design, manufacture, operation and maintenance processes of the whole life cycle. Traditional tools and method cannot complete the data management tasks with huge data due to its defects. In this paper we introduce the k-means algorithm based on MapReduce parallel programming model, which can enhance the data processing and data mining efficiency. Experiments on the Hadoop cluster shows that the proposed algorithm is feasible, stable and efficient.

Keywords: The whole life cycle · K-means · Big data · MapReduce · Data mining

1 Introduction

With the rapid expansion of sensor, network and information technologies [1, 2], all kinds of intelligent terminal, information management system have rapid promotion and application in large complex equipment manufacturing enterprise. Behind the current enterprise of information processing level and the growth of the contradiction between the huge amounts of data that will be more and more obvious. While, for the vast amounts of data, traditional database management systems reflect the low data performance analysis, the lack of advanced data mining methods and other defects which impact the EMU manufacturers which transform from digital to intelligence.

Clustering analysis [3] is widely used as a data mining algorithm which can be acquired at a high processing efficiency in the global scope of distribution and gradually applied to the intelligent manufactures and others fields [4]. Many researchers use clustering analysis as a core algorithm to deal with problems such as data mining and data processing [5–7]. On the basis of above, this paper applies the k-means algorithm based on MapReduce Parallel Programming Model to analyze and research the massive data of the whole life cycle of EMU. The study has a certain reference value to the application of large data processing and data mining on cloud platform in theory.

© Springer International Publishing AG 2017
F. Xhafa et al. (eds.), *Recent Developments in Intelligent Systems and Interactive Applications*, Advances in Intelligent Systems and Computing 541, DOI 10.1007/978-3-319-49568-2_31

2   Related Technology

2.1   Hadoop

Hadoop is an open source project of the Apache software foundation. It supports different hardware architecture for distribution storage and parallel computing so that it can provide the high reliability for the application and the high mobility for data. The entire platform includes Hadoop kernel, HDFS (Hadoop distributed file system), MapReduce [8] parallel computing framework and some related open source projects, such as Hive data warehouse infrastructure, HBase non-relational distributed database, etc.

2.2   MapReduce

Google's MapReduce proposed a simplified and efficient distributed programming model and it is used to handle large amounts of data distributed computation. The main idea of MapReduce model is to calculate the process data, is divided into two phases: Map and Reduce phase stage. The two phases are used function mapper() and reducer (). In the Map stage, the raw data is input mapper () filter and convert the intermediate data acquired in the Reduce stage as a reducer () input through reducer () is processed to obtain the final results.

2.3   k-means Algorithm

k-means algorithm [9] is a classical algorithm of cluster analysis, it generally uses the Euclidean distance as the evaluation index for the similarity degree of two sample. According to the distance of each sample in the data sets to the center of the k and make it to the smallest distance class, then calculate the all average of the samples in each class, update each class center, until it is stable at the minimum square error criterion function.

Assume object SET $M = \{\chi_1, \chi_2, \cdots, \chi_n\}$, $\chi_i = \{\chi_{i1}, \chi_{i2}, \cdots, \chi_{it}\}$, the computational formula for Euclidean distance between Sample χ_i and Sample χ_j is following:

$$d(\chi_i, \chi_j) = \left[(\chi_{i1} = \chi_{i2})^2 + (\chi_{i2} - \chi_{j2})^2 + \ldots + (\chi_{in} - \chi_{jn})^2 \right] \tag{1}$$

The Rule of square error function is:

$$I_C = \sum_{i=1}^{k} \sum_{j=1}^{t_i} \|\chi_j - n_j\|^2 \tag{2}$$

In this formula, k is the number of clustering, t_i is the number of samples in the NO. I class, n_i is the average of samples in the NO. I class.

3 The Overall Structure and Workflow

3.1 The Architecture for Massive Data of the Whole Life Cycle of EMU

This system, which is mainly composed of PDM and the simulation system as the core of innovation platform; ERP, the supply chain management and the knowledge management system as the core management platform; MRO system as the core of operational platform; MES, quality management and digital manufacturing as the core of intelligent manufacturing platform of these four parts. The fact is all the different platform system function are mostly focused on a certain stage of the whole life cycle of products which related to sui generis, independent design and the implementation, all the data type of each system whose format and produce mode is not the same. The main idea of constructing EMU data management system solution is that the system extracts the data of PDM, ERP, MES and MRO business system to the integrated platform which implement the unity of the whole life cycle data management based on big data technology. Framework diagram is shown in Fig. 1.

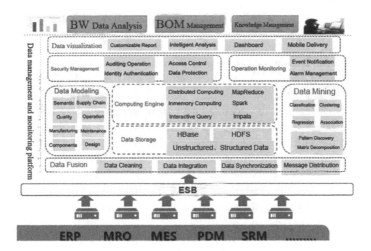

Fig. 1. System Frame

3.2 Analyze Massive Data of EMU PLA Based K-Means Algorithm

In this part, we take the production data as an example, use k-means data mining algorithm to achieve massive data analysis [6].

(I) Choose the initial clustering center

A density-based method is used to choose the initial clustering center, the smaller the Euclidean distance is, the similarity is bigger between the data object. If data object in the more data object area become more and the distance is smaller, that means the object data has big density which can well reflect the characteristics of the data distribution.

According to the formula (1) to calculate the distance between two objects $d(\chi_i, \chi_j)$. Assume the average distance in all objects:

$$\text{AveDis}(M) = \frac{1}{n(n-1)} \sum d(\chi_i, \chi_j) \tag{3}$$

The density of Object χ_i:

$$\text{Den}(\chi_i) = \left(\sum_{j=1}^{n} u\text{AveDis}(M) - d(\chi_i, \chi_j) \right) \tag{4}$$

Density sets:

$$D = \{\text{Den}(\chi_1), (\chi_2), \ldots, (\chi_3)\}. \tag{5}$$

According to max $(\min(d(y_i, O_1)), \min(d(y_i, O_2)), \ldots, \min(d(y_i, O_{n-1})))$. to get the required clustering center (y_i follow the density of object which meet the compare set.

(II) Data Analysis

Use minimum the distance of object to get the best clustering.

At first, the product data are stored in the form of line data sets in a distributed file system, task manager by the above k-means algorithm to choose the number k of clustering and initialization, and then send them to execute Map objection. Then, the Map function record every data set by sent and get the Euclidean distance which calculate the center of Euclidean distance and k, record the center of the smallest distance. The Map function to generate and output intermediate results defined as set <smallDistance,info>, smallDistance means the cluster of the identity of the user belongs to and the info including property, product date, purpose and other information. Partition function will hash the intermediate result and split into different partitions which assigned to specify the Reduce function. Assigned the Reduce task node from the corresponding Map task reads the intermediate results according to sort data, makes the data together with the same smallDistance node which traversal sequence after the middle of the data and list to Reduce function which according to the list values, then calculate with the same vector data and update smallDistance corresponding to the center of the cluster. The work should be repeated until the square error criterion function is stable at the minimum then respectively output the data of clusters. Finally, Map Function scan the input data to get middle result, then Reduce mission read every middle result and merge this data into ordered datasets.

4 Experiment

This demonstration system is deployed on CRRC QINGDAO SIFANG CO., LTD servers, databases, network and other basic platform. The experimental data come from product data in EMU production line and assembly shop.

The cluster building unified used Hadoop2.0 version and the operating system for the Red Hat Linux, which are deployed in 12 xeonE52600 server, including one for the

Table 1. Pilot test result

Data size(10thousand)	Solo(second)	MapReduce(secnd)
1	14	15
15	114	44.1
31	212.7	104.8
200	650.39	343.9

NameNode, 10 servers for DateNote, and one for the client to the server which are named as master, slave0, slave1, slave2, slave3, slave4, slave5 and client can support 8 Map and 8 Reduce mission. Compare the solo server running k-means algorithm with MapReduce programming model k-means, statistics the time-consuming with same work and the test is show in Table 1.

Experimental tests show when faced with massive amounts of data, the MapReduce programming model k-means algorithm is more suitable.

5   Conclusion

This paper analyze and research the k-means algorithm based on MapReduce Parallel Programming Model used in the Hadoop platform and put forward the big data technology application in manufacturing industry. It can practically achieve the exploration and research for big data processing in Enterprises and related departments under the cloud platform.

Acknowledgements. Supported by National High Technology Research and Development Program of China (2015AA043701).

References

1. Cherbakov, L., Galambos, G., Harishankar, R., et al.: Impact of service orientation at the business level. IBM Syst. J. **44**(4), 653–668 (2005)
2. Lee, W.S., Alchanatis, V., Yang, C., et al.: Sensing technologies for precision specialty crop production. Comput. Electron. Agricu. **74**(1), 2–33 (2010)
3. Feng, Z., Xiaoning, Z.: Application of cluster analysis to preventive maintenance scheme design of pavement. J. Harbin Inst. Technol. **16**(4), 581–586 (2009)
4. Hongze, L., Sen, G., Bao, W., et al.: Evaluation on power customer value based on ants colony clustering algorithm optimized by genetic algorithm. Power Syst. Technol. **36**(12), 256–261 (2012). (in Chinese)
5. Li, Z., Wu, J., Wu, W., et al.: Power customers load profile clustering using the SOM neural network. Autom. Electr. Power Syst. **32**(15), 66–70 (2008). (in Chinese)
6. Zhao, L., Hou, X., Hu, J., et al.: Improved k-means algorithm based analysis on massive data of intelligent power utilization. Power Syst. Technol. **38**(10), 2715–2720 (2014). (in Chinese)
7. Youbo, L., Junyong, L., Yan, Z., et al.: Calculation of characteristic attributes of consumer aggregations based on multi-objective clustering. Automa. Electr. Power Syst. **33**(19), 46–51 (2009). (in Chinese)

8. Dean, J., Ghemawat, S.: MapReduce: simplified data processing on large clusters. Commun. ACM **51**(1), 107–113 (2008)
9. de Amorim, R.C., Mirkin, B.: Minkowski metric, feature weighting and anomalous cluster initializing in k-means clustering. Pattern Recognit. **45**(3), 1061–1075 (2012)

Correlation-Merge: A Small Files Performance Optimization for SWIFT Object Storage

Xiguang Wang[1], Xueming Qiao[2], Weiyi Zhu[3], Yang Gao[1],
Dongjie Zhu[4(✉)], and Zhongchuan Fu[1(✉)]

[1] Department of Computer Science and Technology,
Harbin Institute of Technology, Harbin, China
15124585561@163.com
[2] State Grid Shangdong Electric Power Company,
WeiHai Power Supply Company, Beijing, China
[3] State Grid Shangdong Electric Power Company, Beijing, China
[4] Department of Computer Science and Technology,
Harbin Institute of Technology at Weihai, Weihai, China
zhudongjie@163.com

Abstract. In recent years as the internet applications e.g. e-business, web browsing, video website, and social network etc. prevail, data increase at an unprecedented rate. Cloud computing plays an significant role in the big-data era. However, most cloud storage platforms, e.g. SWIFT of OpenStack, aims at optimizing large files performance. Small files with average size less than 15 KB severely hurdles the storage performance. In this paper a framework, namely Correlation-Merge, is proposed to optimize the storage performance of small files in cloud. An optimization layer, namely SFAL (small files additional layer) between Object Server and file system is introduced. Experimental results show that the optimization approach acquires satisfactory efficacy.

Keywords: Cloud computing · Small files · Storage optimization · SWIFT

1 Introduction

As internet applications prevail, such as E-business, Internet Browsing, Video Website and Social Network etc., the amount of data increase at an unprecedented rate. Generally a file with size less than 1 M is regarded as a small file. In big data era, the amount of small files takes a prior dominance. IDC estimates that global data increase at a rate of 58 % per year, meaning that data in the year 2020 will reach 40 ZB, which is 44 times of year 2010 [1]. At that time, one-third of global data is stored in the cloud, and small files such as data, image and audio with an average size of 15 KB, take a prior dominance. For example, nowadays over twenty billion images with an average size of 15 KB are stored in TaoBao e-business platform, while over sixty billion small images are processed in FaceBook and over four thumbnails are generated for each video in Youtube.

However, the distributed and cloud storage platform, e.g. GFS [2], HDFS, and the SWIFT of the OpenStack Object storage, optimizes the performance of large files [3],

© Springer International Publishing AG 2017
F. Xhafa et al. (eds.), *Recent Developments in Intelligent Systems and Interactive Applications*,
Advances in Intelligent Systems and Computing 541, DOI 10.1007/978-3-319-49568-2_32

consequently the large amount of small files incurs a great performance overhead in big data era [4]. In essence, the local file system such as EXT and XFS, are specifically designed to optimize the storage performance for large files, so the random read and write operations for large amount of small files will incur a great overhead.

In this paper Correlation-Merge framework is proposed to optimize the storage performance of small files in cloud platform. In the proposed framework, a SFAL (small files additional layer) layer is introduced to optimize the storage performance of small files by merging related small files into a large file. Then the number of files is decreased. To this end the correlation between a workload and related small files is essential, so an algorithm is elaborately designed and multiple logical small files are mapped into one large physical file. The random small files access is transformed into sequential access, optimizing the file access I/O and read/write performance [5].

2 Design and Implementation of Correlation-Merge

2.1 Correlation-Merge Framework

The principal idea of storage performance optimization for small files is to find the correlations of small files and then related small files are merged into a large file. The overall architecture of Correlation-Merge framework is depicted in Fig. 1. An optimization layer, namely SFAL (small files additional layer) is introduced into the OpenStack platform, lying between Object Server and the underlying OS file system. The correlation of a workload between related small files is modeled by an elaborately designed algorithm, and an indexing service is implemented to manage the mapping between related small files and the physical large file.

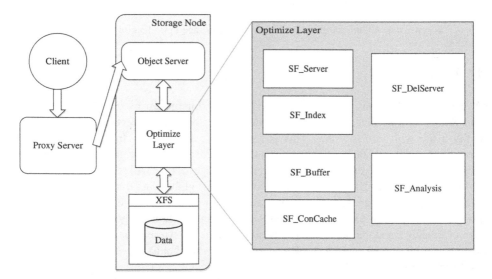

Fig. 1. The *Correlation-Merge* framework for small files storage performance optimization. An optimized Layer is introduced, including the following components.

The SFAL optimization layer includes the following components, such as the indexing service SF_Index, buffering service SF_Buffer, small files deletion service SF_DelServer, files prefetcher SF_ConCache, and correlation modeling service SF_Analysis. In the Correlation-Merge framework the small files are combined and optimized, whilst the large files are processed in the original manner. The correlation between a workload and related small files is analyzed and multiple logical small files are mapped into one large physical file. This transforms the random multiple small access into a sequential access, thus optimizing the file access I/O and read/write performance.

The data structure of indexing service SF_Index is depicted in Fig. 2. The sflag designates if a file is a small file or a normal file, the Bflag indicates the buffering status, cflag represents the correlation between small files, blockID is the block identification of the resulting merged of large file, and the offset depicts the corresponding offset in the merged file. In addition, some other information is needed by the SF_Index service as well, such as the Metadata length, the Datalength, etc. For the SF_Index, the HashTable is used for indexing by using in memory computing [6]. In case of an update, the modified item is saved into a log file, and then it is serialized at a specified time interval. When the indexing service restarts, it is initialized by loading the serialized snapshot.

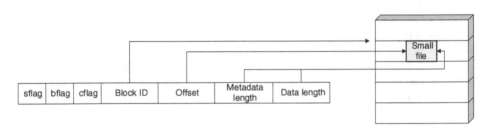

Fig. 2. The metadata of small files index.

SF_Buffer is the buffering optimization of small files. Small files are saved in a buffer. When the buffer is full, it is serialized by using corresponding bitmap ID as a normal block. Through a preliminary experiment, 8 MB is adopted as the buffer size of SF_Buffer. A larger size has modest effect to improve the indexing performance.

SF_DelServer service is in charge of merge file deletion, in the case of a lighter load of the SWIFT storage. The following information is recoded into log file, including BlockID, Offset, and Length etc. By using the SF_DelServer service, the merged file is split into multiple small files according to the log again.

SF_ConCache is the caching mechanism for small files read performance optimization. In the case of a miss, the resulting merged large file is cached. A naive replacement algorithm is designed, where the number of hits is acompanied for each file. And a metric is defined, as the ratio of the number of hits to the size of file, namely Evictbit, which is used for SF_ConCache replacement. This means that when the Evictbit metric is less a threshold, the small file is marked as low correlated and evicted.

SF_Analysis is the vital service, which is used for small files correlation modeling for a specific application. The access correlation between small files is abstracted for a specific application [7], which is used for large file merging. The detailed algorithm is described in the next section.

2.2   Correlation Analysis Algorithm

In a specific time interval, the small files that are accessed frequently are named after a workload correlated files [8]. SF_Analysis is one of the most important services to model small files correlations for a specific workload dynamically, and an algorithm is designed. It is a necessity for small files.

In this paper, a modeling algorithm similar to clustering is designed. First, a triple tuple is defined as follows: (userID, file name, timestamp). The number of accesses of a file is used as dimensionality, so the euclidean distance between two files is used as a metric to quantitatively describe their correlations. Then, the modeling of a correlation between small files is transformed into the clustering problem in a multi-dimensional space. Owing to the difference of time stamps of small files, a calibration algorithm is a necessity as well.

The number of accesses of a file is used as its dimensionality. First, a metric namely Distance is defined to designate the euclidean distance between two nodes. Second, another metric namely Entropy is defined as the ratio of dimensionality of node tNode to that of central node cNode. The fact that related files cannot be accessed simultaneously leads to a non-negligible error. Accordingly, a calibration algorithm is a necessity to adjust the euclidean distance of each node [9]. By aligning the timestamp, the algorithm makes the proposed Correlation Analysis Method conforms to the facts.

Above all, by applying the proposed Correlation Analysis method, the correlation between small files under a specific workload is dynamically sought out. This is of great importance to small files merging and read performance optimizations. Related small files are packed into a single file block with an updated index, together with the SF_ConCache caching the read performance is improved.

3   Experimental Results

3.1   Evaluation Method

To validate the feasibility of the proposed Correlation-Merge framework, a prototype based on SWIFT of OpenStack is implemented. The experimental environment is depicted in Table 1, including a proxy server, three storage nodes, and a benchmark node.

Account server, container server and an object server are deployed on the storage nodes respectively. In present work, COSBench [10] is adopted as workloads. COS-Bench is an opensource benchmark, targeting distributed object storage developed by Intel.

In present work, a file with size larger than 1 M is regarded as a large file. For small files five configurations are adopted, including 1 KB–10 KB, 10 KB–20 KB, 20 KB–50 KB, 50 KB–100 KB, 100 KB–1MKB. In each experiment a total of 320,000 files

Table 1. Experimental environment

CPU	Memory	Disk	Running service
Intel I7-2600 4-core 3.4 GHz	4 GB	1 TB 5400 rpm	Proxy server
Intel E3-1220 2-core 3.10 GHz	4 GB	1 TB 5400 rpm	Storage node
Intel E5500 2-core 2.8 GHz	6 GB	500 GB 7200 rpm	Storage node
Intel Q8300 -core 2.5 GHz	4 GB	500 GB 7200 rpm	Storage node
Intel I7-2630 m 4-core 2.0 GHz	8 GB	750 GB 5400 rpm	Test node

are generated, and sixteen worker threads are used emulating workloads. For read performance experiment, the characteristic of a workload is as follows: 1,000 trails constitute an experiment. For each trail 10–100 correlated files are accessed with a specified probability in a short time interval. In this work, the access probability is set to 80 %. This means that altogether 64,000 files are correlated taking a 20 % of the total files. The performance metrics such as response time, processing time, and bandwidth are evaluated (Figs. 3, 4 and 5).

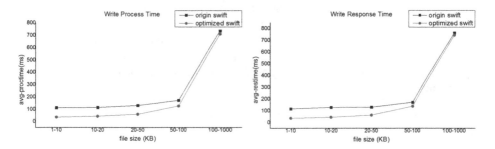

Fig. 3. Write response time   **Fig. 4.** Write process time

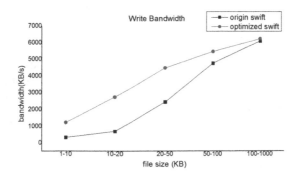

Fig. 5. Write bandwidth

3.2   Experimental Results

An in-depth analysis reveals that the write overhead includes two components, e.g. network transmission overhead and disk I/O overhead. The following factors contributes to disk I/O overhead, including Disk Seek Time Tseek and Disk Write Time Twrite. The smaller the files, the greater overhead it takes for the Tseek component. The proposed optimization mechanism buffers and merges a lot of small files, effectively reducing the Disk Seek Time Tseek. Experimental result demonstrates that the smaller the size of the files, the greater improvements it accquires.

The preliminary experimental results demonstrate that, the read performance, response time, read processing time, and bandwidth improves as well, but they do not improves so remarkably as expected. In-depth analysis shows that traditional caching in SWIFT contributes to this. The detailed evaluation is left for future work.

4   Conclusions

Aiming at storage performance optimization for small files, an optimized framework namely Correlation-Merge is put forward and a prototype basing on OpenStack SWIFT is implemented as a motivating case study. First, the optimized framework is proposed. An optimization layer, namely SFAL (small files additional layer) is introduced into the OpenStack platform, laying between Object Server and the underlying OS file system. And a correlation modeling method is proposed and described in great details. Experimental results show that the proposed optimization approach acquires satisfactory efficacy.

References

1. Turner, V., Gantz, J.F., Reinsel, D., et al.: The digital universe of opportunities: rich data and the increasing value of the internet of things. IDC Analyze the Future (2014)
2. Ghemawat, S., Gobioff, H., Leung, S.T.: The Google file system. ACM SIGOPS Oper. Syst. Rev. **37**(5), 29–43 (2003). ACM
3. Zeng, W., Zhao, Y., Ou, K., et al.: Research on cloud storage architecture and key technologies. In: Proceedings of the 2nd International Conference on Interaction Sciences: Information Technology, Culture and Human, pp. 1044–1048. ACM (2009)
4. Yang, Y., Qi, L., Lu, Y., et al.: A small-file writing acceleration mechanism for distributed file system. In: Cyberspace Technology (CCT 2013), International Conference on. IET, pp. 558–561 (2013)
5. Zhang, W., Lu, G., He, H., et al.: Exploring large-scale small file storage for search engines. J. Supercomput. 1–13 (2015)
6. Lensing, P., Meister, D., Brinkmann, A.: hashfs: applying hashing to optimize file systems for small file reads. In: 2010 International Workshop on Storage Network Architecture and Parallel I/Os (SNAPI), pp. 33–42. IEEE (2010)
7. Wang, T., Yao, S., Xu, Z., et al.: An effective strategy for improving small file problem in distributed file system. In: 2015 2nd International Conference on Information Science and Control Engineering (ICISCE). IEEE, 122–126 (2015)

8. Feng, D., Qin, L.: Adaptive object placement in object-based storage systems with minimal blocking probability. In: AINA 2006. 20th International Conference on Advanced Information Networking and Applications, vol. 1. IEEE, pp. 611–616 (2006)
9. Han, J., Cheng, H., Xin, D., et al.: Frequent pattern mining: current status and future directions. Data Min. Knowl. Discov. **15**(1), 55–86 (2007)
10. Zheng, Q., Chen, H., Wang, Y., et al.: COSBench: cloud object storage benchmark. In: Proceedings of the 4th ACM/SPEC International Conference on Performance Engineering, pp. 199–210. ACM (2013)

The Research of Cyber Situation Awareness Model

Jing An[(✉)], XueHu Li, ChunLan You, and Lei Zhang

Beijing, China
anj21_2000@sina.com

Abstract. Cyber situation awareness is an vital source and prerequisite to carry out the cyber operation. This paper proposed the key problems of model using in cyberspace based on the general reference model of cyber situation awareness. Finally, the cyber situation awareness model for cyber resistance operation is built.

Keywords: Cyberspace · Situation awareness

With the rapid growth of information technology and network technology, the threat and attack of cyberspace has been able to affect the military operation in a multiple ways, the cyberspace has become the fifth battle field, besides the land, the sea, the air, the space. Cyber situation awareness is an important basis and prerequisite to carry out the cyber operation, as the conflict in cyberspace between countries are growing, it is realized that the first combat in "cyberspace" is so essential. Therefore, the development of cybers situation awareness ability is particularly vital.

1 Cyber Situation Awareness

Historically, in the war, commander have to answer three questions "what happened?", "why?" and "what to do?" in the decision-making process.

In cyberspace, it is mainly depended on the following skills to answer these three problems.

(a) Vulnerability analysis, intrusion detection and forensics analysis ability;
(b) Track and analyze of attack;
(c) The information transformation ability.

For a systematic description, the "situation awareness" was introduced into the cyber space security field. In "The US army field operation manual (September 2004)", "situation awareness" is defined as "the knowledge and the understanding of the current situation, which can help to give an accurate, relevant and timely assessment of the actions not only friendly but also competitive in the operation space. The purpose is to assist for decision making. Situation awareness is the ability and skill to observe information. It can be used for promptly determine the developing background of numerous events and event correlation."

Refer to the above definition, cyber situation awareness can be used with the aid of new technologies, such as hardware sensors, intelligent computer program and

© Springer International Publishing AG 2017
F. Xhafa et al. (eds.), *Recent Developments in Intelligent Systems and Interactive Applications*,
Advances in Intelligent Systems and Computing 541, DOI 10.1007/978-3-319-49568-2_33

advanced strategy of human thinking process. It is used to provide accurate, concise index of attack, and reveal the most prominent attack types. So for a security analysts, it can be helped to predict cyberspace development trend in the future, make the effective planning and response to the current and emerging cyber attacks.

Cyber situation awareness can be divided into three stages: situation identification, situation understanding and situation prediction. As shown in Fig. 1.

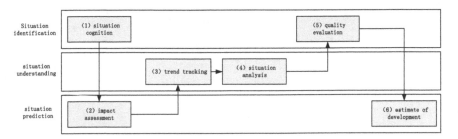

Fig. 1. The stages of cyber situation awareness

Among them, three stages can be divided into the following six steps:

(a) Situation cognition: To understand the current state, including the state identification and confirmation. By detecting, to aware of the attack, at the same time, confirm the attack types, sources, properties and target, etc.
(b) Impact assessment: Understand the impact of attacks, including damage assessment of the current and impact assessment of the future.
(c) Trend tracking: Understanding of evolution, analysis of the attack trend, intention, and methods.
(d) Situation analysis: By the causality analysis, forensic analysis to understand the cause of the current attacks.
(e) Quality evaluation: To analyse the authenticity, integrality and timeliness of the situation awareness information.
(f) Estimate of development: Predict the possible action, activity and path.

Through the above six steps, data is transferred to information, to knowledge, to intelligence. So it is helpful for security analysts to understand the current situation and predict the future trend.

2   The General Reference Model of Cyber Situation Awareness

2.1   The Description of Model

This model mainly from the data stream to describe situational awareness. It combines JDL Data fusion model and Endsley situation awareness model, and it is expanded by McGuinness, Foy, as shown in Fig. 2.

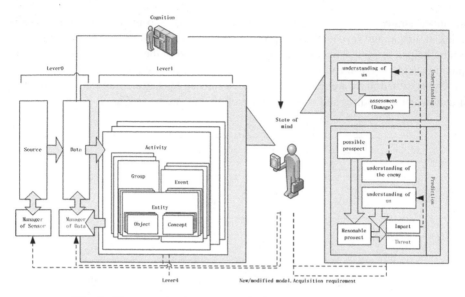

Fig. 2. The general reference model of cyber situation awareness

Level 1. Cognition. Sensor is the source of data. Based on the data, there will be recognition and tracking. In this lever, the model and prior knowledge are require to identify objects, groups and activities. There are two method to get prior knowledge: one is the knowledge discovery technology, the results are verified by the operator. The second is directly provided by the operator. Cognitive can provide status, properties, and dynamic informations of relevant factors, including the information in the form of a variety of understandable process. It is the basis for understanding and forecast. If no basic cognitive for important environmental elements, there will be highly probability of forming error situation.

Level 2. Understanding. If it is answered "how" by cognitive process, so the next level will solve the problem of "what", namely understanding. Understanding of the situation is the process of combination, reading, choosing and storing of information. Integration is the key to the vast amounts of information, determine the relevance of the information, and according to the largest and most threatened to classify activities, deduce the series of conclusions related to focus, to form a structured current situation

Level 3. Prediction. Understanding layer focuses on the current situation, and the level 3 focuses on the forecast and analysis according to the current situation. "The answer for "what might happen". Predict layer clear for the future, refers to the situation of the present state and the change of all the elements.

Level 4. Solution. It is made up of the extension model of McGuinness and Foy. This level is trying to choose a series of actions in a set of available operations and form a plan of action, to answer the problem of "how to". McGuinness and Foy think that, the result in solution is not to tell the decision makers directly what actions or decisions must to do, but ultimately to provide alternative actions and their impact on the environment.

Although our description of model is based on hierarchy. As a matter of fact, model is parallel execution, rather than a serial process. That is, all levels of the model should be operate simultaneously, trigger and interact continuous, form continuous changes and updates. For example, in the level 2, there will be some problems in the process of current situation understanding, so more data is needed to compensate for the lack of items or reduce the uncertainty of the given data. As increased or modified data collection requirements, these data requirements can be feedback to the process of data collection requirements. Level 3 can from another side to provide such data in data collection process. Analysts can predict one or more activities possibility, do the speculation on the basis of the current situation, discover and determine the key events on development. And these key events can also be used as a data collection require-ment for feedback to the data collection process. As new information injection, analysis and forecasting process will form new knowledge, better complete object tracking and confirmation.

2.2   The Key Problems of Model Using in Cyberspace

The general reference model of cyber situation awareness is applied to cyberspace, the key to improve the ability of cyber situation awareness are shown as follow.

2.2.1   Choice of the Influence Domain

The purpose of Situation analysis and evaluation is to determine the impact on focus areas, deduce the impact of reasonable prospect and future attacks. Influence domain is difference, the focus of attack is not the same.

For example, cyber managers might be pay more attention to a strategy change detection, firewall configuration issues, and for the cyber department, are more con-cerned about the impact on sectors of improper use and attack of equipment, etc.

From operation level to strategic level, the situation awareness of interdisciplinary activities will become a kind of demand. For example, the commander of air operations center may pay more attention to what kind of impact to the mission will be bring by cyberspace activities, and logistical support commanders were more concerned that the attacks would affect logistics transport routes.

The choice of the influence domain, therefore, become the basis of analysis and evaluation, is also one of the main requirements of transfer from data to knowledge.

2.2.2   The Acquisition of Attack Behavior Information

To a great extent, efficient cyber situation awareness dependent on the acquisition of accurate, concise, and high quality informations related to attack. Cyber attack behavior mainly comes from intrusion detection system, the log file sensors, anti-virus systems, malicious software test procedure, firewall and other cyber sensors.

Because of the fast evolution of the characteristics of malicious attacks, and the potential threat identification in huge amounts of information is very difficult for cyber security analysts. So it become increasingly difficult to get effective information of cyber malicious activities.

In fact, the acquisition of attack behavior is not only the starting point of the situation awareness, but also one of the determinants of assessment and speculation. For example, it can be speculate whether the attack with pertinence by the acquisition of attack range, which makes the corresponding risk assessment.

3 The Cyber Situation Awareness Model for Cyber Resistance Operation

3.1 Cyber Collapsing Operation

In modern warfare, cyber as ubiquitous are charged with nerve and hub, once the paralysis, consequence is inconceivable. In cyber collapsing operation, attack is the weakness of the network.

Cyber collapsing operation is based on the main and key nodes, using botnets, take "the swarm tactics" attack, lead to network paralysis, can produce a great battle with the small input efficiency. In fact, it is a kind of Internet business flow that programming in advance, it can make the network paralysis or blocking.

These computers were always in the "sleeping", when the attack started, they began to active. The computer is called "zombie", they are in remote control, like robot network composed of zombies. The attack of zombie computer according to the load on them instructions to attack, but nothing to be aware of by computer owners.

At present, the most representative is known as "the first network war" (WW I) in Estonia. From April to May, 2007, Estonia across the country suffered three rounds of large-scale distributed denial of service attack (DDoS). The important goals include Estonia national leadership, congress, government departments, major political parties, Banks and major news media sites, vertical scale widely and deeply.

On August 8, 2008, Russian troops across the Georgia border. At the same time, in Georgia, had also launched a similar cyber collapsing attack. The attack lead to the paralysis of important websites, including the tv media, financial and transportation. Government agencies operation into chaos. The airport, the logistics, the communication and other information network were collapsed. So the supplies of logistics in needed can't timely sent to the specified location, potential ability of war severely weakened, it directly influence the social order of Georgia as well as the army's battle command, logistics.

3.2 The Building of the Model

Based on the above analysis and research, from the perspective of process, building situation awareness model for the cyber collapsing attack, as shown in Fig. 3.

In the model, choosing the outbreak of worm, the spread of the botnet as observation, can help to form the opinion of happening incident. By contrast with various attacks in experience library, to determine whether the ongoing attacks will affect us, such as network interruption, communication etc. To generate "damage" assessment of

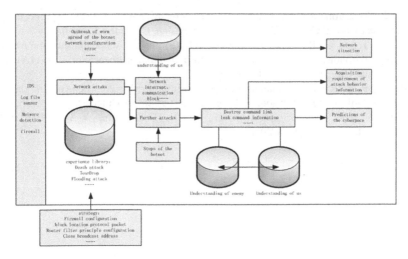

Fig. 3. The situation awareness model for the cyber collapsing attack

the cyber attack situation. More specifically, whether needed to compensate for the damage by recovery plan.

In addition, the model must be focal point on the ongoing or potential attacks. The faster understanding what attackers can be done, the greater the range of decision makers can choose. A priori knowledge, therefore, can according to the experience in the library, namely the harm of various attacks, predict the future state of every important activity.

So far, we only predict the development of the current activity on the basis on the model itself, but did not examine the rationality of these predictions. As a result, other knowledge is also required to consider, including "the understanding of the enemy" and "the understanding of us". That we need to know whether the enemy has the ability and intent to predict the action, and whether they used to show similar behavior. Based on these knowledge, get what we pay special attention to the next development of activities, such as whether there is the risk of damage command link.

Based on this reasonable supposition, not only further predictions of the cyberspace can be concluded, but also the acquisition requirements of future attacks can be formed at the same time.

References

1. U.S Department of Defense, Data Fusion Subpanel fro the Joint Directors of Laboratories, and Technical Panel for C3.Data fusion lexion (1991)
2. Salerno, J.: Measuring situation assessment performance through the activities of interest score. In: Proceedings of the 11th International Conference in Information Fusion, Cologne GE, 30 June – 3 July 2008

3. Tadda, G.: Measuring performance of cyber situation awareness systems. In: Proceedings of the 11th International Conference in Information Fusion, Cologne GE, 30 June – 3 July 2008
4. Salerno, J., Tadda, G., Boulware, D., Hinman, M., Gorton, S.: Achieving situation awareness in a cyber environment. In: Proceedings of the Situation Management Workshop of MILCOM 2005, Atlantic City, NJ, USA, October 2005
5. Tadda, G., et al.: Realizing situation awareness within a cyber environment. In: Dasarathy, B.V. (ed.) Multisensor, Multisource Information Fusion: Architectures, Algorithms, and Aplications 2006, Proceedings of SPIE (SPIE, Bellingham, WA, 2006) 624204, Kissimmee FL, vol. 624, April 2006

High Performance Low Latency Network Address and Port Hopping Mechanism Based on Netfilter

Yue-Bin Luo[✉], Bao-Sheng Wang, Gui-Lin Cai, Xiao-Feng Wang,
and Bo-Feng Zhang

College of Computer, National University
of Defense Technology, Changsha, China
{luoyuebin, bswang, glcai, xf_wang, bfzhang}@nudt.edu.cn

Abstract. Network address and port hopping (NPAH) is an effectual moving target defense tactic that comes from frequency hopping in wireless communication, and it is proposed for host and service hiding and attack resistance. In this paper, we propose a high performance low latency network address and port hopping implementation mechanism, using the netfilter framework inside the Linux kernel. We have conducted experiments and tests to evaluate the performance of our method, and the result shows that the proposed mechanism is efficient in implementing NPAH on Linux platform.

Keywords: Network security · Moving target defense · Address and port hopping · Netfilter

1 Introduction

With the widespread of Internet and the enhance of its significance, the network security demand grows as well. However, in the legacy networks, a server host usually provides a standard network service accessible anytime and anywhere by using a static IP address and a well-known communication port, and the port is kept open for the entire service lifetime, this is known as the static service model. However, the durative accessible port and address also make the service suffer long-term network reconnaissance and attacks. The attackers can perform host discovery and port scanning, obtain detailed information, target the system and find its potential vulnerabilities, and further launch attacks to compromise the target system. However, existing security systems such as intrusion detection/prevention systems (IDS/IPS) [1], firewall, and antivirus software usually detect threats based on pre-defined rules and existing attacks' fingerprints, therefore they can only detect and cope with attacks recognized already. Today, the increasing security affairs such DoS/DDoS [2] and advanced persistent attacks are defeating our faith.

Moving target defense (MTD) [3] is a proactive cyber defense technology, whose basic idea is continuously shift systems' attack surfaces so as to increase attacker's efforts in exploiting target systems' vulnerabilities. Network address and port hopping (NAPH) is a novel MTD technique that comes from frequency hopping in wireless

© Springer International Publishing AG 2017
F. Xhafa et al. (eds.), *Recent Developments in Intelligent Systems and Interactive Applications*,
Advances in Intelligent Systems and Computing 541, DOI 10.1007/978-3-319-49568-2_34

communication, which mutates the server' real communication port and IP address to virtual port and address randomly selected based on pseudo-random function. NAPH constantly mutates communication identities (i.e. the service's port and address) and create high unpredictability, therefore the attackers must pay more efforts in their attack preparations. Consequently, the varying host and service environments force the attackers to re-scan/re-connect the target network and system frequently in different time, this significantly improving its detectability while slowing down the attack progress.

Today, several network address and/or port hopping mechanisms have been proposed, but it is still lack of a high performance and low latency implementation mechanism. Motivated by this, we design an effective implementation mechanism for NAPH based netfilter framework, and we provide the detailed design proposal and conduct experiments and tests to evaluate the performance and overhead of our method.

2 Related Work

In [4], a random port hopping (RPH) technique is proposed based on time-synchronization, where UDP protocol and the setting up TCP communications are implemented through the socket communications. This is a user space model that needs to develop a dedicated software between the applications and the operating system to perform hopping, and the drawback is that TCP protocol will use a given port number for the entire connection.

In [5], a port-based rationing channel and protocol are presented for two-party communication. This is a protocol middle layer model [6] that requires changes to the protocol stack of the end-host operating system. In [7], two designs for port and address hopping are implemented, where hopping delegate and NAT gateway are used for port and address hopping and mapping. It relies on devices special designed for hopping.

In [8], a full service hopping tactic is proposed, and an IP mutation technique called OF-RHM (OpenFlow Random Host Mutation) is developed in [9]. Both of these two mechanisms depend on centralized controllers to control and enforce hopping, which is unavailable on the conventional networks. In [10], a network address space randomization (NASR) scheme is presented. But it implements address mutation based on DHCP update and it provides very limited unpredictability, and its mutation speed is slow (the maximum IP mutation speed is once every 15 min).

In [11], we proposed a universal and multi-platform deployable port and address hopping mechanism based on virtual network kernel driver TAP, called TPAH, which implements hopping relying on a user-space data processing process. Although TPAH applies to various OS platforms, it introduces nonnegligible overheads in the communications.

Although various port and address hopping and implementation mechanisms proposed, a high performance low latency implementation mechanism is also important and in design of need for network address and port hopping. Motivated by this, we proposed a novel and efficient implement mechanism in this paper, providing the following contributions:

- Based on the netfilter framework, a new implementation mechanism is proposed to enforce NAPH, which provides an efficient mechanism that effectively reduces the overhead of NAPH deployment.
- The proposed method performs hopping in the kernel space, which is transparent to the up layer applications. Besides, it doesn't need to modify the OS network stack, which effectively reduces the difficulty of implementing NPAH.
- We provides an economical method, which minimizes the deployment costs since it can be conveniently deployed on Linux platforms such as client hosts, server hosts, and gateways without introducing dedicated devices.

3 System Model

In this section, a novel NAPH implementation mechanism based on the netfilter framework [12] is proposed. Netfilter is a universal packet processing and filtering framework inside the Linux series after version 2.4.x, which provides multiple hooks inside the Linux kernel that allows kernel modules to register callback functions with the network stack, providing network address and port translation, packet filtering and other packet mangling functions. The model of the netfilter framework is shown in Fig. 1, which provides five hook points in the kernel where we can register callback functions to intercept and modify network packets.

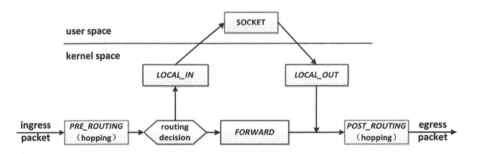

Fig. 1. The model of the netfilter framework

In this paper, we add function handles into netfilter's *PRE_ROUTING* and *POST_ROUTING* chains to hook ingress and egress packets to perform port and address replacement. The system model of the implementation mechanism is shown in Fig. 2, in which service' port and address are randomly selected based on a secret key and a pseudo-random function pre-shared between the client and the server. Although we use NAPH as an example of the hopping implementation, our method can also apply to some other hopping schemes such as port hopping, address hopping, and port and address hopping.

The implementation is accomplished via deployments at three different locations: port and address hopping deployed on the gateway of the client domain that performs port and address hopping and mapping for the client hosts, address hopping deployed

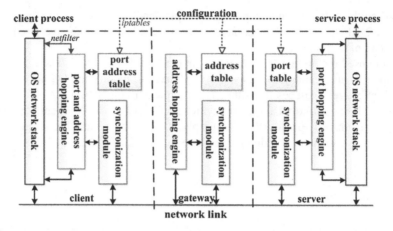

Fig. 2. System model of the proposed method

on the gateway of the server domain that performs address hopping and mapping for the server hosts, and port hopping on the server host that performs port hopping and mapping for the service applications.

Each deployment consists of three components: a port and/or address hopping engine, a port and/or address table, and a synchronization module. The hopping engine is responsible for port and/or address mutation and mapping, which mutates the server's communication identities to/from random identities. In each hopping interval, the valid ports and addresses are stored in the port and/or address table, the hopping engine performs port and address hopping and mapping for the existing connection based on the table, insuring that only the packets whose identities fulfil the NAPH rules can reach the service applications, while attack packets using illegal/invalid communication identities will be effectively detected and denied. The port and/or address table will be updated in the beginning of the each hopping time interval, and the old identities will be deleted and new identities will be added. Furthermore, the system relies on the synchronization module to synchronize communication parties, and the service can be reached only if the port and address hopping is performed based on precise synchronization information. The synchronization can be achieved through time [4] or timestamp-based synchronization [13].

4 Performance Evaluation

In this section, the performance of our method is evaluated through practical implementation on our experimental network. The experiment topology is shown in Fig. 3, which contains two class C subnets, one is the client subnet and the other is the server subnet. The implementation is deployed on Ubuntu Linux (kernel 3.17.3). The client and server connect the same time server to achieve time synchronization using network time protocol (NTP), and perform NAPH with a fixed time period (i.e. 5 s).

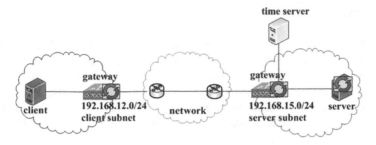

Fig. 3. The experiment topology

We have conducted bandwidth and latency tests with and without port hopping and address and port hopping to measure the influence to the data communication introduced by our method. Firstly, we use Iperf (version 2.05) [14], a network bandwidth benchmark, to measure the bandwidth overhead. Then, round trip time (RTT) tests were performed through using Hping3 [15] to measure the latency overhead. Each of the tests had performed 20 tests, and the results are shown in Figs. 4 and 5.

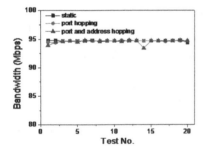

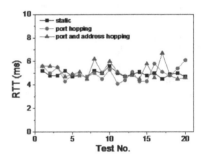

Fig. 4. The bandwidth test result

Fig. 5. The RTT test result

From Fig. 4, we can observe that the costs of introducing netfilter-based port hopping and address and port hopping on bandwidth are 0.02 % and 0.13 %, respectively. From Fig. 5, we can see that the overheads of introducing netfilter-based port hopping and address and port hopping on RTT are 0.4 % and 4.67 %, respectively. Compared to the TPAH method we proposed in [11] that introduces 0.1 % bandwidth overhead and 25.45 % latency overhead, this paper provides a high performance low latency address and port hopping implementation mechanism.

5   Conclusion

In this paper, we have designed and proposed a novel network address and port hopping implementation mechanism based on the universal packet filtering and processing framework, i.e., netfilter. In order to evaluate the network performance and overhead of our method, we conducted network performance tests, and the results illustrate that our netfilter-based method is an efficient mechanism with high performance and low latency for network address and port hopping deployment.

Acknowledgements. This work is funded by Research Fund for the National Key Basic Research Program (973 Program) of China (2012CB315906) and NSF of China (61303264).

References

1. Zhang, K., Cai, Z.P., Zhang, X., Wang, Z.J., et al.: Algorithms to speedup pattern matching for network intrusion detection systems. Comput. Commun. **62**, 47–58 (2015)
2. Wang, F., Wang, H., Wang, X., et al.: A new multistage approach to detect subtle DDoS attacks. Math. Comput. Model. **55**(1), 198–213 (2012)
3. National cyber leap year summit 2009 co-chairs report (2009). http://www.cyber.st.dhs.gov/docs/NationalCyberLeapYearSummit2009Co-ChairsReport.pdf
4. Lee, H.C.J., Thing, V.L.L.: Port hopping for resilient networks. In: Proceedings of IEEE 60th Vehicular Technology Conference, vol. 5, pp. 3291–3295 (2004)
5. Badishi, G., Herzberg, A., Keidar, I.: Keeping denial of service attackers in the dark. IEEE Trans. Dependable Secure Comput. **4**(3), 191–204 (2007)
6. Ma, Q., Dai, H., Zhao, X.L.: Using port hopping to realize information hiding. Comput. Eng. Des. **28**(4), 849–851 (2007)
7. Atighetchi, M., Pal, P., Webber, F., Jones, C.: Adaptive use of network-centric mechanisms in cyber-defense. In: Proceedings of 6th IEEE International Symposium on Object-Oriented Real-Time Distributed Computing, pp. 183–192 (2003)
8. Shi, L., Jia, C., Lu, S.: Dos evading mechanism upon service hopping. In: Proceedings of IFIP International Conference on Network and Parallel Computing Workshop, pp. 119–122 (2007)
9. Jafarian, J.H., Al-Shaer, E., Duan, Q.: Openflow random host mutation: transparent moving target defense using software defined networking. In: Proceedings of HotSDN Workshop at SIGCOMM 2012, pp. 127–132 (2012)
10. Antonatos, S., Akritidis, P., Markatos, E.P., Anagnostakis, K.G.: Defending against hitlist worms using network address space randomization. Comput. Netw. **51**(12), 3471–3490 (2007)
11. Luo, Y.B., Wang, B.S., Wang, X.F., et al.: A universal and multi-platform deployable port and address hopping mechanism. In: Proceedings of 2015 International Conference on Information and Communications Technologies (ICT2015) (2015)
12. The netfilter.org. http://www.netfilter.org/. Accessed 29 Feb 2016
13. Shi, L., Jia, C., Lü, S., Liu, Z.: Port and address hopping for active cyber-defense. In: Yang, C.C., et al. (eds.) PAISI 2007. LNCS, vol. 4430, pp. 295–300. Springer, Heidelberg (2007). doi:10.1007/978-3-540-71549-8_31
14. Iperf. https://iperf.fr/. Accessed 20 Apr 2016
15. Hping. http://hping.org/. Accessed 28 Apr 2016

Sampling Online Social Networks
for Analysis Purpose

Jiajun Zhou$^{(\boxtimes)}$, Bo Liu, Zhefeng Xiao, and Yaofeng Chen

College of Computer Science,
National University of Defense Technology, Changsha, China
{jerome_zjj, fengmail}@163.com,
boliu615@gmail.com, 1124026826@qq.com

Abstract. Online Social Network is a novel communication pattern of great convenience and abundant interest, influencing people's life in extensive regions. A prominent problem in OSN is the access to data form a whole view of network topology from various social network sites. This paper describes the fundamental method to obtain a representative sample of online social network. According to the analysis of user interactions with social networking sites and the comparison of different traversal algorithms, this paper simulates the social network traversal based on the breadth-first algorithm. Then, we implement the algorithm on real OSN by sampling. A considerable data set is available by our sampling method. Finally, several social network analysis are experimented to characterize key properties of OSNs.

Keywords: Online social networks · Graph sampling · Social network analysis

1 Introduction

The popularity of online social networks (OSNs) is constantly rising in the recent years. Hundreds of millions of users are addicted to enjoy the convenient service and share new contents in OSNs. Facebook, in particular, is the largest online social networks (OSNs) today with over half active users logging daily. This success draws lots of attention from researchers and brings about numerous measurement and characterization studies. A few studies pay main attention to community structure. However, the complete dataset are not available due to large scale of users. Therefore, a relatively small but representative dataset is necessary in order to characterize properties and test algorithms.

Our goal of this paper is to compare the pros and cons of various graph-sampling algorithms for generating a representative sample. We finally obtain a representative dataset through web sampling and several data analysis are experiment to our dataset.

We analyze several candidate crawling techniques and design our own traversal process. First, BFS is widely used in crawling because of wide range data covering. One disadvantage of BFS is the bias produced when sampling. DFS is easy to implement but costs highly. Random walk (RW) is also biased but the bias can be revised by adding weighs. Some other algorithms, like re-weighted random walk (RWRW) and metropolis hasting random walk (MHRW), are low efficient, which is

F. Xhafa et al. (eds.), *Recent Developments in Intelligent Systems and Interactive Applications*,
Advances in Intelligent Systems and Computing 541, DOI 10.1007/978-3-319-49568-2_35

not sufficient for our crawling aim. So does random walk as well. According to our analysis, we finally choose BFS as our sampling algorithm balancing the contradiction between bias and efficiency. If we could get a similar global data set, the bias effect can ignored. To get a more complete network in OSN, BFS is the best choice to the best of our knowledge. Moreover, we provide the detailed design alternative of our sampling method in the paper.

In terms of results, statistical method are applied to characterize properties of OSNs. We show the number comparison of male and female people in OSNs. Degree distribution of friend are demonstrate by complementary cumulative distribution function (CCDF). Geographical positions are presented in global world view, thanks to Google Map API. Some influential people, which is significant researching subjects in the future, are observed through our analysis.

This paper is organized as follows. Section 2 introduces several common data collection method. Sections 3 presents social network traversal process. Section 4 describes our data collection prototype system and its implementation. Section 5 provides a characterization of some key OSN properties. Section 6 concludes our work.

2 Related Work

M. Gjoka [4] and others point out that two kinds of crawling algorithm, MHRW and RWRW, will not lead bias while crawling. On the contrary, BFS and RW, which are widely used in crawlers, cannot avoid crawling bias. Unbiased data set is essential to the analysis of any online social network. They obtain unbiased data set by data collection. Convergence analysis of different algorithm and some preliminary analysis of Facebook were also carried out in the paper. Catanese, S. [5] collect data on Facebook with breadth-first and uniform sampling algorithm. Node degree, group closeness and feature vector distribution were then analyzed by them to the data set.

3 Social Network Traversal

Online social network is a self-organization. A connection exists between any two users, constituting a certain relationship. In the sense of figure, each node in the graph represents a user in the online social network. Each edge between two nodes is the relationship between two users. These nodes and edges form a complex network graph mapping the real relationship in social networks.

To learn online social network, the structure of network needs to be acquired. Through online social network sampling, we can obtain a valuable data set and generate a whole view of network topology. Social network traversal is to expand sampling coverage and obtain users as much as possible.

4 Implementation

4.1 Design

As shown in Fig. 1, multi-threads are implemented in the desgin. There are four modules which are briefly described as follows:

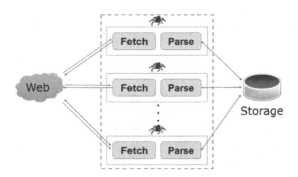

Fig. 1. Architecture design

4.2 System Functional Modules

Fetch Module. Fetch module is the foundation of the system, which is the first step of the whole system. Single collecting behavior will be detected and so sampling cannot continue any more. Accordingly, task scheduling policy is taken by system while assigning tasks to crawler threads. The thread changes crawling behavior with a certain probability to prevent detection from online social networks to some extent.

Parser Module. Web pages are filled with various, discrete and valuable information. These information usually present in various format which increase large amount of crawling cost. Consequently, crawlers require accurate and effective parsing ability extracting data from diversified web sites. To solve this problem, we modify regex pattern to collect different format of data. Regex expressions are a string of characters describing or matching certain rules. It is highly effective when extracting certain words from a sentence for a crawler. However, web page format varies with different kinds of web sites. We have to design specialized regex templates for collection system. Regex templates are a series of regex expressions to gain data to meet our acquisition demands.

Storage Module. After crawlers have collected data from online social networking sites, are gathered and then put into storage module. Storage module consists of deduplication strategy and crawling queue. Deduplication strategy. In collection process, web crawlers fetch a URL as next acquisition task from crawling queues. New URLs are then added to the queue for future scraping according to the crawling mission. Due to six-degree theory, six people in average can connect with two people never recognized before. Web crawler will inevitably encounter some tasks, which are already done or added to crawling queue. Consequently, Deduplication strategy is designed to remove duplication of tasks to improve crawling speed to obtain a whole data set in short time.

Queue. Queue comprises waiting queue, running queue and completion queue. Waiting queue refers to task uncollected. Inside running queue are tasks in the collecting process. Completion queue is the task queue filled with finished tasks.

5 Social Network Analysis

The purpose of the sampling method is to sample social relationships (users and their friends or fans) and social activities (status or comments generated by users). Social relationships help to learn the complete topology of the network, understand the network structure and analyze the network characteristics. Social activities can offer strong support for the online social network data in public opinion research and product marketing. Our collection starts at April 13, 2015 to May 8, 2015. During this time, we got over 1 million users and about 8 million status and comment information. According to our analysis, the average friends' degree is 32.68, while the average fan degree is 19.72. The degree distribution of friends and fans are separately shown in Fig. 2.

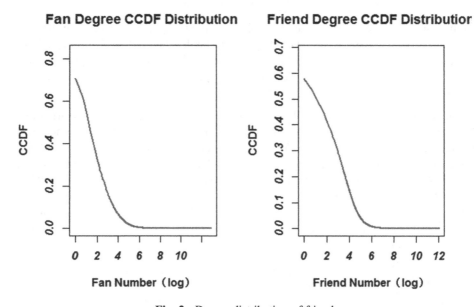

Fig. 2. Degree distribution of friends

6   Conclusion

In this paper, we analyze the data sampling methods of online social networks. Based on these analysis result, we prefer a specific sampling method and implement it on real OSN. Through our sampling method, a relatively complete user relational data sets is accessible. Finally, this paper conducted a preliminary analysis of the data collected. These data provide a solid basis for the future analysis.

References

1. Ellison, N.B.: Social network sites: definition, history, and scholarship. J. Comput. Mediated Commun. **13**(1), 210–230 (2007)
2. Gjoka, M., Kurant, M., Butts, C.T., et al.: Walking in facebook: a case study of unbiased sampling of OSNS. In: 2010 Proceedings IEEE INFOCOM, pp. 1–9. IEEE (2010)
3. Catanese, S.A., De Meo, P., Ferrara, E., et al.: Crawling facebook for social network analysis purposes. In: Proceedings of the International Conference on Web Intelligence, Mining and Semantics, no. 52, ACM (2011)
4. Mislove, A., Marcon, M., Gummadi, K.P., et al.: Measurement and analysis of online social networks. In: Proceedings of the 7th ACM SIGCOMM Conference on Internet Measurement, pp. 29–42. ACM (2007)
5. Ahn, Y.Y., Han, S., Kwak, H., et al.: Analysis of topological characteristics of huge online social networking services. In: Proceedings of the 16th International Conference on World Wide Web, pp. 835–844. ACM (2007)
6. Wilson, C., Boe, B., Sala, A., et al.: User interactions in social networks and their implications. In: Proceedings of the 4th ACM European Conference on Computer Systems, pp. 205–218. ACM (2009)
7. Pearson, K.: The problem of the random walk. Nature **1905**(72), 294 (1865)
8. Rasti, A.H., Torkjazi, M., Rejaie, R., et al.: Respondent-driven sampling for characterizing unstructured overlays. In: INFOCOM 2009, pp. 2701–2705. IEEE (2009)
9. Stutzbach, D., Rejaie, R., Duffield, N., et al.: On unbiased sampling for unstructured peer-to-peer networks. IEEE/ACM Trans. Network. (TON) **17**(2), 377–390 (2009)
10. Fielding, R., Gettys, J., Mogul, J., et al.: Hypertext transfer protocol–HTTP/1.1. RFC2616, **12**(3), 258–273 (1999)

Study Security Technology of Complex Network Based on Trust Data Aggregation

Yuan Chunlong[⊠] and Zhao Hai

College of Information Science and Engineering, Northeastern University,
Boston, USA
yuanchunlong@189.com

Abstract. With the rising number of wireless sensor networks arranged in harsh environments, the sensor nodes are likely to be captured by the enemy, so that the data gathered under great security threat. Therefore, to deal with the attacks in the captured data aggregation node process, to ensure reliable data aggregation has become an important part of wireless sensor network security research.

Keywords: Component · Traffic · Complex networks · Optimization · Routing

1 Introduction

In the process of gathering information, the use of each individual sensor nodes to transmit data aggregation method node communication bandwidth exhausts energy and reduce the efficiency of information gathering. To avoid these problems, sensor networks in the process of collecting the data require the use of data aggregation technology [1].

As an important supplement means cryptography-based security, trust management in wireless sensor networks to solve internal attacks, to identify the malicious node, the node selfishness and low competitiveness node, to improve system security, reliability and fairness and other aspects have a significant advantage [2].

Wireless sensor network node has limited resources, relatively simple network application features, and based on signed authorization ticket public key algorithm of the public key certificate authorizing the use of limited resources, which is not suitable for sensor networks. The present study sensor networks trust management system focused on a node trust value evaluated by means of enhanced trust value assessments sensor network security, robustness, and so on [3].

2 Related Work

Sensor networks consist of a large number of sensor nodes to monitor the coverage area. With the limited scope of monitoring and reliability of the individual sensor nodes, in the network deployment, sensor nodes require to reach a certain density in order to enhance the accuracy and robustness of the entire network monitoring

© Springer International Publishing AG 2017
F. Xhafa et al. (eds.), *Recent Developments in Intelligent Systems and Interactive Applications*,
Advances in Intelligent Systems and Computing 541, DOI 10.1007/978-3-319-49568-2_36

information, and sometimes even essential to the scope of monitoring a plurality of nodes overlap [4]. This overlap results in a monitored area adjacent node information reported some degree of redundancy. This redundancy for the data which is gathered within the data processing network, i.e., the intermediate node before forwarding the sensor data, the first data is integrated, redundant information is removed, under the premise of application needs the data amount to be transmitted should be minimized. Network processing makes use of computing resources and storage resources node, its energy consumption is much smaller as compared to the data transmission.

Sensor network consists of a large number of inexpensive sensor nodes deployed in a variety of environment [5], the information obtained from the sensor nodes there is a high unreliability. These factors are not reliable, mainly from the following aspects: cost and size restrictions, the sensor node configuration accuracy is generally low; a wireless communication mechanism makes data transfer more easy, disturbed by who were destroyed; in addition to poor working conditions outside influence data transmission, but also undermine the feature node, so that it is not working, report wrong data. From this data, only a few distributed collections of sensor nodes is difficult to ensure the accuracy of the information obtained, the data required for monitoring the same object by a plurality of sensors integrated collected, improve the accuracy of the information obtained and credibility. Further, since the adjacent sensor nodes monitoring the same area, the differences between the information available is small, if an individual node reports inaccurate information or error is superior, it is easy to be handled locally by a simple comparison algorithm excluded.

In addition to these several factors [6], depending on the trust management system design goals, there are other factors that affect the node trust value. For example, the higher the value of the trust node is selected to perform the task, the higher the probability, the faster the power consumption, one of the considerations of power as a trust value calculation to prolong the entire network lifecycle. Another example is the availability of the node trust value calculation and as one of the factors in the sensor network node localization based anchor, the anchor node state can survive periodic detection node to be periodically sends Hello messages to determine whether the node alive. However, the nodes periodically sleep will affect its availability. Redundant increase the accuracy of the judgment is a common means, so many systems require the deployment of a network with sufficient density.

3   Wireless Sensor Network Trust Management

3.1   Wireless Sensor Networks Trust Management Categories

Hierarchical trust management is the assessment of the trust and other values [7], and pass a hierarchical storage management features, often with the application of network topology and confidence values are closely linked. Data security; if the base station and sensor nodes based on natural hierarchy formed by the base station as the center of trust management; - - cluster head node ordinary nodes of layer 3 of trust in management there is a cluster structure of a sensor network, the base station will form aggregation applications, often based on trust management aggregation tree hierarchy. In the hierarchical trust management, the trust can pass layer by layer, the higher the value of the trust store all subordinate or subordinates adjacent. Trust can also be progressively converge, forming different levels of trust value. Planar trust management is the process of trust management, the network status of all nodes and base stations are equal, take the same model and management strategies, there is no obvious or central level.

(1) Global trust management and trust management local

Global trust management means that a node has a unique trust value across the entire network [8], the general common sensor network cluster structure. Local trust management refers to the node being evaluated to assess different values may be inconsistent trust at the node in the node to make decisions based on trust value stored locally or transmitted credibility comprehensive decision based on local trust value and neighbors.

(2) Trust management based on trust and credibility of local management based on information collection

When conducting trust evaluation because of incomplete information may cause deviation of the assessed value [9], in order to obtain a more accurate value of the trust, the trust often need to consider the assessed value of the other node to correct local assessment results, which is based on the credibility of the trust management the basic idea. But in the sensor network, node due to limited resources, to reduce the cost of communication and computing, and some trust management system in the assessment of the value of trust, simply consider the node itself to be observed and the results of the assessment node interactions evaluation local information, save other transport node reputation value of energy consumption.

(3) General trust management and trust management related applications

General Trust management is considering all aspects of trust defined element of the definition of a complete trust management framework, including information collection, transmission, storage trust management, calculation, update other aspects of the design [10]. Calculating trust value untargeted, is a comprehensive assessment of the credibility of the nodes, sensor networks can be used to run applications and all related technologies. And the application of the relevant trust management are well targeted, trust management of all aspects of the design process are closely associated with a particular application, such as trust management often need to identify safe routes selfish nodes, low competitiveness node, and for secure data fusion trust management selfish nodes generally do not need treatment, even in some systems, when a malicious node does not send the error data will not affect the value of their trust.

3.2 Wireless Sensor Network Trust Management Framework

Due to the limited resources of wireless sensor network nodes, trust management framework in the sensor network environment need to optimize all aspects of transport, computing [11], storage and other characteristics of the sensor according to the network environment.

(1) Element of trust

Establishment of the constituent elements of a trust management system should first clear the trust, that trust what factors mainly include. This is directly related to the definition of trust, but also the entire trust management framework based on the basic design and implementation. Trust form a trusted, define different elements under very different. In general, the lower sensor network environment of trust from all major comprehensive assessment of the following factors.

In sensor networks, control commands, or application data transmission is the main node behavior, which can be observed. Malicious nodes may be manifested as discarded packets and other acts of tampering, selfish nodes may also be discarded because of energy saving needs to forward a packet to identify malicious nodes or selfish behavior observation node via the communication nodes is a common mechanism for trust management system.

(2) Factors of cryptography

In most applications, the trust management system [12] is a complement to password-based security mechanisms, to improve network security. At the same time, the password mechanism has become one of the main considerations for many trust management systems assessment. Password mechanisms can be used to initialize the value of trust - which have 1 as the key neighbor node, and 0 otherwise. It can also be used to update the trust value - if a node cannot decrypt the packets meaningful plaintext, encrypted data packet node trust value of the increase, decrease or else, if the message authentication code is verified, then increase the value of the node trust, otherwise reduced. Hash chain mechanism in the presence of a system, if the hash values are not derived, trust value corresponding node is reduced; if the current hash value can be derived, but with the previous hash value far apart, it is considered that the transfer process packet loss occurs, the corresponding node trust value due to lower interval based on hash chain.

4 Energy Calculation of Routing Path

In our design of trust model, we establish and record the trust value of each other by actively listening to the communication behavior of the neighbor nodes. Nodes that often drop out, start an attack, or work in a selfish way can easily be detected by neighbors. Each node stores a trust table for a neighbor node, which stores the trust value of the other neighbor nodes, as shown in Table 1.

Table 1. Table of trust value of neighbor nodes

ID of node	csi	isi	ssi	sfi
i	nonce	nonce	nonce	nonce

csi: This node is consistent with the neighbor node i acquisition data.

Isj: This node and neighbor node i data collection is not consistent with the number of times.

Ssi: This node and neighbor node i aware of the number of the same event.
Sfi: This node and neighbor node i do not perceive the number of the same event.
Ti: The trust value of the first i neighbor node of this node.

$$C_i = \frac{cs_i - is_i}{cs_i + is_i} \quad -1 \le C_i \le 1$$

$$S_i = \frac{ss_i - sf_i}{ss_i + sf_i} \quad -1 \le S_i \le 1$$

Routing path of delay and energy consumption, and the number of AP on the routing path, each AP asked to lie off the tight phase, in fact, is how to set the distance between the various sectors of the base station, that is, how to select the ring size. The energy consumption of the routing is also related to the sector size and the distance from the base station of each sector. The nearest ring starting from the tomb station is in turn: r1, r2, ..., rk. For r1, for the R1 assuming a bit transmission of this distance of energy consumption for the Cr12, then, for the Aj to Aj-1, it is easy to get its value of c (rj-rj-1)2. This set of r0 = 0, the path from the sector Ai to the base station for the energy consumption of the routing path:

$$E_i = c \sum_{j=1}^{i} (r_j - r_{j-1})^2 \tag{4.1}$$

introduced the following:

$$\sum_{1 \le p < q \le i} (a_p b_q - a_q b_p)^2 = \sum_{p=1}^{i} a_p^2 \sum_{p=1}^{i} b_p^2 - \left(\sum_{p=1}^{i} a_p b_p \right)^2 \tag{4.2}$$

4.1 into the 4.2 type, you can get:

$$E_i = \frac{c}{i} \left(r_i^2 + \sum_{1 \le p < q \le i} (a_p - a_q)^2 \right) \tag{4.3}$$

It is obvious that the (4.3) type left is the smallest and must be made:

$$\sum_{1 \le p < q \le i} (a_p - a_q)^2 = 0 \tag{4.4}$$

For (4.4) - type when and only if:

$$a_1 = a_2 = a_3 = \ldots = a_4 \tag{4.5}$$

Form 4.3 and 4.4, we can get:

$$E_i = icd^2 \tag{4.6}$$

Let tx for the maximum transmission radius of AP, and set up d = tx:

$$E_i = i(ct_x^2) \tag{4.7}$$

From the (4.7) type, it is known that the minimum energy consumption of the base station must be equal to the distance between each ring, and the distance is the maximum transmission distance of AP.

5 Secure Cluster Head Election

In the following experiments, we randomly placed 100 nodes to monitor the temperature at 140 m * 40 m in the sensing area, and investigate the accuracy of the probability and data acquisition results of the safety election cluster heads.

Neighbor nodes monitor each other, determine each other's trust value, and submit their most trusted neighbor nodes as candidate cluster heads in each cluster head election. In the presence of malicious nodes, they may cheat the neighbor's trust to become the candidate cluster head, Fig. 1 shows the probability that the node in our program will be malicious neighbors selected as candidate cluster heads. As shown in the graph, the probability of a malicious node to become the candidate cluster heads is very small, and it has little effect on the correct candidate cluster head, and decreases with the decrease of the captured rate.

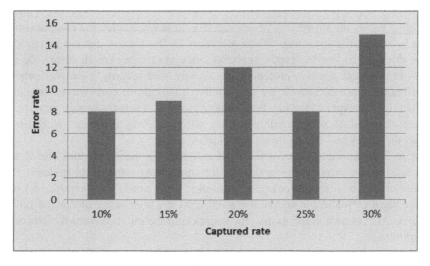

Fig. 1. Error rate of node

Table 2. Probability of selecting compromised nodes as cluster head

Compromised node ratio(%)	Probability(Using trust grid head election)	Probability(without trust mechanism)
0	0	0
15	0	0.2
50	0.1	0.4
85	0.2	0.8
100	1	1

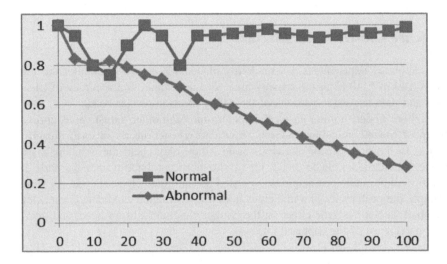

Fig. 2. Trust value of node

Table 2 shows the superiority of our security cluster head election mechanism with respect to no security measures. When less than 15 % of the nodes are captured, our scheme does not choose the captive node as the cluster head, which shows the effectiveness of our mechanism in protecting the cluster head security. However, when the captured nodes more than 85 %, the probability of the captive node was elected cluster head. This can be explained, when the captured nodes more than a certain amount, with the increase of the packet loss rate and false voting, making it difficult to distinguish between the nodes and the captured non captive node.

Figure 2 shows the change of the trust value of the member nodes stored in the cluster head. In the protocol of our design, the initial trust value of the nodes is 1, and the value of trust is gradually changing with the cluster head. The overall trend is that the trust value of malicious nodes decreases, and the trust value of normal nodes is close to 1. As is shown in the graph, the variation of the curve verifies the correctness of the protocol.

6   Conclusion

This paper introduces the neighbor node trust evaluation mechanism by neighbor nodes monitor each data aggregation behavior, according to monitoring results, the use of trust mechanism to conduct the evaluation nodes is to identify the captured node, based on trust and safety assessment of the elections cluster head. Security cluster head node based on past members of the aggregation behavior trust evaluation by reliable data aggregation. Experimental result illustrates that trust mechanism designed in this paper, is possible to deal with captured node attack, to ensure the safety of the data gathered.

Acknowledgements. I would like to express my gratitude to all those who helped me during the writing of this thesis. I gratefully acknowledge the help of my supervisor Professor Zhao Hai. I do appreciate his patience, encouragement, and professional instructions during my thesis writing. Also, I would like to thank Mr Chengdw who kindly gave me a hand in the preparation of the thesis.

References

1. Zhang, X.: Heterogeneous statistical QoS provisioning over 5G mobile wireless networks. IEEE Netw. **11**(6), 46–53 (2014)
2. Gabale, V.: A classification framework for scheduling algorithms in wireless mesh networks. IEEE Commun. Surv. Tutorials **5**(1), 199–222 (2012)
3. Bukhari, S.H.R.: A survey of channel bonding for wireless networks and guidelines of channel bonding for futuristic cognitive radio sensor networks. IEEE Commun. Surv. Tutorials **18**, 924–948 (2015)
4. Wang, X.: Scalable routing modeling for wireless ad hoc networks by using polychromatic sets. IEEE Syst. J. **7**, 50–58 (2012)
5. Yulei, W.: Performance analysis of hybrid wireless networks under bursty and correlated traffic. IEEE Trans. Veh. Technol. **62**, 449–454 (2012)
6. Shen, H.: A scalable and mobility-resilient data search system for large-scale mobile wireless networks. IEEE Trans. Parall. Distrib. Syst. **25**, 1124–1134 (2013)
7. Fotouhi, H.: Reliable and fast hand-offs in low-power wireless networks. IEEE Trans. Mobile Comput. **13**, 2620–2633 (2014)
8. Zhang, R.: A hybrid reservation/contention-based MAC for video streaming over wireless networks. IEEE J. Sel. Areas Commun. **28**, 389–398 (2010)
9. Ahmed, A.: Enabling vertical handover decisions in heterogeneous wireless networks: a state-of-the-art and a classification. IEEE Commun. Surv. Tutorials **16**, 776–811 (2013)
10. Park, J.-H.: All-terminal reliability analysis of wireless networks of redundant radio modules. IEEE Internet Things J. **3**, 219–230 (2016)
11. Niati, R.: Throughput and energy optimization in wireless networks: joint mac scheduling and network coding. IEEE Trans. Veh. Technolo. **61**, 1372–1382 (2012)

The Design and Implementation of Medical Wisdom Based on Internet of Things Technology

Jie-Min Yang$^{(\boxtimes)}$ and En-Hong Dong

Shanghai University of Medicine and Health Sciences, Shanghai, China
yjm0212@163.com

Abstract. The paper proposes a medical wisdom conceptual model composed of regional health platform, hospital wisdom system and family health management system. In the system, all equipment will connect with basic database by standardizes interface. Other users can obtain the personal health data by logging in resident electronic health records based on the user's authorization, so all medical resources can be shared in the system. The paper defines the function and the composition of each subsystem, in which the regional health platform takes resident electronic health records as the core and hospital wisdom system takes information as the center and family health management system bases on intelligent network equipment. The paper also introduces the model implementation of the system. The system is a scientific, standardized and opening system, which provides significant reference for top designers and builders.

Keywords: Medical wisdom · Regional health platform · Hospital wisdom · Family health management system

1 Introduction

Medical wisdom uses networking and cloud computing and other information technology to attach with physics, information, social and commercial infrastructure associated with the health, and intelligently meet the needs in medical and health ecosystem. Medical wisdom will be more comprehensive interaction and more intelligent insight, and realize self-management and optimization to meet personalized medical service experience.

There is a big gap between Chinese medical services with the developed countries, and medical resources are in short and the doctor-patient contradiction is gradually more tense. The medical resources are imbalance in different regions, and east resource is significantly richer than the western resource. All these restrict the improvement of our medical treatment. Medical wisdom can meet the personalized medical experience because it bases on network, which not only solve conflicts between doctors and patients because of personalized demand but also connect the different medical resources in other regions, so it realizes the interaction between the patients, medical personnel, medical institutions and medical equipment. Medical wisdom is coming into the lives of ordinary people.

F. Xhafa et al. (eds.), *Recent Developments in Intelligent Systems and Interactive Applications*,
Advances in Intelligent Systems and Computing 541, DOI 10.1007/978-3-319-49568-2_37

There is no complete set of system specification for Medical wisdom, and it is difficult to realize, so it is necessary to design a set of scientific and open system architecture, interconnecting with other systems.

2 The Model of Medical Wisdom

Medical wisdom will consist of regional health platform, hospital wisdom system, and family health management system. It contains disease prevention, health clinics, emergency, physical rehabilitation etc. The regional health platform takes resident electronic health records as the core, which establishes health information sharing platform by unified information interface. Hospital wisdom system takes patients as the center, which can realize the information on internet of the patient treatment through the electronic medical record and improve efficiency of the hospital. Family health management system bases on intelligent network equipment, and timely detects health problem.

As shown in Fig. 1, medical wisdom system consist of basic hardware layer, standard interface, basic database, cloud computing layer, comprehensive application and service system, user experience layer. All kinds of medical equipment acquisition, self-service terminals, electronic tag, card and smart sensing devices can be seen as basic hardware support, which can access to the basic database through the Internet, wireless network, health network, and constitute the basis data of personal health information. It will manage and analyze these data in the cloud layer to get data used for the comprehensive application and service system, and they will meet the needs of personalized medicine. The user can enjoy convenient services in the hospital or by remote medical treatment at home by using resident electronic health records, also there is no need to rebuild medical records in other areas. The system will open the data resources to the third party based on the user's authorization, which will actively lead the company to develop more quality products, improving the medical level.

Regional health platform is an authority management of medical wisdom, which is responsible for management of resident electronic health records, and supervises the system operation. People can log into health information platform by residents electronic health records to query all kinds of their health data including inspection data and medical examination reports. Hospital is the main place of diagnosis and treatment, so hospital wisdom system is responsible for collection of physiological data and the expert consultation results and treatment effect. All of these data will be uploaded to the regional health platform. Hospital wisdom system can also obtain these data to let doctor understand patient history for more comprehensive and scientific judgment. Family health management system can meet the individual needs of patients. Through the intelligent equipment various types of health data collected at home can be uploaded to resident electronic health records according to the open interface. The experts in the hospital wisdom system can access to the data in electronic health records to understand the disease, and realize remote treatment, and generate health reports.

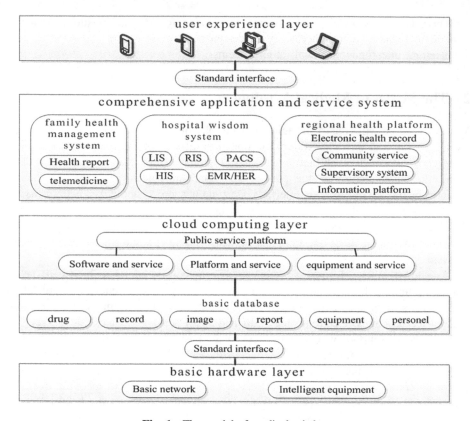

Fig. 1. The model of medical wisdom

3   The Architecture Design of Wisdom Medical System

Medical wisdom is composed of regional health platform, hospital wisdom system, and family health management system. It must guarantee the uniqueness of the personal data when users enjoy the service in the three subsystems, so it is necessary to establish the real-name system based on resident electronic health records to avoid conflict of personal data.

3.1   Hospital Wisdom System

The hospital wisdom system directly to the patient takes the main role in the medical wisdom system. All medical information can connect with each other through intelligent identification technology, network technology, cloud computing technology, which will optimize the management process and health care process to improve work efficiency and prevent medical errors. As shown in Fig. 2, the system will integrate the clinical medical service system, the hospital comprehensive management system, the clinical service assistant system and the external interface system. The medical wisdom system will build hospital intelligent system as the core of system integration and hospital information system as the core of information integration and clinical information system as the center of the electronic medical record and data integration platform system as the core of data sharing and information security system based on data security.

Fig. 2. The structure of medical wisdom

The system will conduct data integration and data exchange to construct an open system with the standard data format, so that it can realize data fusion from all platforms.

Medical personnel can work without paper through the hospital information system, medical image storage and transmission system, hospital laboratory system, hospital management system, office automation system and teleconferencing system. Patients can enjoy high-efficient registration and treatment through the external system interface by the resident electronic health records, and it can backup the health data and access to all kinds of inspection report.

3.2 Family Health Management System

Family health management system bases on intelligent networking monitoring equipment. It is an application of Internet of things, which is consist of the perception layer, network layer, application layer.

The architecture diagram of family health management system is shown in Fig. 3. It consist of physiological detection equipment in perception layer, including the computer for manual input information and telemedicine video equipment. All equipments directly or indirectly capture physiological parameters in perception layer and connect with internet equipment such as mobile phones and routers by the Bluetooth and wifi and cable network. The captured data will be sent to the application layer through wireless or wired network. Application layer based on third-party cloud services will generate a series of the user's health evaluation report and expert guidance, at the same time, the system can be authorized by user to transfer these data to their electronic health records, so other doctors can access to electronic health records in the regional

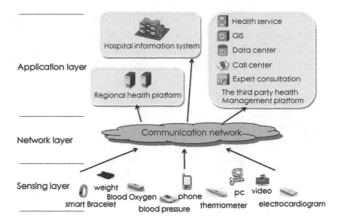

Fig. 3. The structure of family health management system

health platform to obtain a more comprehensive health data for more accurately determine and treatment in the future.

3.3   Regional Health Platform

The regional health platform is the central nervous system of medical wisdom system, responsible for the management and supervision of all kinds of medical resources. The resident electronic health records build in this system, which needs real name certification. The medical institutions and individuals can be authorized to access to electronic health records and upload the treatment results of data through the health information exchange layer (Fig. 4).

Regional health platform need to check any request for access, and the system will automatically reject the false request to protect user's privacy. Information security is

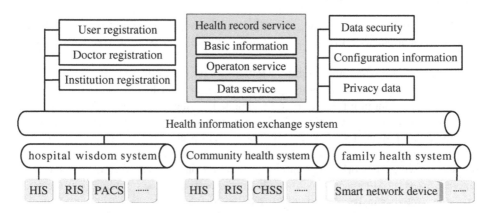

Fig. 4. The structure of regional health platform

the focus in the platform construction, so the system uses a hierarchical access method, and the different permission can access to the different content. As the data exchange center of the medical wisdom system, regional health platform connect hospital and patients, and it needs to formulate a series of information exchange standard to support equipment manufacturers to develop compatibility products for personalized needs.

4 Summary

Medical wisdom aims to establish an open and shared medical mode. It integrates the current decentralized medical resources into a standardized platform to share all resources, and provides personalized service for the residents. It is a huge project that the whole society should participate in the system construction. The system uses a hierarchical mode. Government and medical institutions build the regional health platform, and establish electronic health records and make all standard data interfaces. The hospital is responsible for the construction of hospital wisdom system which can connect with the regional health platform. In family health management system, enterprise will be encouraged to develop new products which can be communicated with the regional health platform. The enterprise also can use the personal health data opened by user to develop more personalized service.

Acknowledgements. This research was financially supported by the Research Fund of Shanghai Municipal Commission of Health and Family Planning (NO. 20144Y0117).

References

1. Hao, X., Chen, L., Wei, L., Li, J.: Study on the virtual care system for the elderly. In: Proceedings of the 2015 2nd International Conference on Engineering Technology and Application, vol. 1 (2015)
2. Zhao, L., Ma, Y., Yang, G., Meng, Q.: Research on development and application of tele-medicine. In: 2010 International Conference on Computer and Communication Technologies in Agriculture Engineering(CCTAE 2010), vol. 2 (2010)
3. Hanfeng, B.: Hospital medicine management information system design and implementation (2012)
4. Xiantao, L.: Design and implementation of speech recognition client system based on android platform. China Acad. J. Electron. Publishing House (2013)
5. Li, J., Zhang, H., Bai, Y.: Research on information sharing between the hospital and the community based on the regional health information platform. In: Proceedings of 2015 International Conference on Education Technology, Management and Humanities Science (ETMHS 2015) (2015)
6. Chen, D.: Design of a system circuit for intelligent terminal nursing home applications. J. Hangzhou Dianzi Univ. (2012)
7. Al-Taei, M.H., Abdul-Mehdi, Z.T., Hamdoon, S.H.: The need for teleradiology system in medical remote-diagnosis process. In: Ma, M. (ed.) Communication Systems and Information Technology (EEIC 2011, Volume 4). LNEE, vol. 100, pp. 1021–1026. Springer, Heidelberg (2011)

Pattern Recognition and Vision Systems

Fast Gesture Recognition Algorithm Based on Superpixel Distribution and EMD Metric

Cheng Ming and Xu Jianbo[(⊠)]

Hunan University of Science and Technology, Xiangtan, China
jbxu@hnust.edu.cn

Abstract. This paper presents an efficient gesture recognition algorithm based on superpixel distribution and EMD (Earth Mover's instance) metric with the help of the Kinect depth camera. In the first step we make full use of the depth and skeleton information from Kinect to accurately locate and segment hands from cluttered backgrounds. In the second step we adopt superpixel distribution to describe gestures and the SLIC algorithm works out superpixels. At last EMD metric is applied to measure the distance between two superpixel distributions. And the FC-EMD algorithm is proposed to calculate EMD distance immediately. Experimental results show that the proposed FSP-EMD can speed up and accurately detect hands, extract gesture features and calculate EMD distance. The running time is smaller when compared to the F-EMD and SP-EMD algorithms.

Keywords: Gesture recognition · Superpixel · EMD · Kinect · SLIC

1 Introduction

Gesture recognition is primarily divided into three parts. First of all, hand detection is the foundation of gesture recognition algorithms. The problem of gesture segmentation based on depth information is essentially a depth clustering problem, where the pixels are divided into different depth levels in [1]. One of the most important jobs is to determine a threshold indicating at which depth level the hand is located. But it is difficult to get the proper depth threshold, because the threshold value is affected by complicated backgrounds. Secondly, various hand features can be extracted from either the depth maps e.g. Histogram of 3D Facets [2] and 3D point distribution histogram [3] or the corresponding color images such as Histogram of Oriented Gradients (HOG) [4] and contours [5]. Finally, an appropriate algorithm is selected to identify gestures. F-EMD is proposed in [6] to measure the dissimilarities between different hand shapes. The performance of F-EMD depends on the accuracy of finger detection. In practice, the results of finger detection are not satisfactory due to distortion. SP-EMD is proposed in [7] to avoid the difficult of finger detection. Based on F-EMD and SP-EMD, we propose FSP-EMD to speed up gesture recognition.

© Springer International Publishing AG 2017
F. Xhafa et al. (eds.), *Recent Developments in Intelligent Systems and Interactive Applications*,
Advances in Intelligent Systems and Computing 541, DOI 10.1007/978-3-319-49568-2_38

2 Hand Detection

2.1 Hand Localization

It's the first step of hand detection. The work of hand segmentation would be greatly reduced if the hands are accurately located. In this paper, we use the Kinect's skeleton tracking function to accurately and immediately locate the hands. With the help of joint points of hands, the hands could be directly located. We can get a square in the center of joint point of left or right hand as shown in Fig. 1.

Fig. 1. Hand localization. (a) in the color image. (b) in the depth image. (c) hand segmentation.

2.2 Hand Segmentation

Hand segmentation refers to separating the hands from the background in order to obtain a binary image containing hand shapes, as shown in Fig. 1. In the process of hand segmentation, the critical part is to set a proper threshold. We select the depth value of hand joint point as a center to form a small interval to segment hands. The hand shape could be segmented quickly by simply judging whether the depth value of a pixel is in the interval or not. Moreover we don't need to consider the background outside the square. Compared to dealing with all pixels in the image, it greatly improves the speed of hand segmentation.

3 Feature Extraction

Superpixel refers to an image block which is comprised of adjacent pixels which are similar in color, texture, brightness etc. Superpixel segmentation algorithms are based on the similarity between pixels to divide an image into many superpixels. We adopt and modify the SLIC algorithm [8] in our system. We perform the clustering in a six dimensional space including the CIELAB color space and the (x,y,d) pixel coordinates, where d is the depth value at the pixel location (x,y). We define the following metric to measure the distance between pixels.

$$D_{ij} = d_c + c \times d_x, \qquad d_c = \sqrt{(l_i - l_j)^2 + (a_i - a_j)^2 + (b_i - b_j)^2} \qquad (1)$$

$$d_x = \sqrt{(x_i - x_j)^2 + (y_i - y_j)^2 + \theta(d_i - d_j)^2} \qquad (2)$$

Fig. 2. Superpixel segmentation in the color image.

Where θ is the compactness coefficient of supper pixels. Using the modified SLIC, gesture is divided into a number of clusters, i.e. superpixels, as shown in Fig. 2.

We express superpixels in the form of (l,a,b,x,y,d). The (x,y) is the coordinate of superpixel's center. d is the average depth value of superpixel. (l,a,b) is the average color value of superpixel. The distance between superpixels is defined as D_{ij}, which will be used in the calculation of the distance between superpixel distributions.

4 Gesture Recognition

In this paper, the superpixel distribution is used to describe and summarize the features of the hand shape. Inspired by the work of F-EMD and SP-EMD, we use the EMD metric to measure the distance between two superpixel distributions. The method of calculating the EMD distance is derived from the classical transportation problem. Now we talk about transportation problem.

4.1 EMD Mathematical Model

Given a set of suppliers $I = \{A_1, A_2, \cdots, A_m\}$, a set of consumers $J = \{B_1, B_2, \cdots, B_n\}$ and the cost C_{ij} to ship a unit of supply from $i \in I$ to $j \in J$, the aim is to find an optimal set F of flow f_{ij} which means the amount of supply shipped from A_i to B_j, to minimize the overall cost. The objective function is

$$\sum_{i \in I} \sum_{j \in J} C_{ij} f_{ij} \qquad (3)$$

subject to the constraints $f_{ij} \geq 0$, $\quad \sum_{i \in I} f_{ij} \leq t_j$, $\quad \sum_{j \in J} f_{ij} \leq s_i$, $\quad i \in I, j \in J$ where s_i is the supply of A_i and t_j is the capacity of B_j.

4.2   EMD Between Superpixel Distributions

How can we transform the problem of measuring the distance between two superpixel distributions into the transportation problem? Firstly, see Fig. 3 below.

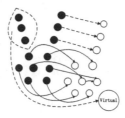

Fig. 3. An example of gesture matching.

Now we assume that the black part (gesture) is matching a gesture in the template library indicated by the white part. We use the model of transportation problem to describe the matching. Then, a black circle corresponds to a suppler, the supply of a black circle is set as the total amount of pixels in the black circle (superpixel). And a white circle means a consumer, the capacity of a white circle is set as the number of pixels in the white circle. The cost to ship a unit of supply from a black circle to a white circle is set as the distance between the corresponding superpixels which is defined in the Formula 1. Obviously the number of pixels in the black part is greater than that in the white part. To solve the partial match, we adopt a novel virtual superpixel. The virtual superpixel is in the center of the palm because it represents the folded fingers. The virtual superpixel will take part in the white part as showed in Fig. 3. The virtual superpixel represents a consumer. The capacity of the special consumer is

$$t_{virtual} = \sum_{i \in I} s_i - \sum_{j \in J} t_j \tag{4}$$

On the contrary, if the black part is a gesture in the template library and the white part is matching to the black part, the virtual superpixel would be considered as a supplier. The supply of the special supplier is

$$s_{virtual} = \sum_{j \in J} t_j - \sum_{i \in I} s_i \tag{5}$$

Calculating the EMD distance between two superpixel distributions is a linear programming problem, the objective function is:

$$minZ = \frac{\sum_{i \in I} \sum_{j \in J} C_{ij} f_{ij}}{\sum_{i \in I} \sum_{j \in J} f_{ij}} \tag{6}$$

As we can see, the objective function in Formula 6 has a denominator compared to that in Formula 3. The denominator represents a normalized operation. It can't remove because the greater the number of supply have, the greater the cost become.

4.3   FC-EMD

We can solve the linear programming problem (Formula 6) by using the simplex method [9]. The time complexity of the simplex method is $O(n^3 \log n)$. The proposed FC-EMD algorithm adopts the dynamic programming method to avoid the iterative in complex method. The time complexity of the FC-EMD algorithm is $O(n^3)$.

Firstly, we give some data structures used in the algorithm. The supply of the i-th supplier is stored in S[i]. The capacity of the j-th consumer is stored in T[j]. C[i][j] is used to store the cost to ship a unit of supply from i to j. An auxiliary array C_1 is initialized to C. We use F[i][j] to store the amount of supply shipped from i-th supplier to j-th consumer which will be calculated in the FC-EMD algorithm. The FC-EMD algorithm steps are as follows

Step1,   initialize the array F and sum. $F[i][j] \leftarrow 0$, $sum \leftarrow \sum S[i]$.

Step2,   select the minimum element from C_1. It means choose the least cost route to ship. Assume it's $C_1[i][j]$, then $row \leftarrow i, col \leftarrow j$.

Step3, $F[row][col] \leftarrow \min\{S[row], T[col]\}$. If S[row] is greater than T[col], T[col] would be assigned to F[row][col]. It means that the supply of row-th supplier is greater than the capacity of col-th consumer and the number of supply equaled to T[col] would be shipped from row-th supplier to col-th consumer. So the capacity of col-th consumer will be zero. So $T[col] \leftarrow 0, C_1[i][col] \leftarrow \infty$. If S[row] is smaller than T[col], S[row] would be assigned to F[row][col]. It means the remaining supply of row-th supplier is zero. So $S[row] \leftarrow 0, C_1[row][j] \leftarrow \infty$. In a word, a supplier or consumer would be reduced. The total amount of supply would reduced by F[row][col], $sum \leftarrow sum - F[row][col]$.

Step 4,   determine whether sum is greater than zero. If yes, go to step 2. If no, go to the next step.

Step 5,   calculate the EMD distance between two superpixel distributions according to the Formula 6.

4.4   Proof of the Time Complexity $O(N^3)$

Now we analyze the time complexity of FC-EMD algorithm according to the algorithm steps above. We assume n (the length of T) is greater than m (the length of S). Obviously, the time complexity of the first step is $O(n^2)$. The second step is to select the minimum element from C_1. If we adopt methods based on comparison, the selecting operation needs $m \times n-1$ comparison operations. In the third step, we need to update the i-th row or j-th col of the array C_1. So the time complexity of the third step is $O(n)$. The second step, third step and fourth step constitute a loop. The time complexity of the loop body is $O(n^2)$. Then, we analyze the execution times of the loop body. At the end of each loop, the variable sum will subtract F[row][col]. The value of F[row][col] is S [row] or T[col]. It means that every execution of the loop body will lead to an element in the array S or T reduced to zero. So the maximum number of loop times is m + n. So we can get that the time complexity of the loop is $O(n^3)$. Obviously, the last step needs m × n basic operations to calculate the EMD distance. So the time complexity of FC-EMD algorithm is $O(n^3)$. Moreover, if we sort the two dimensional array C_1 beforehand, the time complexity of the optimized FC-EMD algorithm can drop to O $(n^2 \log n)$.

5   Experimental Evaluations

We evaluate and compare the proposed gesture recognition system with F-EMD and SP-EMD algorithms, using two different datasets, namely self-built dataset and SP-EMD dataset. We collect a joint color-depth hand gesture dataset using Kinect. It contains 5 gestures with 8 different poses from 5 subjects. So, there are a total of 200 cases for testing, each of which is comprised of color texture and depth information with corresponding skeleton information used in our system. 5 gestures are showed in Fig. 4. The SP-EMD dataset contains 10 gestures [7] with 20 different poses from 5 subjects. In the experiments, LOOCV is conducted to evaluate the recognition performance.

5.1   Performance Evaluation

Robustness of complex environments: First, hand localization makes a substantial reduction in the background area. Second, we use a small depth interval to segment the hand. So the hand could be detected accurately and efficiently as shown in Fig. 4.

Mean accuracy: The confusion matrix for LOO CV on our dataset and SP-EMD dataset is shown in Fig. 4. The mean accuracy of the proposed system on our dataset is 98 %. As we can see, the most confusing case is between gestures 2 and 3, and 4 and 5.

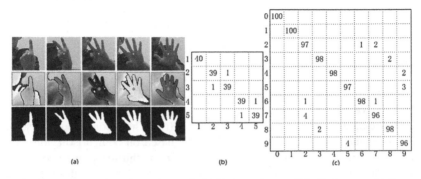

Fig. 4.   (a) Five gestures have been detected. (b) Test result on self-built data set. (c) Test results on SP-EMD dataset.

Table 1. Comparison among three algorithms on mean accuracy and running time.

Algorithm	F-EMD	SP-EMD	QSP-EMD
Mean accuracy	93.9 %	99.6 %	98.0 %
Running time	0.085 s	0.074 s	0.068 s

Because they are similar in the hand shapes. The only difference is on little finger or thumb. We also applied the proposed system to the SP-EMD dataset. The mean accuracy on SP-EMD dataset is slightly degraded to 97.8 %. It can be seen that the most confusing cases are 2, 6 and 7, 5and 9, and 3 and 8. Because they have the same number of fingers stretched out. We also applied the F-EMD and SP-EMD algorithms to our dataset. The results are shown in Table 1. On the average accuracy, the best is SP-EMD, the second is FSP-EMD, F-EMD is the worst. Because the F-EMD algorithm is difficult to correctly segment the fingers due to distortion.

Average running time: As presented in Table 1, the proposed algorithm FSP-EMD, is superior to the classical F-EMD and SP-EMD algorithm on the running time.

6   Conclusion

A fast gesture recognition algorithm FSP-EMD based on superpixel distribution and EMD matric has been proposed. It adopts a compact and efficient representation in the form of superpixel. The virtual superpixel is introduced to solve the partial matching issue and the EMD metric is adopted as the dissimilarity measurement for gesture recognition. To speed up calculate the EMD distance, the FC-EMD algorithm is proposed. The time complexity of the optimized FC-EMD algorithm is $O(n^2 log n)$. Experimental results show that the proposed FSP-EMD algorithm has high mean accuracy (98 %, 97.8 %). And it has faster recognition speed compared to F-EMD and SP-EMD algorithms.

To further improve the speed of gesture recognition, the next thing is to adopt a better superpixel segmentation algorithm.

This work is supported by Project supported by Hunan Provincial Natural Science Foundation of China, 2016JJ2058.

References

1. Rayi, Y., et al.: Hand segmentation from depth image using anthropometric approach in natural interface development. Int. J. Sci. Eng. Res (2012)
2. Zhang, C., Yang, X., Tian, Y.L.: Histogram of 3D facets: a characteristic descriptor for hand gesture recognition. In: 2013 10th IEEE International Conference and Workshops on Automatic Face and Gesture Recognition (FG), pp. 1–8 (2013)
3. Mihail, R.P., Jacobs, N., Goldsmith, J.: Real time gesture recognition with 2 kinect sensors. In: International Conference on Image Processing, Computer Vision, and Pattern Recognition (IPCV) (2012)

4. Dalal, N., Triggs, B.: Histograms of oriented gradients for human detection. IEEE Conf. Comput. Vis. Pattern Recogn. **2005**, 886–893 (2005)
5. Yao, Y., Fu, Y.: Contour model-based hand-gesture recognition using the kinect sensor. IEEE Trans. Circ. Syst. Video Technol. **24**(11), 1935–1944 (2014)
6. Ren, Z., et al.: Robust part-based hand gesture recognition using kinect sensor. IEEE Trans. Multimed. **15**(5), 1110–1120 (2013)
7. Wang, C., Liu, Z., Chan, S.C.: Superpixel-based hand gesture recognition with kinect depth camera. IEEE Trans. Multimedia **17**(1), 29–39 (2015)
8. Radhakrishna, A., et al.: SLIC superpixels compared to state-of-the-art superpixel methods. IEEE Trans. Pattern Anal. Mach. Intell. **34**(11), 2274–2282 (2012)
9. Aiyer, A., et al.: Lloyd clustering of Gauss mixture models for image compression and classification ☆. Signal Process. Image Commun. **20**(5), 459–485 (2005)

Comparative Analysis of Different Techniques for Image Fusion

Shigang Hu[1], Huiyi Cao[1], Xiaofeng Wu[1(✉)], Qingyang Wu[1],
Zhijun Tang[1], and Yunxin Liu[2]

[1] School of Information and Electrical Engineering,
Hunan University of Science and Technology, Xiangtan 411201, China
xfwuvip@126.com
[2] Department of Physics and Electronic Science, Hunan University of Science
and Technology, Xiangtan 411201, China

Abstract. Image fusion technology is an emerging field of research which is gaining currency in the recent years, Such technology focuses on extracting more concise and useful information with higher quality from source images. This paper deduces and generalizes the algorithms of the existing mainstream pixel level image fusion in details and realizes the Matlab experiment and adopts the contrast analysis to analyze the research findings. Wavelet analysis theory and fusion methods based on wavelet are introduced in detail. The techniques are compared by using different objective criteria. The feasibility and effectiveness of the image fusion technology based on wavelet transform are verified by experiments.

Keywords: Image fusion · Wavelet transform · Multi-resolution · DWT

1 Introduction

Currently due to the rapid development of multi-media technology and the wide application of image sensor, the scope of image application is expanding. The traditional image processing technology cannot meet the need for increasing accuracy. Hence new image fusion technology is effective on improving image quality and obtains more information.

There exists a lot of image fusion algorithms, but they will more or less bring negative influence to fusion images. Wavelet transform has been rapidly developing in the past few years for its good frequency characteristics [2]. It has the characteristics of the low entropy, the de-correlation and the flexibility of choice of the base. With good localized time domain and frequency domain and multi-resolution nature, wavelet has become a mainstream technology in the image fusion application.

The article first introduces some theories of image fusion, introduces both algorithms of traditional image fusion and multi-resolution wavelet-based image fusion, and put emphases on multi-resolution wavelet-based image fusion.

© Springer International Publishing AG 2017
F. Xhafa et al. (eds.), *Recent Developments in Intelligent Systems and Interactive Applications*,
Advances in Intelligent Systems and Computing 541, DOI 10.1007/978-3-319-49568-2_39

2 Theory of Image Fusion and Wavelet Image Fusion Technique

Image fusion is a kind of image processing techniques, which is the fusion of multi-sensor information from vision. The aim of image fusion is to realize the detection, feature extraction and recognition of the target by making full use of the complementary and redundant information provided by different image sensors. Image fusion generally consists of three levels: pixel level fusion, feature level fusion and decision level fusion [3]. Pixel level image fusion has been widely used at present.

Image fusion method can be roughly divided into two categories: spatial domain airspace fusion and transform domain fusion. The fusion methods such as averaging, maximal pixel selection, minimal pixel selection and principal component analysis (PCA) belong to spatial domain approaches [4]. For these spatial image fusion methods, the pixel values coming from two or more images are combined in a linear or non-linear way. Spatial domain methods will lead to spatial distortion of fused images [5]. This spatial distortion will adversely affect the further processing of the fused image. Transform domain fusion methods will be able to overcome such shortcoming [6].

Wavelets are mathematical functions, which are used for signal analysis at different resolutions. The irregularity of wavelets makes them suitable for discontinuous signal analysis of the real world. Excellent localization performance in frequency domain and spatial domain is the significant characteristics of wavelets.

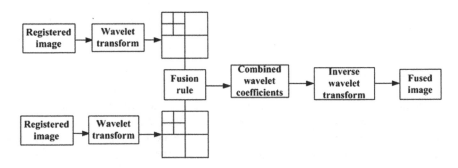

Fig. 1. Wavelet based image fusion

In general, the basic idea of wavelet transform image fusion is to execute a multiresolution decomposition for each source image, and then to deal with coefficient according to some fusion rules, as shown in the middle block of Fig. 1 [7]. After that, the IDWT of the corresponding combined wavelet coefficients is performed in order to obtain the fused image.

Coefficient combination is the key step of wavelet image fusion. This can be achieved by a set of fusion rules.

3   Evaluation Standard of Image Fusion

Generally, image assessment methods can be divided into two classes: firstly qualitative (or subjective) methods and secondly quantitative (or objective) methods [8]. Qualitative measures fundamentally rely on the vision, and can not be represented by a rigorous mathematical model. In view of the shortcomings of subjective quality evaluation methods, various efforts have been committed to the development of objective image quality assessment methods.

For measuring the effect of image fusion with different methods, some objective criteria such as Mean (μ), Standard Deviation (σ), Root Mean Square Error (RMSE), Peak Signal to Noise Ratio (PSNR), Entropy (EN), Average gradient (AG), Spatial frequency (SF), and Correlation coefficient (r) are used to compare the fusion results.

4   Results and Discussion

Experiments were performed on a standard image of size 256×256 as shown in Fig. 2(a), which served as ideal reference image here. The ideal image was blurred twice and after that, we got an image blurred on the left as shown in Fig. 2(b) and the other image blurred on the right as shown in Fig. 2(c).

(a)   (b)   (c)

Fig. 2. Ideal reference image and two blurred images

Firstly, image fusion algorithms including spatial domain methods and transform fusion methods were compared. The fused images with different algorithms are shown in Fig. 3. The fusion results were evaluated as shown in Table 1. The experimental results show that the clarity of fusion images by transform domain methods is far better than that of fusion images by spatial domain, and the comprehensive performances of fusion images by transform domain methods are better. The fusion effect using Laplace pyramid is the best, the indicators are relatively good. The fusion effect of wavelet transform fusion method is close to that of Laplace pyramid fusion method. By using the wavelet transform, a more compact representation is given, spatial orientation in different bands is separated, and interesting properties in the original image are efficiently de-correlated. The wavelet transform can overcome the instability of Laplacian pyramid scheme. In addition, wavelet transformation also provides other advantages, such as directional information, computational efficiency and so on.

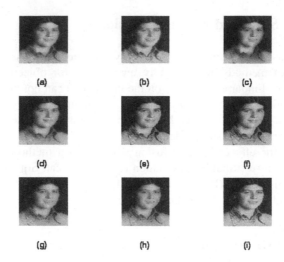

Fig. 3. Fusion images with different fusion methods (a) averaging, (b) maximal pixel selection, (c) inimal pixel selection, (d) PCA method, (e) laplacian pyramid, (f) contrast pyramid, (g) gradient pyramid, (h) FSD pyramid, (i) wavelet transform

Table 1. The fusion results with different algorithms

IMAGE	μ	σ	EN	AG	SF	r	RMSE	PSNR
Reference image	104.8458	76.8805	7.2158	7.1621	21.4278			
Image blurred on the left	99.2038	77.3057	7.0063	5.4223	17.7890	0.9842	14.8410	24.7015
Image blurred on the right	95.0710	69.8224	7.2644	5.3111	17.4432	0.9648	22.8639	20.9478
Averaging	97.3119	72.4379	7.3878	6.2180	15.3311	0.9905	13.5013	25.5233
Maximal pixel selection	105.6270	76.4244	6.9108	4.4459	18.6613	0.9969	6.1373	32.3712
Minimal pixel selection	88.6478	69.8215	7.2841	5.0806	15.1205	0.9634	26.5584	19.6468
PCA method	97.2500	72.6326	7.3835	5.0806	15.3518	0.9908	13.3563	25.6171
Laplacian pyramid	97.4549	74.9998	7.3984	7.1785	21.4206	0.9928	11.8808	26.6339
Contrast pyramid	97.2211	74.6243	7.3967	6.6653	19.9102	0.9923	12.3260	26.3143
Gradient pyramid	97.1658	71.7872	7.4678	6.1170	19.2647	0.9927	12.8827	25.9307
fsd pyramid	97.1929	71.7891	7.4685	6.1204	19.2776	0.9927	12.8670	25.9413
Wavelet based mean-max fusion	97.1602	73.5160	7.4610	7.1052	21.1963	0.9921	12.6151	26.1130

The selection of wavelet basis has a certain effect on the fusion result. For the selection of wavelet basis, the properties including orthogonality, compact support, and symmetry and regularity msut be considered. Different wavelet basis in accordance with the above conditions were selected in the experiment. The decomposition level was 3. The fusion results were evaluated as shown in Table 2. As can be seen from the evaluation, using different wavelet, there is little change in image fusion effect.

Table 2. The fusion results with different wavelet basis

wavelet	μ	σ	EN	AG	SF	r	RMSE	PSNR
Reference image	104.8458	76.8805	7.2158	7.1621	21.4278			
Image blurred on the left	99.2038	77.3057	7.0063	5.4223	17.7890	0.9842	14.8410	24.7015
Image blurred on the right	95.0710	69.8224	7.2644	5.3111	17.4432	0.9648	22.8639	20.9478
haar	97.1374	73.8889	7.4040	7.2730	21.5883	0.9926	12.3545	26.2943
db2	97.0458	73.5492	7.3927	7.1642	21.2578	0.9915	12.9529	25.8835
db5	97.1005	73.5376	7.4472	7.1265	21.2572	0.9920	12.7077	26.0494
db8	97.1286	73.4121	7.4637	7.1162	21.2151	0.9920	12.7204	26.0408
sym4	97.1602	73.5160	7.4610	7.1052	21.1963	0.9921	12.6151	26.1130
sym8	97.1557	73.4123	7.4628	7.1022	21.2036	0.9920	12.6870	26.0636
coif2	97.1851	73.4591	7.4621	7.1060	21.2056	0.9921	12.6410	26.0952
coif5	97.1303	73.5321	7.4336	7.1189	21.2384	0.9920	12.7084	26.0490
bior2.2	97.1181	73.5911	7.4478	7.1046	21.2035	0.9923	12.5491	26.1585
bior6.8	97.1052	73.5100	7.4135	7.1121	21.2180	0.9918	12.7904	25.9931
rbic1.3	97.1431	73.4266	7.4342	7.0824	21.1374	0.9919	12.7527	26.0188
rbic3.9	97.1424	73.4725	7.4914	7.1916	21.2979	0.9918	12.8107	25.9794
dmey	97.1331	73.5024	7.4563	7.1261	21.2494	0.9920	12.7210	26.0404

Wavelet decomposition level number is an important factor to influence the quality of the fused image. The more the number of wavelet decomposition level, the more abundant the details of fused image, but more levels are not always better. On the other hand, boundary extension should be carried out during wavelet decomposition and synthesis, the more the level number, the greater the border distortion. Therefore, wavelet decomposition level number should not be too large. Here wavelet bases functions 'db5' was used. The mean fusion rule is applied to fuse the approximation image. The detail images are fused using maximum fusion rule. The fusion results are shown in Table 3. AG and SF increases with the increase of decomposition level. EN, r and PSNR increase with the increasing of level number firstly, when the number reaches a certain value, they then decrease with increasing of level number. The RMSE variation is just the reverse of PSNR variation. It can be seen from the fusion results, 3 or 4 levels wavelet decomposition is more appropriate.

In the process of wavelet image fusion, the original images are first transformed into their multi-resolution representations. A new composite multi-resolution

Table 3. Effect of wavelet decomposition level number on fusion result

Decomposition level	μ	σ	EN	AG	SF	r	RMSE	PSNR
Reference image	104.8458	76.8805	7.2158	7.1621	21.4278			
Image blurred on the left	99.2038	77.3057	7.0063	5.4223	17.7890	0.9842	14.8410	24.7015
Image blurred cm the right	95.0710	69.8224	7.2644	5.3111	17.4432	0.9648	22.8639	20.9478
1	97.1375	72.6454	7.3948	6.0971	19.2361	0.9911	13.2800	25.6669
2	97.1257	72.9585	7.4299	6.7500	20.5727	0.9919	12.8661	25.9419
3	97.1005	73.5376	7.4472	7.1265	21.2572	0.9920	12.7077	26.0494
4	96.9410	74,1405	7.4601	7,2561	21.5118	0,9914	12.9725	25.8703
5	96.9293	74.8572	7.4559	7.2928	21.5770	0.9913	12.9333	25.8966
6	96.9501	75.4900	7.4523	7.3077	21.5937	0.9906	13.1572	25.7476
7	96.9059	77.3792	7.4466	7.3155	21.6012	0.9899	13.5349	25.5017
8	96.6312	78.2245	7.4831	7.3181	21.6026	0.9895	13.9838	25.2183

representation is then created from those inputs according to a certain fusion rule. Therefore, fusion rule is the core of image fusion, which directly affects the speed and quality of image fusion. Generally, the approximations or details structures can be merged by taking maximum, minimum, the mean rules. Wavelet base function 'db5' was used. The wavelet decomposition level number is set to 3. Nine kinds of fusion operators were used. The fusion results are shown in Table 4. Using different fusion rules, fusion effect is different. Mean and maximum fusion rules are usually used. Many more sophisticated fusion rules need to be developed to improve the quality of the fused image.

Table 4. Effect of fusion rule on fusion results

Fusion rule	μ	σ	EN	AG	SF	r	RMSE	PSNR
Reference image	104.8458	76.8805	7.2158	7.1621	21.4278			
Image blurred cm the left	99.2038	77.3057	7.0063	5.4223	17.7890	0.9842	14.8410	24.7015
Image blurred cm the right.	95.0710	69.8224	7.2644	5.3111	17.4432	0.9648	22.8639	20.9478
Max-max	104.4548	76.9452	7.2768	7.2111	21.4478	0.9974	5.5471	33.2494
Max-mean	104.4917	75.8845	7.4324	5.3368	15.6094	0.9952	7.5549	30.5662
Max-min	104.5287	75.5118	7.4809	4.3663	13.0911	0.9839	13.7499	25.3648
Mean-max	97.1005	73.5376	7.4472	7.1265	21.2572	0.9920	12.7077	26.0494
Mean-mean	97.1374	72.4189	7.4237	5.0715	15.3255	0.9904	13.6408	25.4340
Mean-min	97.1743	72.0201	7.4619	3.8989	12.7258	0.9789	17.7857	23.1294
Min-max	89.7461	71.1064	7.4582	7.2211	21.2170	0.9725	23.6959	20.6373
Min-mean	89.7831	69.9405	7.5010	5.0907	15.2484	0.9712	24.1736	20.4640
Min-min	89.8200	69.5190	7.4515	3.7161	12.6074	0.9595	26.7002	19.6005

5 Conclusions

This paper introduces the principle of image fusion and the theory of wavelet. Different image fusion techniques are simulated. The fusion techniques are compared by using various objective criteria. Different wavelet bases, several kinds of decomposition levels and nine kinds of fusion rules are compared. The results show that Transform domain fusion methods are significantly better than spatial domain fusion methods. Wavelet transform fusion method has pretty good performance. Selection of wavelet type, decomposition level and fusion rule are key problems for wavelet transform image fusion.

Acknowledgments. This research was financially supported by the National Natural Science Foundation of China (Grant Nos 61376076, 21301058, 61274026, 61575062 and 61377024); supported by the Scientific Research Fund of Hunan Provincial Education Department (Grant No. 14B060); supported by the Science and Technology Plan Foundation of Hunan Province (Grant Nos 2014FJ2017).

References

1. Shu-tao, L., Hai-tao, Y., Le-yuan, F.: Remote sensing image fusion via sparse representations over learned dictionaries. IEEE Geosci. Remote Sens. Lett. **51**(9), 4779–4789 (2013)
2. Auli-Llinas, F.: General embedded quantization for wavelet-based lossy image coding. IEEE Trans. Sig. Process. **61**(6), 1561–1574 (2013)
3. Zhaoxia, L., Jubai, A., Jing, Yu.: A simple and robust feature point matching algorithm based on restricted spatial order constraints for aerial image registration. IEEE Geosci. Remote Sens. Lett. **50**(2), 514–527 (2012)
4. Paul, P.P., Gavrilova, M.L., Alhajj, R.: Decision fusion for multimodal biometrics using social network analysis. IEEE Trans. Syst. Man Cybern. Syst. **44**(11), 1522–1533 (2014)
5. Joshi, M., Jalobeanu, A.: MAP estimation for multiresolution fusion in remotely sensed images using an IGMRF prior model. IEEE Trans. Geosci. Remote Sens. **48**(3), 1245–1255 (2010)
6. Javidi, B., Do, C.M., Hong, S.H., Nomura, T.: Multi-spectral holographic three-dimensional image fusion using discrete wavelet transform. J. Disp. Technol. **2**(4), 411–417 (2006)
7. Ellmauthaler, A., Pagliari, C.L., Da Silva, E.A.B.: Multiscale image fusion using the undecimated wavelet transform with spectral factorization and nonorthogonal filter banks. IEEE Trans. Image Process. **22**(3), 1005–1017 (2013)
8. Liu, T.J., Lin, W.S., Kuo, C.C.J.: Image quality assessment using multi-method fusion. IEEE Trans. Image Process. **22**(5), 1793–1807 (2013)

Mickey Mouse 3D CAD Model Reconstruction Based on Reverse Engineering

Zhang Mei[(✉)] and Wen Jinghua

Working with School of Informatics,
Guizhou University of Finance and Economics, Guiyang, China
zm_gy@sina.com

Abstract. This paper aims at reconstruction of 3D CAD model by using Reverse Engineering (RE) which is the key-technology emphasized in this context of research. First the Mickey was scanned by active range laser scanning system (Konica Minolta VIVID910), and its data of points cloud was obtained. Then the data were preprocessed in RE software of Imageware12, also it was segmented with curvature analytical method and its contour line was created. Finally the 3D model reconstruct was made with the function of high-level surface building and quality evaluating. The novelty of the study is that the model presented here is the fast design of CAD model of the Mickey Mouse.

Keywords: Reverse engineering · Laser scan · 3D reconstruct · CAD · Imageware12

1 Introduction

During the process of design, study and development of products when users do not provide blueprint or CAD data file instead they only provide sample-piece of the products. Since most of these kinds of products are composed of free-surface, their size can not be measured and the graphics can not be drawn with the routine method. If the traditional design method is used, the design time is long and it will fall short of precision and in some cases it will be inextricable. Hence it needs a kind of bran-new and high-effective resolved scheme of products exploitation, which is sample-piece or model of a real object → 3D surveying data → 3D product digital model → product, namely Reverse Engineering. The Reverse Engineering is a process which starts with an exist real object or prototype and first deal it with digital method, then the curved-surfaces are reconstructed and their CAD model is also constructed. finally the product is produced [1] along with Reverse Engineering technologies which are experiencing wider applications. Their corresponding software and hardware system also get development. This paper takes real object model of a Mickey Mouse as example and it adopts initiative range laser scan system produced by Konica Minolta company as well as Imageware12 which is a professional Reverse Engineering software to complete its process of digital survey and CAD model reconstruction.

Reverse Engineering is a advanced idea and method of product design which develops at the beginning of 90 years during 20th century and it is the total of correlative digital technologies which translate real object to CAD model. Geometrical

F. Xhafa et al. (eds.), *Recent Developments in Intelligent Systems and Interactive Applications,*
Advances in Intelligent Systems and Computing 541, DOI 10.1007/978-3-319-49568-2_40

model reconstruction technologies and products manufacture technologies At present, whether interiorly or overseas the research related to Reverse Engineering mainly focus on converse of geometrical shape and structure which is related to function factor of a product [2]. Its application fields are quite broad, and it can be used in mould manufacturing, plaything industry, game trade, electron occupation, shoes industry, art occupation, medical engineering and industrial design and so on. The process of Reverse Engineering is first using laser scanner to survey the sample or model known in order to obtain its 3D contour points cloud data; then making data process and CAD model reconstruction cooperating with software of Reverse Engineering, moreover making precision analysis, appraising construct effect and redesigning for the curved surface reconstructed. Finally creating data with format of IGES or STL, and it can be machining by numerical control or rapid molding to obtain entity model [3].

2   Modeling Foundation of Imageware12

2.1   Curve and Curved Surface in Imageware12

InImageware12, we use Bezier curve, B-spline curve, NURBS curve to define free curve and curve close to taper such as ellipse, hyperbola and parabola. The main information of these kinds of curves is as follow:

Direction: When the curve is used to built mold curved-surface, the direction of a curve is very important

Nodes: They are the positions in the curve which the span is joined by them

Segment: The cycle part which pertain to the nodes of a curve

Control Points: The mathematical points which can affect and restrict the shape of a curve in little area

Order: The shapes of curves mostly are determined by their orders, on the default instance, the 4-order curve formula is used to describe a curve in the software of imageware12

The Interrelation among Parameters of a Curve: Number of Control Points = Number of nodes inside +degree of freeness +1 = Number of nodes inside +Order = Number of Segments−1 + Order.

The basic elements of Bezier curved-surface, B-spline curved surface and NURBS curved-surface include the following:

Normal: There is a normal direction of each curved surface, and the positive normal direction displays the color of the curve, while the negative normal direction displays gray;

Nodes and Spans: When the curved-surface is created by curves, the curved-surface will have the same nodes and spans as the curves;

Control Points: Like the Control Points of a curve, the control points of a curved-surface are the points which can affect and restrict the shape of the curved-surface in little area;

Order: The shapes of curved-surfaces mostly are determined by their orders, on default instance, the 4-order curved-surface formula is used to describe a curved-surface in the software of imageware12;

Direction of U and V: Each NURBS curved-surface has 4 borders, the position is disparted into U, V direction where the two borders are vertical each other.

2.2   Curved-Surface Reconstruction in Imageware12

Curved-surface reconstruction adopts proper algorithm to fit into mathematical model for the curved-surface with the discrete 3D-surveyed data which has been preprocessed and it is the key technology of Reverse Engineering. In Reverse Engineering, a curved-surface object sometimes is not simply consisted of single curved-surface, but it is shaped by the mix of extension, transition and cut-out of many curved-surfaces, so the construction should be completed one block by one block. For the limit of digital technology, the problem of "multi-view data" (namely which is the data block surveyed from different direction or position) still exists in Reverse Engineering [4].

In a general way, in order to ensure integrality of digitization, there is some overlap among each-view data, so it brings multi-view align and merge. Although the process, operation steps and detail are not all same to realize curved-surface reconstruction in different CAD system.

2.3   Smoothness Evaluate of Curves and Curved-Surface in Imageware12

Geometrical Continuity of Parameter Curve: When a complex curve is designed, it always is combined by multi-segment curve, so it needs to resolve the question which is how to realize smooth connection among curve segment. There are two types of measurement of smoothness of connection among curve [5]: One is differentiability of a function, if we can build the combination parameter curve which has continuous derivative vector till n-order at the point of joint, namely it is n-order continuous differentiability, and this kind of smoothness is called C_n or n-order parameter continuity. The other is called as geometrical continuity, which is the combination curves satisfy one group restrict condition different from C_n at the point of joint, and the combined curve is called as possessing n-order geometrical continuity, and it is simply noted as G_n. The two measurement methods of smoothness are no inconsistent, C_n continuity is included in G_n continuity. They are showed in Fig. 1, for the two curves P (t) and Q (t), $t \in [0, 1]$.

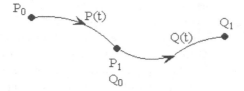

Fig. 1. Continuity of two curves

If it is require achieving G^0 continuity or C^0 continuity, namely the two curves can achieve position continuity at the joint, Eq. 1 should be satisfied:

$$P(1) = Q(0) \qquad (1)$$

If it is require achieving G^1 continuity, namely the two curves except being required to satisfy the condition of G^0 continuity at the joint, moreover it is require having common tangent vector, Eq. 2 should be satisfied:

$$Q'(0) = P'(1) \qquad (2)$$

If it is require achieving G^2 continuity, namely the two curves except needing to satisfy the condition of G^1 continuity at the joint, moreover it is require having common curvature vector, Eq. 3 should be satisfied:

$$Q''(0) = P''(1) \qquad (3)$$

Geometrical Continuity of Parameter Curved-Surface: For the shape of entity surface is complex in the world, it is difficult to describe its shape with a single curved-surface. It is easier to build surface model of complex entity if we first decompose its surface into many smaller curved-patches then fit these curved-patches with NURBS curved-surface. Commonly there is some smoothness such as position continuity, tangent plane continuity and curvature continuity on the surface of an entity. When making sculpt with disported curved-patches, we should assure smoothness of the result curve-surface, and it is necessary to adjust continuity between two coterminous curved-patches. The continuity between two coterminous curved-patches takes very important action in the fields of CAD/CAM, geometrical sculpt and reversing engineering etc. This is not only because of Geometrical continuity providing the free parameter which may be used to construct and modify very complex geometrical entity but also because of this kind of continuity reflect essential continuity between two curved-patches in practice, namely it does not depend on parameterization of curved-surface. Hence it is widely used in theory research and engineering practice [6].

Continuity of curved-surface is smoothness degree of connection between two curved-patches. Similar to parameter curve, parameter continuity also cannot measure smoothness of connection between two curved-patches exactly. Continuity of parameter curved-surface also needs to be evaluated by gather continuity. As it is showed in Fig. 2, if two curved-surfaces $P(s,t), Q(u,v)$ have common connection-line they are called as position continuity or G^0, the 0-order geometrical continuity of curved-surface also is consistent with its 0-order parameter continuity C^0. The G^1 continuity of two

Fig. 2. Continuity between two curved-patches

parameter curved-surfaces is called as well as tangent-plane continuity, its definition is as following: if the two curved-surfaces have common tangent plane or common normal at each joint in the common connection line of them.

The G^2 continuity of two parameter curved-surfaces is called as well as curvature continuity, its definition is as following: if the two curved-surfaces have common tangent plane and common main curvature moreover common main direction if main curvature is not equal at each joint in the common connection line of them [7].

3   Digitization of Product Surface

3.1   3D Data Acquirement of Surface

The shape of "Mickey Mouse" is complex as it consists of many curved-surfaces, it can not be surveyed by routine method. So the original digitization of part prototype is always completed by obtaining 3D coordinate value of surface points with measurement device such as 3-COMERO or laser scanner. The common surveying method in reversing engineering is disparted into two kinds of way of touch and way of not touch According to the different surveying principle, the un-touch surveying is approximately disparted into optical surveying, ultrasonic surveying, electromagnetism surveying etc. Among them the most used measurement method is laser scanning which adopting optics triangle principle and belonging to optical measurement method [8], this paper adopts this method to make 3D measure for "Mickey Mouse" to gather data, it is showed in Fig. 3 that the data points of cloud gathered by 3D laser scanning.

Fig. 3. Points of cloud of "Mickey Mouse"

3.2 Preprocess of Points of Cloud Data

For the character of reverse modeling, forward CAD software such as Pro/E and UG can not satisfy the need of rapid and correct modeling, the points of cloud data should be processed by special reverse engineering software such as Imageware12 and Geomagic studio, the penman adopt the Imageware12 to preprocess the points of cloud, there are main three steps as follow: (1) Read in of points of cloud data, it is imported into Imageware12 that the points of cloud data obtained by 3D laser scanner; (2) Creating color character, the points of cloud will be drawn different color according to curvature of each point after it has been computed; (3) Points of cloud segmentation, we use curvature analytical method to disport cloud data, and the cloud of "Mickey Mouse" is segmented into 5 parts which is ear, face, eyes, nose and mouse.

4 Reconstruction of CAD Model

4.1 Building of Contour Line

In order to reconstruct curved-surface, it is needed to obtain the contour line which can reflect character fracture first. The building of contour line is completed by adopting 4-order B-Spline interpolation for cloud data. The curve can be adjusted by changing the number of control points in such manner that if the control points are added the degree of shape inosculating is good and if the control points are reduced, the curve is smoother. We can estimate smoothness of curve by its curvature, and we can inspect degree of inosculate between a curve and its corresponding cloud, also we can change continuity between one curve and another curve which can be position continuity, tangent continuity and curvature continuity. The whole contour line of "Mickey Mouse" is showed in Fig. 4.

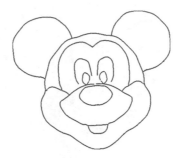

Fig. 4. Contour line of "Mickey Mouse"

4.2   Reconstruction of Curved-Surface

The building of ear part may be completed by first fitting 5*5-order well-proportioned top surface with cloud data directly, then tensioning contour line to create side, and spilling cyclo-angle between the top surface and the side. The fitting error between the top surface and its corresponding cloud is showed in Table 1, from this table we know the total max error is 0.0707 mm which can satisfy precision requirement of A-level curved-surface. The execution method of other part curved-surface of "Mickey Mouse" is relatively similar to the curved-surface of ear part. The whole 3D CAD model of the "Mickey Mouse" reconstructed is showed in Fig. 5.

Table 1. Fitting error between the top surface and its cloud of the ear part

	Max (mm)	Average (mm)	Std.Dev
Euclidean	0.0707	0.0108	0.0080
Lateral	0.0002	0.0000	0.0000
Neg.Norm	−0.0707	−0.0111	0.0090
Pos.Norm	0.0322	0.0105	0.0070

Fig. 5. The whole 3D CAD model of the Mickey Mouse reconstructed

5 Conclusion

For the sample piece with complex shape, it can enhance precision of product design and manufacture by adopting technology of laser scanning to complete the collecting of model data rapidly. Then using software of reverse engineering to complete process, curved-surface reconstruction and entity sculpt of cloud data, at the same time the blueprint does not need to be drawn during the whole process (also it is difficult to be drawn), it realizes process of no paper, and shortens development period of the product. Hence it has important meaning for enhancing competition ability of product economy.

Acknowledgment. Thank Project Supported by Regional Science Fund of National Natural Science Foundation of China (Project approval number: 41261094).

References

1. Wang-Xiao, Liu-Hui: Applied tutorial of CATIA reverse engineering. Publishing company of chemic industry, Beijing (2006)
2. Shan-Yan, Xie-Binfei: Technologies foundations of imageware reverse sculpt. Publishing company of Tsinghua University, Beijing (2006)
3. Chen-Xuefang, Sun-Chunhua: The application of reverse engineering and rapid molding technologies. Publishing company of mechanical industry, Beijing (2009)
4. Ji-xiaogang: Key technologies research of reverse engineering, Xiandai Zhizao Gongcheng **7**, 29–33 (2008)
5. Chang-Ming: Computer Graphics, 3rd edn. Publishing company of Huazhong university of science and technology, Wuhan (2009)
6. Li-Peng, Li-Yuan, Liu-Ping, Zhang-Kaifu: On continuity conditions for C-B-Spline curves and surfaces. J. Northwest. Polytech. Univ. **25**(6), 890–895 (2007)
7. Zhang-Xiaoxiang: Encyclopedia of Computer Science and Technologies, 2nd edn. Publishing company of Tsinghua University, Beijing (2005)
8. Xie-Longhan: Basic Tutorial of UGNX Surface Sculp. Publishing company of demotic post and communication, Beijing (2006)

Age Estimation of Asian Face Based on Feature Map of Texture Difference Model

Jie Yang, Peng Liu, Yuxin Jiang, and Songbin Li$^{(\boxtimes)}$

Haikou Laboratory, Institute of Acoustics,
Chinese Academy of Sciences, Haikou, China
yangjie13@mails.ucas.ac.cn, liup@dsp.ac.cn,
anakin320@126.com, lisongbin_work@126.com

Abstract. Combining with Support Vector Regression (SVR), this paper proposes a feature map of texture difference (FMTD) model for Asian face age estimation. The FMTD model is based on the standard bio-inspired feature model and can learn the feature information of the important face organs area as well as the wrinkles area. The learnt feature is strengthened by image processing, including image difference, down scaling, dividing and max-pooling. The resulting feature is sensitive to age estimation. Experimental results on the Asian face dataset and two public datasets prove that the proposed method reduces the mean absolute error (MAE) of age estimation comparing with other current methods and improves the degree of accuracy, which results in effective age estimation for Asian face.

Keywords: Age estimation · Feature map of texture difference model · Mean absolute error · Support vector regression

1 Introduction

Over the years, with the development of computer vision and machine learning, the recognition and the analysis based on face have drawn scholarly attention for research. Face image-based age estimation as one part of face analysis techniques, has been widely studied by researchers. Age estimation study can be approximately divided into three categories: the first category is based on the regression solving strategy, which implements the regression prediction by treating age as a positive real number [1–3], the second category is based on the multi-classification solving strategies which implements the multi-classification treating age or age-group as a category [4–6] and the third category is based on the combination of classification and regression solving strategies, which implements multi-classification before regression [7].

Recently, some new methods different from the above three categories were proposed. Chang et al. [8] proposed a novel algorithm of age estimation based on ranking (Rank). Then, they proposed a method of ordinal hyperplanes ranker based on Rank (OHRank) [9]. Lately, they combined the latest scattering transform and the cost-sensitive technology to optimize the OHRank algorithm and proposed a new cost-sensitive ordinal hyperplanes ranker (CSOHR) model [10]. This approach can

F. Xhafa et al. (eds.), *Recent Developments in Intelligent Systems and Interactive Applications*,
Advances in Intelligent Systems and Computing 541, DOI 10.1007/978-3-319-49568-2_41

achieve the mean absolute error (MAE) equal to 4.48 years on the FG-NET [12] age dataset, representing the state-of-the-art level of the existing researches.

However, the current methods of age estimation almost all study on the FG-NET age dataset or the MORPH [13] age dataset which are composed of Europeans and Americans. There is lack of the research for Asian face age estimation. In practical scenarios, we need to deal with Asian face in many cases. Compared with the Europeans and Americans, Asians look younger at the same age. Therefore, the existing methods directly used for Asian face age estimation will lead to a large error. This paper implements age estimation by in-depth analysis for Asian facial feature, and experiments on Asian face image dataset and the public age datasets.

2   Feature Map of Texture Difference

The important organs of face such as eyes, nose, eyebrows and the surrounding area contain abundant information which are used for face recognition, face gender recognition, facial expression recognition and face age estimation. Daily experience-based humans also exploit facial wrinkles which express rich texture feature to judge a person's age. This paper proposes the feature map of texture difference (FMTD) model based on the standard bio-inspired feature model. The FMTD can not only learn the feature information of the important organs and the surrounding area, but also learn and strengthen the feature information of facial wrinkles area.

2.1   The Feature Map of Texture Difference Model

The bio-inspired feature model is originated in the biologist's study on the human visual cortex system. Riesenhuber and Poggio [11] proposed a feed-forward mode "HMAX" model for visual object recognition. The model includes the processing layer S1 with simple cells and the processing layer C1 with complex cells.

Figure 1 illustrates the implementation process of the FMTD model. Compared with the standard bio-inspired feature model, the FMTD model has been improved from three aspects. Firstly, the processing layer S0 is added before S1 to obtain the local binary pattern (LBP) image of the original image. Then the processing layer C0 is added between S1 and C1 to implement image difference computation. Finally the FMTD is generated through max-operation and down scaling and the corresponding feature vectors are output.

LBP is an effective description method for face texture feature. Since the LBP image of face can distinctly describe the important organs and the wrinkles, the LBP image is introduced to the FMTD model.

This paper captures the most prominent change in the LBP image based on the Gabor wavelet transform. The idea of difference of Gaussian (DoG) is utilized to further strengthen the change. DoG is an important approach of corner enhancement in image processing and can strengthen the part contours of the image. In essential, the Gabor transform is also Gaussian filtering processing with multiple different filters. The same Gabor wavelet transform is implemented for both the original image and the LBP

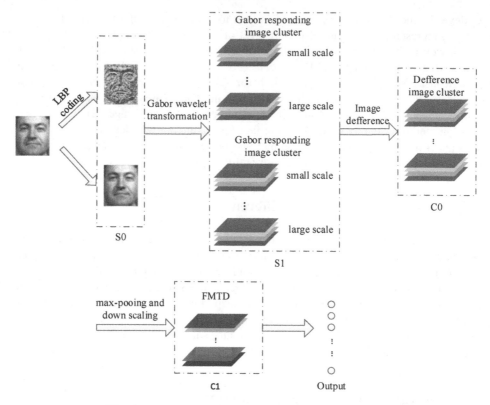

Fig. 1. The implementation process of the FMTD model

image to obtain two image clusters corresponding to the original image and the LBP image respectively. Then a difference image cluster with part contours of critical regions enhanced can be obtained through computing the difference of the two image clusters.

The difference image cluster contains the enhanced feature information of the important face organs area as well as the wrinkles area. However, the difference image cluster is a large set of images, which also contains some redundant information. We implement the max-pooling operation and down scaling operation for the difference image cluster to extract the most sensitive information for age estimation.

2.2 Feature Dimensionality Reduction

The FMTD model obtains feature vector based on the Gabor feature space, in which there is a problem of high dimensionality. Suppose that the original face image is $m \times n$ pixels and suffers Gabor filtering with p different scales and q different directions, and then the image cluster contains $p \times q$ images with the size of $m \times n$. The corresponding feature vector of the image cluster has $(p \times q) \times (m \times n)$ dimensions, which easily leads to the curse of dimensionality. This paper adopts the principal component analysis (PCA) to reduce the feature dimensionality. Suppose there are N data samples, and the corresponding feature vector of each sample is $X_i = (f_1^i, f_2^i, \ldots, f_d^i)^T$, where $1 \le i \le N$ and d denotes the feature vector dimensionality. Through the PCA, the feature vector dimension reduces from d to m. In this paper, m is decided by the experimental effects on the MAE of age estimation.

2.3   Age Estimation Based on FMTD

In this paper, support vector regression (SVR) is selected as the regression algorithm. Through dimensionality reduction, the feature is used for SVR model training and regression prediction.

3   Experiments

This paper implements the experiments of age estimation based on the regression model and adopts the common evaluation criteria MAE to assess the results. The regression result for each face image is the estimated age.

3.1   Experimental Dataset

Experimental dataset has an important influence on the performance of the age estimation algorithm. Two public age datasets FG-NET and MORPH both consist of European and American face, while this paper aims to study the age estimation for Asian face. Therefore, this paper collects abundant network face images with Asian face and selects part images from FG-NET and MORPH. These images make up a new face age estimation dataset with 2700 images called HL-AE. In addition, since each person in HL-AE corresponds to multiple images, this paper experiments with another dataset HL-AE+ for reducing the influence of individual difference on the age estimation. HL-AE+ is composed of 3000 Asian face images with corresponding to 3000 people under various conditions, including the strong sunlight, profile, partial occlusion, etc.

3.2   Experimental Results and Analysis

In this paper, the aforementioned dimensionality m is decided by the experimental effects on the MAE of age estimation. The experimental results show that the MAE reaches the minimum when m is equal to 1600.

After m is determined, 5-fold cross validation is applied to assess the proposed method FMTD+SVR with the evaluation criteria MAE on the HL-AE dataset. This paper also reproduces five another age estimation methods for comparison under the identical experimental conditions, including AAM+CSOHR [10], BIF+SVR [2], LARR [3], HMM [7] and AGES [5]. Table 1 shows the results.

From the data in Table 1, the proposed age estimation method FMTD+SVR has achieved the lowest MAE = 4.25 in HL-AE, which performs better than the other

Table 1. MAE of different face age estimation methods on the HL-AE dataset

Methods	AAM +CSOHR [10]	BIF +SVR [2]	LARR [3]	HMM [7]	AGES [5]	FMTD +SVR
MAE	4.82	4.89	5.39	7.33	9.63	4.25

methods. The MAE of FMTD+SVR reduces 0.57 years compared with the other best method AAM+CSORH, reduces 0.64 years compared with BIF+SVR, and reduces more than 1.0 years compared with the others. The results show that the proposed method greatly improves the accuracy of age estimation.

In order to further verify the generalization ability of the method, this paper selects three best methods on the HL-AE dataset, FMTD+SVR, AAM+CSOHR and BIF +SVR. We use the model trained on HL-AE dataset to test the images in the HL-AE+ dataset. The MAE of BIF+SVR is 5.83, the MAE of AAM+CSOHR is 6.12, and the MAE of FMTD+SVR is the lowest, which is 4.96. The performances of the three methods on the HL-AE+ dataset all reduce compared with those on the HL-AE dataset. This is because the image quality of HL-AE+ is poorer than that of HL-AE. The experimental results demonstrate that the proposed method FMTD+SVR has excellent generalization ability.

4   Conclusion

This paper proposes a new feature extraction model FMTD based on the standard bio-inspired feature model. The FMTD model can learn the abundant feature information of the important organs area as well as the wrinkle area of face, and enhance the information through a series of image processing. The dimensionality of the extracted feature is reduced with PCA, and SVR is employed to implement age estimation. The experimental results demonstrate that the proposed method has better performance of age estimation than the current methods and excellent generalization ability for Asian face.

Acknowledgments. This work is supported by Application Technology Research & Demonstration Promotion Project of Hainan Province of China under grant ZDXM2015103 and JDJS2013006.

References

1. Lanitis, A., Taylor, C.J., Cootes, T.F.: Modeling the process of aging in face images. In: 7th IEEE International Conference on Computer Vision, vol. 1, pp. 131–136 (1999)
2. Guo, G.D., Mu, G.W., Fu, Y., Huang, T.S.: Human age estimation using bio-inspired features. In: IEEE Conference on Computer Vision and Pattern Recognition, pp. 112–119 (2009)

3. Guo, G.D., Fu, Y., Dyer, C.R., Huang, T.S.: Image-based human age estimation by manifold learning and locally adjusted robust regression. IEEE Trans. Image Process. **17**(7), 1178–1188 (2008)
4. Lanitis, A., Draganova, C., Christodoulou, C.: Comparing different classifiers for automatic age estimation. IEEE Trans. Syst. Man Cybern. Part B (Cybern.) **34**(1), 621–628 (2004)
5. Geng, X., Zhou, Z.H., Zhang, Y., Li, G., Dai, H.H.: Learning from facial aging patterns for automatic age estimation. In: 14th ACM International Conference on Multimedia Proceeding, pp. 307–316 (2006)
6. Wang, C.C., Su, Y.C., Hsu, C.T., Lin, C.W., Liao, H.M.: Bayesian age estimation on face images. In: IEEE International Conference on Multimedia and Expo, pp. 282–285 (2009)
7. Zhuang, X.D., Zhou, X., Hasegawa-Johnson, M., Huang, T.: Face age estimation using patch-based hidden markov model supervectors. In: 19th International Conference on Pattern Recognition, pp. 1–4 (2008)
8. Chang, K.Y., Chen, C.S., Huang, Y.P.: A ranking approach for human ages estimation based on face images. In: 20th IEEE International Conference on Pattern Recognition, pp. 3396–3399 (2010)
9. Chang, K.Y., Chen, C.S., Huang, Y.P.: Ordinal hyperplanes ranker with cost sensitivities for age estimation. In: IEEE Conference on Computer Vision and Pattern Recognition, pp. 585–592 (2011)
10. Chang, K.Y., Chen, C.S.: A learning framework for age rank estimation based on face images with scattering transform. IEEE Trans. Image Process. **24**(3), 785–798 (2015)
11. Riesenhuber, M., Poggio, T.: Hierarchical models of object recognition in cortex. Nat. Neurosci. **2**(11), 1019–1025 (1999)
12. The FG-NET Aging Database [DB/OL] (2011). http://www-prima.inrialpes.fr/FGnet/html/benchmarks.html
13. MORPH Face Database [DB/OL] (2012). http://faceaginggroup.com/

User Intent for Virtual Environments

Wei Ge$^{(\boxtimes)}$, Cheng Cheng, Ting Zhang, Jing Zhang, and Hong Zhu

Beijing Laboratory of Intelligent Information Technology,
School of Computer Science, Beijing Institute of Technology,
Beijing 100081, People's Republic of China
gwei_hello@163.com, zhang_ting0402@163.com,
cailingjingjing@163.com, guoguocheng@vip.sina.com,
zhuhong0204@yeah.net

Abstract. In the virtual environments, multimodal interaction not only makes the traditional event-driven system complicated but also increases the task of the users to handle the heavy loads of cognitive and operational complexity. The authors extracted users' operation intents and used intents to trigger system state transition. This paper investigated establishing operators' intent set and intent experiment in a virtual assembly system. By observing actual operation processes and analyzing the expression of intent, the authors constructed an intent-driven system. The research laid a solid foundation for capturing and reasoning intents. The comparison of experimental results show that the intents make the intent-driven system more efficient than the traditional event-driven system and they are able to express users' ideas accurately.

Keywords: Intent · Human-computer interaction · Multimodal · Virtual environment · Direct manipulation · Virtual assembly

1 Introduction

Virtual design technology can demonstrate the feature and shape of products clearly in intuitively three-dimensional environment [1]. We apply virtual design technology to mechanical part assembly and develop a 3D interactive virtual assembly system. Multimodal human-computer interaction system needs to integrate information which acts as input by a variety of interactive devices [2]. This kind of system makes interactive operation in a concurrent and collaborative manner, which can improve efficiency and naturalness. Vidakis et al. presented a natural user interface system, in which users could interact with the desktop application through the face, object, voice and gesture [3].

In the last few years, intent recognition widely aroused the interest because it can make interaction convenient and make intelligent analysis coordinated [4, 5]. There are many methods of constructing and expressing intent. Sergey Sosnovsky [6] presented an approach that combines the logic-based Event Calculus (EC) and probabilistic modeling. Zhang Yingzhong et al. [7] used semantics to describe intent, which is based on the shareable and inferential expression pattern. Song Wei [8] classified the net surfing intents into subject categories, topics and the point that users interested in. Mukherjee S [9] used features to identify intent and constructed an intent graph.

© Springer International Publishing AG 2017
F. Xhafa et al. (eds.), *Recent Developments in Intelligent Systems and Interactive Applications*,
Advances in Intelligent Systems and Computing 541, DOI 10.1007/978-3-319-49568-2_42

The authors analyze the intent and use it instead of event to trigger the system state transition. This paper explores the construction of intent, including two aspects: constructing intent set by observing and analyzing the user's operation process and building intent expression pattern and doing experiments.

2 Intent Analysis and Design

2.1 Intent Set Investigation

Virtual assembly system is a 3D interactive system. To assemble parts we construct some specific perceptions to cater to particular interactive scenarios such as FeatureMatching, Aligning and FaceMating [10]. Previous virtual assembly system was event-driven and input event triggers the system state to change. This system has many complex virtual scenes and a variety of 3D objects, which makes event-driven system complicated on aspects of design and operation. Constructing an intent-driven system can solve the above question.

Intent is a mental activity, which assumes that the virtual environment reach a given state in a short term during the interactive process. It depends on user's intellectual state and current working scene [11]. We need to investigate what intents exist in assembly system for building an intent set. In this assembly system, the typical intents include Picking, FeatureMatching and so on. Picking is to pick up a part and the part will be together with 3D mouse. FeatureMatching automatically judges the assembly feature pairs that the user intends to fit together [10]. FaceMating classifies the parallel faces belonging to the currently assembled parts and calculates the potential set of matching pairs [10]. The authors collect the intents through investigating many mechanical designers, researchers, and experiments. The intent set is shown in Eq. (1).

$$\text{ISet} = \{\text{Picking, MultiPicking, FeatureMatching, Coincidence, FaceMating, DisableCN, Browse, Disassembly, MultiDisassembly, AutoAssembly.MultiAutoAssembly, TechPlanning, InteractiveSimulation, AutoSimulation}\} \tag{1}$$

2.2 Intent Expression

We construct the intent expression pattern, with which virtual assembly system can reason the intent. The current situation and intent pattern have a mapping relationship. The system processes input information and recognizes intent. In our virtual system, eyes and hands can express the user's thought, namely, intent. User's intent has relation with the multimodal input. We use state vector to describe this relation. The state vector is a multi-dimensional vector structure that depicts a state of multimodal inputs at a time point [11]. It is expressed as <T, Sce, Obj1, Obj2, Obj3, Tsk, Stt, Mod1, Mod2, Mod3>. 'T' is the logic time point, 'Sce' is the current scene, 'Obji' is the virtual object related, 'Tsk' is the user's interactive task, 'Stt' is the state of system, and 'Modi' is the input state of multimodal. Every channel has a set of the channel state. For example, the eye channel's states include: gazing at static object (GazeStatic), gazing at dynamic object (GazeDynamic), and so on.

Intent's expression pattern is a sequence of successive state vectors and it implies an interactive mode. Based on a long-term observation, we discovered that intents could be expressed by within five successive state vectors. We can use temporal logic language XYZ to describe and analyze state vectors, which has the semantic accuracy and practicability [12]. The temporal descriptions of intents are shown in Fig. 1, and they are state vectors of intents from left to right, Picking intent, FeatureMatching and FaceMating. Here we just give three channels' states, eye (Eye) and right hand (RHand) and left hand channel (LHand). Every intent uses less than five time points. The flow of information starts from left to right. New state vector will come into being if any channel's state changes. Symbolic semantics is as follows: ◯Scan ◎ Gaze △ freeStatic △ graspStatic ▢ freeTrans ▣ graspTrans ⟩ handGesture.

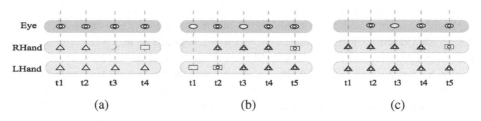

Fig. 1. The state vectors for intents. (a) Picking (b) FeatureMatching (c) FaceMating

3   Experiment and Results

To evaluate the effect of intent and the performance of intent pattern, the authors organized a comparative experiment between event-driven and intent-driven system.

3.1   Participants and Apparatus

Twenty students (10 were male.) from the local university participated in the experiment, with ages ranging from 19 to 23. All had no experience in virtual environments and 3D operation. The experiment included two groups (event-driven system and intent-driven system.) and each group was ten (5 male and 5 female).

The computer was based on the desktop with 4 GB RAM and USB 3.0 interface. The interactive devices included Eye Tribe that got user's fixation point and 3D mouse which was SpaceMouse Plus of 3Dconnexion Company. The operators could control the object and roam in 3D scene. Eye Tribe and computer are connected via USB3.0. Assembly system builds on Open Inventor 5.0 platforms.

3.2   Experiment Design and Procedure

The aim of the experiment is to study the intent's effectiveness and performance. The experiment includes two aspects: to compare the time of tasks between event-driven and intent-driven system; to study intent's accuracy and instantaneity in the intent-driven system. It includes three tasks: pick the 10 certain parts from one specific area to another area; do feature matching; do face mating. The indicators include time, accuracy and instantaneity. Accuracy is used to show if the intent identified is consistent with the operator's thought. Instantaneity describes system's reaction speed. Here we just introduce the operations in intent-driven system.

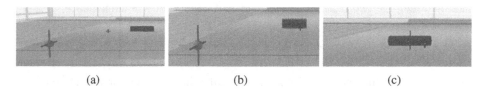

(a)   (b)   (c)

Fig. 2. Picking (a) initialization condition (b) eye lighten the bounding box (c) finish picking

Picking: In Fig. 2(a), The ball and perpendicular bars is 3D mouse and black cross is eye's fixation direction. In Fig. 2(b), the blue bounding box is shown when eyes stare at the part. At the same time, as long as the 3D mouse starts to move to the part the system captures the picking intent and picks up the part automatically. Results is shown in Fig. 2(c), 3D mouse and part are bound together.

(a)   (b)

Fig. 3. FeatureMatching (a) initialization condition (b) eye lightens feature bounding box

FeatureMatching: In Fig. 3(a), left big box has many holes ant it can match many other parts. In Fig. 3(b), 3D mouse is carrying a part and the operators need to find target part to assemble. Eyes scan to find a hole on left box to match the feature of the part, if they can match together the hole's feature box will be highlighted. System identifies FeatureMatching intent and moves the 3D mouse near to the big box.

FaceMating: Coincidence constraint is enforced on current part that is manipulated, the part can only move on the specific axis. When the eyes stare at the part, the current and target part's faces (whether left or right based on the stare point) can be shown. The left face is shown in Fig. 4(a) and the right face is shown in Fig. 4(b). At the same time,

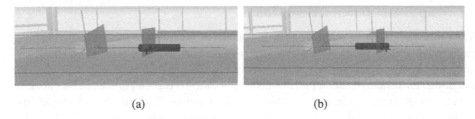

(a)   (b)

Fig. 4. FaceMating. (a) left face (b) right face

as long as the 3D mouse starts to move to the part the system captures the FaceMating intent and then the two faces of current part and target part win coincidence.

3.3   Experiment Results

Table 1 presents the mean task completion time in the event-driven system and intent-driven system. We can see that every task in event-driven system takes more time than in intent-driven system. The intent-driven system can capture intent and implement the intent automatically, so it can save time and improve the efficiency.

Table 1. Mean time of every intent (second)

System	Intent		
	Picking	FeatureMatching	FaceMating
Event-driven system	54.42	23.02	11.02
Intent-driven system	38.75	17.32	10.89

Now we analyze the time differences between these two systems. The difference of Picking intent between two systems is 16.57 s, FeatureMatching intent is 5.7 s, and FaceMating intent is 0.13 s. The longer the operation distance is, the bigger the advantage of event-driven system is. The operation distance of FaceMating is short. So the time difference of FaceMating is smallest.

Table 2. The performance of intent (the full mark is 10)

Intent	Accuracy	Instantaneity
Picking	9.8	8.3
FeatureMatching	9.4	7.9
FaceMating	6.1	9.8

Table 2 presents the performance of three intents in intent-driven system. The instantaneity is quite good and the accuracy on Picking and FeatureMatching is high. However, the accuracy on FaceMating is not good. This problem is related to the eye's threshold value. Later, narrowing the value will solve this problem.

The experiment results show that intent-driven system can save operation time and simplify the operations. Intent provides a significant performance improvement especially on the big scene. The accuracy of FaceMating intent needs to be improved and the authors will continue work on it.

4 Conclusions and Future Work

The authors have constructed the intent's expression pattern and check it in the intent-driven system and such a system has a low design complexity. By using the intent, system can understand user's thought and implement the intent automatically, which can reduce operation load and improve interactive efficiency. In this intent-driven system, the instantaneity and accuracy level increases substantially.

Acknowledgement. This paper is funded by the National Natural Science Foundation of China (Grant No.61370135). This paper is partially supported by Beijing Key Discipline Program.

References

1. Zhang, W.: Research on environment-based intend-driven product form design. Nanjing University of Aeronautics and Astronautics (2006). (in Chinese)
2. Mu, Y.: Research on design strategies of multimodal interaction. Beijing University of Posts and Telecommunications (2008). (in Chinese)
3. Vidakis N, Vlasopoulos A, Kounalakis T, et al.: Multimodal desktop interaction: the face-object-gesture-voice example. In: 2013 18th International Conference on Digital Signal Processing (DSP), pp. 1–8. IEEE (2013)
4. Wang, M., Maeda, Y., Takahashi, Y.: Human intention recognition via eye tracking based on fuzzy inference. In: 2012 Joint 6th International Conference on Soft Computing and Intelligent Systems (SCIS) and 13th International Symposium on Advanced Intelligent Systems (ISIS), pp. 846–851. IEEE (2012)
5. Santos, E., Nguyen, H., Wilkinson, J., Yu, F., Li, D., Kim, K., Russell, J., Olson, A.: Capturing user intent for analytic process. In: Houben, G.-J., McCalla, G., Pianesi, F., Zancanaro, M. (eds.) UMAP 2009. LNCS, vol. 5535, pp. 349–354. Springer, Heidelberg (2009). doi:10.1007/978-3-642-02247-0_35
6. Gu, Y., Sosnovsky, S.: Recognition of student intentions in a virtual reality training environment. In: Proceedings of the Companion Publication of the 19th International Conference on Intelligent User Interfaces (IUI Companion 2014), pp. 69–72. ACM (2014)
7. Zhang, Y.Z., Luo, X.F., Fan, C.: Semantic representation for assembly design intent. Comput. Integr. Manuf. Syst. **17**(2), 248–255 (2011). (in Chinese)
8. Song, W.: Research on topic based query intent identification. Harbin Institute of Technology (2013). (in Chinese)
9. Mukherjee, S., Joshi, S.: Help yourself: a virtual self-assist system. In: Proceedings of the 23rd International Conference on World Wide Web, pp. 171–174 (2014)
10. Cheng, C., Jiang, R., Dong, X.M.: Human knowledge acquisition from 3D interaction in virtual environments. Sci. Chin. Inf. Sci. **55**(7), 1528–1540 (2012)

11. Cheng, C., Zhao, D.P., Lu, B.A.: Intent understanding for virtual environments. J. Graph. **20** (2), 271–279 (2015). (in Chinese)
12. Tang, C.-S.: Temporal Logic Programming and Software Engineering. Science Press, Beijing (1999)

An Image Based on SVM Classification Technique in Image Retrieval

Jiang Qianyi[✉], Zhong Shaohong, and Yang Yuwei

College of Computer Science and Information Technology,
Central South University of Forestry and Technology, Changsha, China
jiangqy226@163.com

Abstract. On the frontier of Image Processing, researchers are encountering the challenge of effectively retrieving and using the information contained in the image. As per the prevailing research after the feature extraction of the relevant properties of a high-level image, the resulting image does not add too many features, When operating directly on the image, because of the high Witte sexual performance data are relatively poor, resulting in the traditional classification method does not apply. So this paper uses support vector machine (SVM) image classification techniques which can overcome this defect. This paper makes the use of Dense SIFT algorithm to obtain image feature and then build Bag of words model. Subsequently establishing training dictionary database and finally, the test set of images SVM classification test. Experimental results show that the use of SVM classification accuracy of image retrieval technology enables greatly increased.

Keywords: Image retrieval · Image classification techniques · Feature extraction · SVM algorithm

1 Preface

It is widely seen that the development of multimedia technology and storage technology have found its applicability in the corridors of medicine, engineering, science, photography, advertising and in many more spheres of our socio-economic life every day such fields of human endeavour will require a huge amount of digital image information. How to effectively manage and use digital image information has become a major problem facing modern society. The traditional method is through the image text annotations, and these methods have been unable to meet the rapidly increasing demands of digital image processing. So this paper presents an image retrieval method based on support vector machine image content [1] which makes the establishment of a database for the image of the test image retrieval prototype system based on support vector machine classification techniques [2, 3], making retrieval speed to suit the actual application requirements.

© Springer International Publishing AG 2017
F. Xhafa et al. (eds.), *Recent Developments in Intelligent Systems and Interactive Applications*,
Advances in Intelligent Systems and Computing 541, DOI 10.1007/978-3-319-49568-2_43

2 Image Feature Extraction

Content-based image retrieval key is to use computers to automatically understand the content of the image and quantitatively expressed, that we call the feature extraction, following the color, shape, texture feature extraction method of image analysis.

2.1 Color Feature

Color image feature is a feature of the most widely used in image retrieval method [4], it has good robustness to image the scale, translation, rotation, translation is not sensitive, while the color feature the calculation is very simple and easy to understand the use of color features.

2.2 Shape Feature

Shape is used to describe one of the most direct features of an object, shape features an image containing some semantic information, which we study the shape feature is particularly important. Shape characteristics of the target translation, rotation and scaling has strong robustness [5, 6], and therefore, the shape feature is an image feature image retrieval technique frequently used to improve the image retrieval efficiency plays an important role.

2.3 Texture

Texture is CBIR retrieval of another common description of the image content features [7]. Its Objective is to describe texture feature is the use of image processing and pixel image analysis to obtain recurring arrangement rule and local mode, these rules and patterns of statistical calculations to distinguish between different textures, so as to realize the distinction between images. Commonly used texture feature extraction methods can be divided into three categories: statistics, structured approach, the model law.

3 Content-Based Image Retrieval Feature (CBIR) Related Technologies

Content-based image retrieval is an integrated research area, which combines the existing pattern recognition, artificial intelligence, image processing, computer vision, information retrieval and other fields of knowledge, image feature extraction, similarity measure, relevance feedback, retrieval performance evaluation criteria and other technology is the key technology of CBIR.

3.1 Similarity Measure

After the image feature extraction, CBIR system, the need for image feature matching, and this process is called matching similarity measure. The similarity measure of accuracy and speed of retrieval CBIR has an important influence. Since the image feature extraction mostly in vector form is saved, in the multi-dimensional vector space in which the image is seen as a point, so the similarity measure between the images is transformed into multi-dimensional vector space between the calculated distance between two points.

3.2 Relevance Feedback

Often there will be some differences due to the presence of ordinary people to understand when the "semantic gap" and therefore in the use of the image of the shape, color, texture and other characteristics of the search results, you can use relevance feedback, RF method to narrow this difference. Relevance feedback method is through repeated interaction between users to find the user's query intent relevance feedback between man and machine through a number of "feedback – Search" process in order to achieve better results for each a search result.

3.3 Retrieval Performance Evaluation Criteria

Evaluate the retrieval performance for content based image retrieval image classification technology plays an important role. It includes two questions: First, select the test database, and second, the evaluation index system.

For the evaluation of the retrieval performance, the most commonly used is recall and precision, recall means is retrieved related image occupies all the relevant proportion of the number of images, precision is It refers to the total return of the image to retrieve relevant images percentage of all images. In search engines, often using normalize Discounted Cumulative Gain, nDCG to measure the quality of search results sorted, detailed definitions and refer to the derivation of reference [8–10].

4 Based on Support Vector Machine Classification Technology

In front of the color, shape, texture features of the image analysis and research carried out, which are characterized by a high level of image extracted image related properties without adding extra features. When the operator directly on the image, because of the high performance data Witte relatively poor, resulting in the traditional classification method is not as useful, but the support vector machine (SVM) for use in image classification can overcome this defect. We first use Dense SIFT algorithm to obtain image feature; then build Bag of words model, a training dictionary database; and finally the use of SVM to classify the test image collection tests, the experimental results.

4.1   Local Image Feature Extraction

Scale-Invariant Feature Transform, SIFT, the core idea of the algorithm is to use different scales of Gaussian smoothing the image, the image smoother difference before and after the big difference pixels are extracted as feature points feature descriptor, and then to match the feature descriptor. The algorithm is able to increase image brightness, scale, rotation, translation changes simultaneously maintained, used to describe the local characteristics of the image.

This article is extracted using image local feature dense Scale Invariant Feature Transform (Dense SIFT), feature extraction using descriptors, not build Gaussian scale space, extract SIFT features only on a single scale, the extraction of interest patches each SIFT feature locations.

Let the image of f(x, y), computing (x0, y0) points Dense SIFT features, Dense SIFT calculated as follows:

1. Calculated with (x0, y0) as the center of the amplitude and direction of 9 * 9 pixel neighborhood gradients. Formula 1 for each pixel difference in x and y directions of:

$$d_x = f(x_0 + i + 1, y_0 + j) - f(x_0 + i, y_0 + j), i, j \in \{-4, -3, \ldots, 3, 4\}$$
$$d_y = f(x_0 + i, y_0 + j + 1) - f(x_0 + i, y_0 + j), i, j \in \{-4, -3, \ldots, 3, 4\} \quad (1)$$

Gradient magnitude and direction are as follows:

$$m(x_0 + i, y_0 + j) = \sqrt{(dx)^2 + (dy)^2}$$
$$\theta(x_0 + i, y_0 + j) = \arctan\left(\frac{dy}{dx}\right) \quad (2)$$

2. Thus with (x0, y0) as the center and form 4 * 4 grid of cells, each cell covering 3 * 3 = 9 pixels, 81 pixels of the difference grid.
3. The 3 * 3 = 9 points in each cell within the pixels using a gradient amplitude weighting, quantization to eight directions, so that each cell can form an 8-dimensional vector, i.e., $v1 = [c_1, c_2, \ldots, c_8]$.
4. The 4 * 4 all cell histograms vector v i are connected in series, to obtain (x0, y0) Point 4 * 4 * 8 = 128-dimensional descriptors C. Wherein, $C = [c_1, c_2, \ldots, c_{128}]$.

4.2   Construction of Bag of Words Model

Bag-of-words model is often used in information retrieval, it is a document that all statements as an assemblage composed by a number of vocabulary, syntax and word order without considering other factors appear, each words are treated as independent occurrences, not to consider the relationship between it and other words. Semantic affect any document that is a word appears is independently selected at any location, regardless of the document.

Bag of words model image Representation of the three steps:

1. Dense SIFT algorithm to use for each type of image extracted "visual vocabulary", without considering the mutual order between the visual vocabulary of the visual vocabulary all together.
2. The use of K-means algorithm to construct the dictionary. K-means algorithm K as a parameter, the N objects into K clusters, such that the higher the similarity within a cluster, a lower degree of similarity between clusters. The use of K-means algorithm Dense SIFT extracted visual vocabulary, according to the distances are combined to obtain a dictionary of basic vocabulary.
3. Use dictionary words an image, based on Dense SIFT algorithm to dictionary words an image, then extract the word list of words close substitutes of feature points from the image, the statistical number of times it appears in the image, so that K dimensional vector values can be used to represent the image.

4.3 SVM Classification Algorithm Utilizing Image

Support vector machine is the mid-1990s by the Vladimir N. Vapnik put forward, and now has been successfully used in the regression analysis, pattern recognition and other fields. This machine learning method is theoretically based on statistical learning theory through structural risk minimization to optimize the experience of risk, mainly used to solve problems in the field of neural networks.

4.4 Results

Experimental data selected a total of six categories (respectively, build, bus, dinosaur, elephant, face, flower), every class a total of 60 images, 40 images as training samples, 20 samples as a test image, the total sample test the number is 120. Bag of words model pyramid matching principle Bag of words model. Select the histogram intersection, HI as nuclear SVM kernel function. Each classification (recognition) in Table 1.

Period, the test correctly identified class build rate of 90 %, 5 % became a bus class identification, identifying five percent became elephant class. Build test class has two image including a plurality of sculpture, be mistaken for elephant classes. Test class

Table 1. Each test class recognition rate

	Build	Bus	Dinosaur	Elephant	Face	Flower
Build	0.90	0.05	0.00	0.05	0.00	0.00
Bus	0.15	0.80	0.00	0.05	0.00	0.00
Dinosaur	0.00	0.00	0.85	0.00	0.15	0.00
Elephant	0.10	0.00	0.00	0.90	0.00	0.00
Face	0.00	0.00	0.00	0.00	1.00	0.00
Flower	0.00	0.00	0.00	0.00	0.00	1.00

bus to identify the correct rate of 80 %, 15 % to be identified would be build, 5 % recognition became elephant. Test class bus image in a plurality of images of the building contains a lot of information, it was mistaken for build. Dinosaur class test correctly identified 85 %, 15 % is recognized as a face. Class elephant test correctly identified 90 %, 10 % to be identified would be build. Test class flower and face recognition rate was 100 %. SVM total accuracy rate of 90.8333 % (109/120).

5 Conclusion

Based on Support Vector Machine classification image retrieval is a blend of computer vision, image processing, information retrieval, pattern recognition, data mining, comprehensive study of artificial intelligence and many other disciplines. Hence it is an important research direction future digital multimedia technology development. Based on the underlying visual characteristics of image retrieval technology this method uses an image, such as shape, color, texture and spatial relationships, to manage and index the image information, and with a certain degree of human-computer interaction through similarity measurement methods to achieve image retrieval.

Acknowledgements. The research work was supported by 2015 Year College Students in Research Learning and innovative experiment project under Grant No. (201510538007).

References

1. Li, L.J., Su, H., Lim, Y., Li, F.F.: Object bank: an object-level image representation for high-level visual recognition. Int. J. Comput. Vis. **107**(1), 20–39 (2014)
2. Bharath, R., Cheng, X., Tong Heng, L.: Shape classification using invariant features and contextual information in the bag-of-words model. Pattern Recogn. **48**, 894–906 (2015)
3. Azad, S.A., Ali, A.B.M.S., Wolfs, P.: Identification of typical load profiles using K-means clustering algorithm. In: 2014 Asia-Pacific World Congress on Computer Science and Engineering (APWC on CSE), pp. 1–6 (2014)
4. Cao, J., Mao, D., Cai, Q., Li, H., Du, J.: A review of object representation based on local features. J. Zhejiang Univ. Sci. C **14**(7), 495–504 (2013)
5. Subrahmanyam, M., Wu, Q.M.J., Maheshwari, R.P., et al.: Modified color motif co-occurrence matrix for image indexing and retrieval. Comput. Electr. Eng. **39**(3), 762–774 (2013)
6. Wang, X., Wang, Z.: A novel method for image retrieval based on structure elements' descriptor. J. Vis. Commun. Image Represent. **24**(1), 63–74 (2013)
7. Lina, C.H., Chen, C.C., Lee, H.L., et al.: Fast K-means algorithm based on a level histogram for image retrieval. Expert Syst. Appl. **41**(7), 3276–3283 (2014)
8. Walia, E., Pal, A.: Fusion framework for effective color image retrieval. J. Vis. Commun. Image Represent. **25**(6), 1335–1348 (2014)
9. Zhang, S., Huang, J., Li, H., et al.: Automatic image annotation and retrieval using group spasity. IEEE Trans. Syst. Man Cybern. Part B **42**(3), 838–849 (2012)
10. http://www.cnblogs.com/uniquews/archive/2012/12/27/2835923.html

Constructing Kinematic Animation of Products in Virtual Assembly Environments

Jing Zhang, Cheng Cheng$^{(\boxtimes)}$, Wei Ge, Ting Zhang, Hong Zhu,
and Sebai Mounir

Beijing Laboratory of Intelligent Information Technology,
School of Computer Science, Beijing Institute of Technology,
Beijing 100081, People's Republic of China
cailingjingjing@163.com, gewei_hello@163.com,
zhang_ting0402@163.com, guoguocheng@vip.sina.com,
zhuhong0204@yeah.net, moumou2788@yahoo.fr

Abstract. The traditional assembly simulation is designed for product design system which doesn't support 3D interaction, and the design is very complicated. This paper proposes a new strategy and an algorithm to construct an assembly animation in VE. Engine components are used to realize animation as building blocks. Algorithm and process about how to construct engine components as well as parameters' transfer between components are presented. The method will decrease designers' burden and improve the production efficiency. At last, two examples are given to verify the feasibility of this strategy.

Keywords: Engine component · Animation · Virtual assembly · Human-computer interaction · Direct manipulation · Virtual environment

1 Introduction

Product assembly analysis and planning evaluation in Virtual Environment (VE) are important in future product design. Simulation of virtual assembly mechanism is necessary part to verify products [1]. Visual assembly mechanism simulation is simulating the motion animation of assembly. However, at present, virtual assembly mechanism are not fully adaptable animations [2].

There are two forms of animation construction: behavior construction in WIMP interface system or in VE [3]. In VE, programs are needed to construct complex behavior mechanisms which depend on concrete animation engine mechanisms. Another traditional way is designed in WIMP interface system. Animations designed through these two ways can't import to virtual environment directly.

Researches about virtual assembly and virtual animation have been carried out for a long time, especially populate in recent years. Pan Wang et al. [4] present a way building assembly model based on semantics and geometric constraints which will optimize the assembly simulation. In the development of virtual animation, interactivity has been gained attention, Stelian Coros, Bernhard Thomaszewski et al. [5] have researched an interactive simulation system. Animation can be constructed through depicting motion curves repeatedly. Most animation constructions need people to fill

© Springer International Publishing AG 2017
F. Xhafa et al. (eds.), *Recent Developments in Intelligent Systems and Interactive Applications*,
Advances in Intelligent Systems and Computing 541, DOI 10.1007/978-3-319-49568-2_44

parameters in forms. Researches about automatic construction algorithms of animation are relatively less. Duygu Ceylan, Wilmot Li et al. [6] proposed an automation algorithm making mechanical figures to simulate input motions. Dance Lessin et al. [7] put forward an open-ended method to complicate the behavior of virtual figures in virtual environment.

In spite of having so many researches on virtual assembly and animation, most animations are complexity and low reusability. This paper presents a strategy to construct virtual animation based on engine components to solve these problems.

2   Automatic Construction Strategy of Virtual Assembly Animation

Assembly animation design in VE is a difficult task. This paper proposes an assembly animation strategy based on engine components which needn't designers to offer kinematic parameters. Parameters are automatically generated through user's direct manipulations.

2.1   Mechanical Motion Specification

Before the simulation stage is the virtual assembly stage. The assembly structure and kinematic pair composition have determined the motion forms of assembly. All kinds of mechanical motions are made up of basic kinematic pairs [8].

The concept of Temporal Assembly Model is introduced for assembly. It combines dynamic and static descriptions to express the relationships between objects. Relationships include temporal aggregation relationship (Tar), temporal constraint dependency relationship (Tcd), constraint (CN) and Trace etc. 'Tar' embodies the assemble intention. 'Tcd' reflects the location relationships between assembly objects. 'CN' is the inner property of part object which is the foundation of 'Tar' and 'Tcd'. Object Trace is a sequence of short behavior sequences (sbs). Its form is: sbs1, sbs2,..., sbsn. Object Trace descripts the motion and behavior of object. Trace file is generated after assemble phase. It records relationships and positions. Short behavior sequence (sbs) is composed of temporal behavior segments (tbs). The form of sbs is: tbs1, tbs2,..., tbsm. Sbs is a piece of Object Trace. The bigger the number n of sbs, the more accurate the description of Object Trace will be. Temporal behavior segment (tbs) is the infrastructure of object behavior. Complex behavior can be decomposed into sequences of tbs. The concept of 'tbs' can reference paper [1]. 'Tbs' describes basic motions of different motion traces.

The authors construct basic animation engine components for every part object, which include a calculation engine, a rotation engine and a translation engine. Calculation engine is important part. It has a set of inputs and outputs and receives parameters through interfaces. Expressions are the core of calculation engine. Reusability is a great advantage of component. Same mechanism will share same component. The basic components can be used to build the 'bigger' components. For example, in spite of different cars, as long as they have same automobile engine, the 'bigger' component, the automobile engine, can be shared. The automobile engine consists of small components such as crank-link mechanism component.

In virtual assembly animation, there are automatic and interactive animations. The driver of automatic simulation is a time engine, while the driver of interactive animation is human being who control with 3D devices such as 3D mouse. The virtual simulation structure of assembly based on engine components is shown in Fig. 1. Simulative inputs are constructed for automatic animation. The event process unit is shared with both user's input and simulative input. Event process translates user's continuous inputs to a sequence of discrete events. They are translation and rotation increments and correspond to tbs. The motion increments of tbs are output from calculation engines solving constraints.

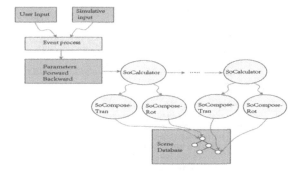

Fig. 1. Virtual simulation construction based on engine components

Construct engine components is core problem of animation. Algorithm of constructing engine components is shown in Fig. 2. In this algorithm, the trace file is first analyzed to recover relationships. Step 4 is a key step of algorithm. Assume that there exists a start and end tbs, information like constraints, rotation axis and central point can be obtained from tbs analysis. Inputs are handed in process unit. Output parameters are sent to the engine component of active part in kinematic chain. Active part is the part that driving other parts to move. The calculation engine receives parameters and calculates constraints. The output parameters are delivered to next engine component in the kinematic chain. Translation and rotation increments are delivered to translation and rotation engines which are connect to the part object in scene database. This is a process of tbs execution. It will not stop until it meets the end tbs. Sbs are executed sequentially. A whole trace simulation cycle is finished when all sbs executed.

The main content of this algorithm is building calculation engines to calculate constraints. Take cam as an example to explain the constraints solving process. Assume cam is the active part. O_1 is the origin point of cam's feature–Cam.fea_1. O_1's constraint type is circle. Another cam's feature is Cam.fea_2 connecting with the roll. Its origin point is O_3. 'Pi' is the position of O_3. The roller also has two features. One is roller.fea_1 connecting with the linkage. O_2 is the origin point of roller.fea_1. Q_i is O_2's position. Another is roller.fea_2 connecting with the cam. Geometric of cam-linkage is shown in Fig. 3.

The contour line of cam is divided into three parts. The top and bottom of cam are semi-circles A and B. R_1 and R_2 are their radiuses. Their centers are A_1 and B_1 respectively. A_1's position is Pa_i. B_1 and O_1 are coincident points. The middle of cam is a isosceles trapezoid. Its height is d. As the symmetry of cam, the authors just analyze half of cam's trace. Assume rotation angle is α. α_1 and α_2 are rotation angles when the

Fig. 2. Engine components construction algorithm

Fig. 3. Geometric of cam-linkage mechanism

cam rotates to the junction of semi-circle and trapezoid. α_1 and α_2 can calculated according R_1 and R_2.

Roller makes same motion with linkage. O_3's position becomes P_{i+1} when cam rotates angle α. The translation amount of O_3 along the direction of $O_1 O_2$ is $(P_{i+1} - P_i)$ O_1O_2. O_2's position is Q_{i+1}. Set $\Delta\text{Trans}_1 = (P_{i+1} - P_i)O_1O_2$. ΔTrans_1 can be calculated according R_1, R_2, d and α:

$$\Delta\text{Trans}_1 = F_1(R_1, R_2, d, \alpha); \tag{1}$$

As $\|Q_{i+1} - Q_i\| = \|(P_{i+1} - P_i)O_1O_2\|$, translation amount of roller:

$$\Delta\text{Trans}_2 = Q_{i+1} - Q_i = F_2(R1, R2, d, \alpha); \tag{2}$$

Functions F_1 and F_2 are two concrete algebraic equation expressions.

3 Experiments and Analyses of Virtual Assembly Animation

Two examples are used to analyze the feasibility and efficiency of this animation algorithm. One is cam-linkage mechanism animation, shown in a and b of Fig. 4. Another is single cylinder engine animation, shown in c and d of Fig. 4.

Fig. 4. Animation states. (a) the start state of cam-linkage (b) one state of cam-linkage animation (c) the start state of single cylinder engine (d) one state of single cylinder animation

Cam is selected as the active part. Initial position is determined through tbs analysis. Constraint expressions in calculation engine are designed according to the structure of assembly. Calculation engine will calculate constraints according the embedded expressions. Translation and rotation value are output to translation and rotation engines to control the cam's motion. Meanwhile, the cam's engine component outputs parameters to engine components of roller and linkage. Roller and linkage calculate their motions according to themselves' constraint expressions. One segment's simulation of cam-linkage mechanism is implemented. The trace's next segment goes on until meet the end position. A cycle of cam-linkage mechanism's animation is finished. The single cylinder engine simulation is driven by the crank. The simulation is the same as the cam-linkage mechanism (Table 1).

Table 1. Performances of different animation design modes

Cam-linkage	CN	Interactive mode	Reusability	Operation step
General software	6	ScrollBar & Fill forms	Low	26
VE	6	3D device	Low	4
Direct manipulation	3	3D device	High	4

A form is given to show performances of different methods to construct assembly animations. General Software needs professional designers to design constraints and fill parameters forms, which will depend on designers' skill and experience. In VE, the reusability is still low. It is obvious that Direct Manipulation has better performances than General software and VE.

4 Conclusion

This paper proposed an animation construction strategy based on engine components for virtual assembly. Experiments show it is feasible. More kinds of kinematic mechanisms should be researched overall so that the algorithm can solve all kinds of automatic animation problems.

Acknowledgements. This paper is funded by the National Natural Science Foundation of China (Grant NO. 61370135). This paper is partially supported by Beijing Key Discipline Program.

References

1. Cheng, C., Lu, B.A.: Interactive animation software construction in virtual environments. Appl. Mech. Mater. **635–637**, 515–518 (2014)
2. Cheng, C., Li, L.Y.: Adaptive animation design method for virtual environments. In: Proceedings of the 10th International Conference on Computer Graphics Theory and Applications (GRAPP 2015), INSTICC, pp. 356–361 (2015)
3. Navarre, D., Palanque, P., Bastide, R., Schyn, A., Winckler, M., Nedel, L.P., Freitas, C.M.D. S.: A formal description of multimodal interaction techniques for immersive virtual reality applications. In: Costabile, M.F., Paternò, F. (eds.) INTERACT 2005. LNCS, vol. 3585, pp. 170–183. Springer, Heidelberg (2005). doi:10.1007/11555261_17
4. Wang, P., Li, Y., Yu, L., Zhang, J.: A novel assembly simulation method based on semantics and geometric constraint. Assembly Autom. **36**, 34–50 (2016)
5. Coros, S., Thomaszewski, B., Noris, G., Sueda, S., Forberg, M., Summer, R.W., Matusik, W., Bickel, B.: Computational design of mechanical characters. ACM Trans. Graph. **32**(4), 83:1–83:12 (2013)
6. Ceylan, D., Li, W., Mitra, N.J., Agrawala, M., Pauly, M.: Designing and fabricating mechanical automata from mocap sequences. ACM Trans. Graph. **32**(6), 186:1–186:11 (2013)
7. Lessin, D., Fussell, D., Miikkulainen, R.: Open-ended behavioral complexity for evolved virtual creatures. In: Proceedings of the 2013 Genetic and Evolutionary Computation Conference, pp. 335–342. ACM (2013)

8. Liu, H.X.: CATIA Digital Prototype Motion Simulation Explanation, 2nd edn. China Machine Press, Beijing (2013). (in Chinese)
9. Cheng, C., Deng, J., Chen, D.: Temporal assembly model for virtual environment. In: Human System Interactions, pp. 948–953. IEEE (2008)

Null Space Diversity Fisher Discriminant Analysis for Face Recognition

Xingzhu Liang$^{(\boxtimes)}$, Yu'e Lin, Gaoming Yang, and Guangyu Xu

School of Computer Science and Engineering,
Anhui University of Science and Technology, Huainan 232001, China
`lxz9117@126.com`, `linyu_e@126.com`,
`{gmyang,gyxu}@aust.edu.cn`

Abstract. The feature extraction algorithms, which attempt to project the original data contained in a lower dimensional feature space, have drawn much attention. In this paper, based on enhanced fisher discriminant criterion (EFDC), a new feature extraction method called Null Space Diversity Fisher Discriminant Analysis (NSDFDA) is proposed for face recognition. NSDFDA based on a new optimization criterion is presented, which signifies that all the discriminant vectors can be calculated in the null space of the within-class scatter. Moreover, the proposed algorithm is able to extract the orthogonal discriminant vectors in the feature space and simultaneously does not suffer from the small sample size problem, which is desirable for many pattern analysis applications. Experimental results on the Yale database show the effectiveness of the proposed method.

Keywords: Feature extraction · Enhanced fisher discriminant criterion · Null space · The within-class scatter · The small sample size problems

1 Introduction

The feature extraction is a critical issue in face recognition activity. The goal of feature extraction is to map high dimensional data samples to a lower dimensional space such that certain properties are preserved. Among all the dimensionality reduction methods, Fisher linear discriminant analysis (FLDA) [1] is the most popular method and has been widely used in many classification applications. FLDA seeks to find directions on which the ratio of the trace of the between-class matrix and the trace of the within-class matrix is maximized. However, some recent research shows that the samples may reside on a nonlinear submanifold. FLDA fails to discover the underlying submanifold structure, due to the fact that it aims only to preserve the global structures of the samples. Another technique called Locality Preserving Projections (LPP) [2] have been proposed for dimensionality reduction, which can preserve the intrinsic geometry of data. However, in some applications, LPP has no direct relationship to classification so that it could not make sure to generate a set of good projections for classification purpose. Yan et al. [3] reformulated a variant of FLDA using graph embedding framework and proposed Marginal Fisher Analysis (MFA). MFA can be viewed as a supervised variant of LPP since it focuses on the characterization of intra-class locality

© Springer International Publishing AG 2017
F. Xhafa et al. (eds.), *Recent Developments in Intelligent Systems and Interactive Applications*,
Advances in Intelligent Systems and Computing 541, DOI 10.1007/978-3-319-49568-2_45

and inter-class locality. However, MFA only pays attention to the intra-class compactness via minimizing the distance among data points from the same class, which will impair the variation of the values of data from the same class and lead to over-fitting problem. In order to overcome the over-fitting problem, Gao et al. [4] propose a new approach, called enhanced fisher discriminant criterion (EFDC), which explicitly considers the variation of the values among nearby data belonging to the same class and the discriminating information. However, the basis vectors of EFDC are nonorthogonal and then the extracted features contain redundancy, which may dramatically degrade performance. Several orthogonal manifold learning algorithms [5–7] were proposed to pursue orthogonal bases, which are believed to preserve the metric structure of the original vector space and have more locality preserving power and discriminating power. Furthermore, when EFDC is applied to face recognition, it meets the small sample size problems because the within-class matrix is singular. In order to address the problems mentioned above, a new feature extraction algorithm, called null space diversity fisher discriminant analysis (NSDFDA) is proposed. NSDFDA based on a new optimization criterion is presented which can directly overcome the small sample size problem and at the same time derive all the orthogonal optimal discriminant vectors. The effectiveness of the proposed algorithm is demonstrated by experiments on the Yale face database.

2   Related Work Enhanced Fisher Discriminant Criterion

Firstly, we give the objective of FLDA, which is defined as the following optimization problem:

$$J(w) = \frac{w^T S_b w}{w^T S_w w} \tag{1}$$

Where S_b the between-class is scatter matrix and S_w is the within-class scatter matrixes, which are defined as the followings:

$$S_w = \frac{1}{n} \sum_{i=1}^{C} \left(\sum_{j=1}^{n_i} (x_j^{(i)} - m^{(i)})(x_j^{(i)} - m^{(i)})^T \right) \tag{2}$$

$$S_b = \frac{1}{n} \sum_{i=1}^{C} n_i (m^{(i)} - m)(m^{(i)} - m)^T \tag{3}$$

Where $x_j^{(i)}$ is the jth image of the class i, $m^{(i)}$ is the mean of the class i and m is the mean of all the samples.

In order to build a stable discriminant criterion, EFDC takes the variation into Fisher discriminant criterion. EFDC models the variation among nearby data from the same class by adjacency graph, which measures the amount of variation of the values of data from this adjacency graph, and then combines the variation and Fisher linear discriminant analysis criterion to build a stable discriminant criterion. Let B denote the local weighted matrix with elements characterizing the diversity of two close data points with same labels. The elements of the diversity weighted matrix B are defined as follows:

$$
B_{ij} = \begin{cases} \exp\left(\frac{-t}{\|x_i - x_j\|^2}\right) & \begin{array}{l} \text{if } x_i \in N_k(x_j) \text{ or} \\ x_j \in N_k(x_i) \text{ and } l_i = l_j \end{array} \\ 0, & \text{otherwise} \end{cases} \tag{4}
$$

Where $N_k(x_i)$ denotes the set of k nearest neighbors of x_i, l_i and l_j denote the class label of data x_i and x_j, respectively. $t \in (0, +\infty)$ is a suitable constant. B_{ij} measures the contribution of x_i relative to x_j to the diversity information. Using the local diversity weighted matrix B, the local diversity scatter S_d can be expressed as:

$$
\begin{aligned}
S_d &= \frac{1}{2} \sum_{i=1}^{N} \sum_{j=1}^{N} B_{ij}(x_i - x_j)(x_i - x_j)^T \\
&= X(Q - B)X^T \\
&= XL_BX^T
\end{aligned} \tag{5}
$$

Where Q is a diagonal matrix whose elements on diagonal are column sum of B, i.e. $Q_{ii} = \sum_j B_{ij}$.

The goal of EFDC is to map data points from different classes to a subspace in which they are as distant as possible. Moreover, the data points from the same class to a subspace in which both the intra-class compactness and intra-class variation can be well preserved. The objective function of EFDA can be expressed as follows:

$$
J(W) = \max_{W} \frac{W^T(aS_b + (1-a)S_d)W}{(W^T S_w W)} \tag{6}
$$

Equation 6 can be solved by generalized eigenvalue decomposition

$$
(aS_b + (1-a)S_d)w_i = \lambda_i S_w w_i \quad i = 1, 2, \cdots, l \tag{7}
$$

3   Null Space Diversity Fisher Discriminant Analysis

Assume that X is a d dimensional face sample set with N elements belonging to C classes. Denote x_i the jth image, and N_l number sample in the class i. Thus, we have

$\sum_{i=1}^{C} n_i = N$ and $X = [X^1, X^2, \cdots, X^c]$. In the face recognition, it is almost impossible to make S_w invertible because of the limited amount of training samples, which is the small sample size problem. In order to solve the small sample size problem, the objection function of null space diversity fisher discriminant analysis (NSDFDA) is defined as follows:

$$J(W) = \max_{|W^T S_w W| = 0} \text{tr}(W^T [aS_b + (1-a)S_d]W) \tag{8}$$

Equation 8 means that all the discriminant vectors can be chosen from the null space of S_w. In order to calculate the orthogonal basis in the null space of S_w, we redefine the S_w as follows:

$$S_w = H_w H_w^T \tag{9}$$

$$H_w = \frac{1}{\sqrt{n}} \left[(X^1 - m^{(1)}e^T), \cdots, (X^C - m^{(C)}e^T) \right] \tag{10}$$

Where $m^{(i)}$ is the mean of the class i and e is a $1 \times N$ vector with all terms equal to 1. As to resolve Eq. 8, we firstly calculate the orthogonal basis in the null space of S_w. Denote $r = \text{rank}(H_w)$ and $H_w = U\Sigma E$, $\Sigma = \begin{pmatrix} \Sigma_{r \times r} & 0_{r \times (N-r)} \\ 0_{(d-r) \times r} & 0_{(d-r) \times (N-r)} \end{pmatrix}$ and $U = (U_1, U_2)$. As to $U_1 \in R^{d \times r}$, $U_2 \in R^{d \times (d-r)}$. Then, we know that U_2 is the null space of S_w. Denote $P = U_2$, then Eq. 8 is transformed into Eq. 11

$$J(W) = \max_{V^T V = I} \text{tr}(V^T [aS_b^P + (1-a)S_d^P]V) \tag{11}$$

Where $S_b^P = P^T S_b P = U_2^T S_b U_2$ and $S_d^P = P^T S_d P = U_2^T S_d U_2$. Then, Eq. 11 can be solved by generalized eigenvalue decomposition

$$(aS_b^P + (1-a)S_d^P)v_i = \lambda_i v_i \quad i = 1, 2, \cdots, l \tag{12}$$

Let the column vectors $v_1, v_2, \cdots v_d$ be the solutions of Eq. 11 ordered according to their eigenvalues $\lambda_1, \lambda_2, \cdots \lambda_d$. Then, we have the optimal matrix $W = PV$.

NSDFDA for computing orthogonal discriminant vectors can be described in the following steps.

Step 1. Compute the diversity weighted matrix B and the local diversity scatter S_d according to Eqs. 4 and 5;

Step 2. compute the between-class scatter matrix S_b according to Eq. 3;

Step 3. Compute the within-class scatter matrix S_w according to Eq. 10, then we calculate the orthogonal basis in the null space of S_w, which is $P = U_2$;

Step 4. Compute $S_b^P = P^T S_b P = U_2^T S_b U_2$ and $S_d^P = P^T S_d P = U_2^T S_d U_2$

Step 5. Solve Eq. 12 by generalized eigenvalue decomposition, then we can obtain the basis vectors $\alpha_1, \alpha_2, \cdots, \alpha_r$. Denote $Q = (\alpha_1, \alpha_2 \cdots \alpha_r)$, the matrix $W = QP$ is the optimal matrix. Obviously, $W^T W = I$, W is the optimal orthogonal matrix.

4 Experimental Results

In order to test performance of our proposed method, the Yale database is used. The Yale face database contains 165 grayscale images of 15 individuals. The images demonstrate variations in lighting conditions (left-light, center light, right-light) and facial expression (normal, happy, sad, sleep, surprised, and wink). Figure 1 shows the sample images of one person. In the experiments, each image is manually cropped and resized to 32×32 pixels. We evaluate the performance of NSDFDA and compare it with two dimensionality reduction methods including EFDC, and LFDA. Ten tests are performed and these results are averaged. The recognition rates for the three methods are listed in Table 1.

Table 1. Recognition rates on the Yale face database (%)

Methods	Training samples/class				
	3	4	5	6	7
NSDFDA	79.7	82.7	83.4	87.8	88.4
EFDC	79.2	81.1	82.2	86.2	87.6
LFDA	78.3	80.5	80.7	85.6	87.2

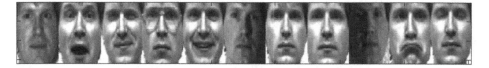

Fig. 1. Images of one person in YALE database

From Table 1, we find that NSDFDA is the most efficient dimensionality reduction method, and is much more efficient than EFDC and LFDA. The one reason is that the NSDFDA not only exploits the more useful bases in the null space of within-class scatter matrix but also preserves the variation among nearby data from the same class, which avoids the over learning problem. The other reason is that the NSDFDA overcomes the small sample size problems, which is very suitable for face recognition.

5 Conclusions

In this paper, we present the algorithm called NSDFDA. In comparison with EFDC and LFDA, NSDFDA has two prominent characteristics. First, it is designed to achieve good discrimination ability by explicitly considering the within-class information, the between-class information and the variation among nearby data from the same class. Second, NSDFDA not only exploits the orthogonal discriminant vectors in the feature space but also overcomes the small sample size problems. Finally, experimental results on the Yale face database show that the proposed method is effective and feasible.

Acknowledgment. This article is supported by the Nature Science Foundation of China (No. 61471004), the Key Project of Higher Education Natural Science Foundation of Anhui Province (No. KJ2016A203, No. KJ2014A061) and the Master and Doctor Foundation of Anhui University Of Science and Technology (No. 2010yb026).

References

1. Belhumeur, P., Hespanha, J., et al.: Eigenfaces vs. fisherfaces: recognition using class specific linear projection. IEEE Trans. Pattern Anal. Mach. Intell. **19**, 711–720 (1997)
2. He, X.F., Yan, S.C., Hu, Y., et al.: Face recognition using laplacianfaces. IEEE Trans. Pattern Anal. Mach. Intell. **27**, 328–340 (2005)
3. Xu, D., Yan, S., Tao, D., et al.: Marginal fisher analysis and its variants for human gait recognition and content- based image retrieval. IEEE Trans. Image Process. **16**, 2811–2821 (2007)
4. Gao, Q.X., Liu, J.J., Zhang, H.J., et al.: Enhanced fisher discriminant criterion for image recognition. Pattern Recogn. **45**(10), 3717–3724 (2012)
5. Cai, D., He, X.F., Han, J.W.: Orthogonal laplacianfaces for face recognition. IEEE Trans. Image Process. **15**, 3608–3614 (2006)
6. Hu, H.F.: Orthogonal neighborhood preserving discriminant analysis for face recognition. Pattern Recogn. **41**, 2045–2054 (2008)
7. Zhu, L., Zhu, S.N.: Face recognition based on orthogonal discriminant locality preserving projections. Neurocomputing **70**, 1543–1546 (2007)

Research on 3D Imaging Based on Linear Ultrasonic Phased Array

Zhiwei Han, Chao Lu$^{(\boxtimes)}$, and Zhihao Liu

Key Laboratory of NDT, Ministry of Education,
Nanchang Hangkong University, Nanchang, China
luchaoniat@163.com

Abstract. In order to further enhance the detection capability of ultrasonic phased array, in this paper an attempt has been made to make up for the singleness of the spatial information of the defect image and obtain more benefit evaluation and analysis 3D image, 1D linear array probe carried an encoder was applied to obtain a complete A-scan data and form B-scan image through the secondary development module of Phascan platform. The 3D reconstruction was realized through hybrid rendering method based on region growing technology. Most of the measurement error is less than 5 %. The experiment research shows that this method and system can form a more intuitive defect performance in space and can accurately reflect the location, shape and size information of the defect. It establishes foundation for the succeeding experiential researches.

Keywords: Ultrasonic phased array · 1D linear array · Region growing · Hybrid rendering · 3D reconstruction

1 Introduction

In recent years, ultrasonic phased array technology [1] in industrial non-destructive testing field has been used more widely, and its features make it possible to meet a detecting under a variety of complex conditions and complex shape of the workpiece, which includes controllable beam deflection, dynamic focusing, wide detection range, high spatial resolution and a variety of scanning images [2, 3]. Currently the industrial application of ultrasonic phased array is mainly based on 1D (one-dimensional) linear array. Some factors that large number of array elements, more independent transmit and receive channels, complex crafts, expensive, and low echo signal to noise ratio (SNR) [4, 5] limit the application of 2D (two-dimensional) array transducer in the actual testing. Fixed probe also limits the range of its detection. With the improvement of testing standards and requirements, just obtaining a simple 2D images have not been satisfied, while 3D (three-dimensional) images, especially in real-time 3D images will become the new requirements [6]. Currently 3D ultrasonic imaging has been very widely used in medicine. Since the reasons of industrial detection environment, technology, material properties and complex characteristics of defects, the 3D imaging is still in the development stage [7].

In the study, by phased array scanning device with encoder, the data of distance configuration information is collected and imaged, high precision 3D imaging of defect

© Springer International Publishing AG 2017
F. Xhafa et al. (eds.), *Recent Developments in Intelligent Systems and Interactive Applications*,
Advances in Intelligent Systems and Computing 541, DOI 10.1007/978-3-319-49568-2_46

of controllable scanning has finally been realized. These establish foundation for the subsequent detection researches of 3D imaging.

2 2D Slice Obtainment

3D image is generally composed of a series of two-dimensional image slices by a certain stack. The obtainment of slices is the first step of 3D imaging. The research selected a mechanical support with the encoder, to achieve linear scan imaging based on mechanical positioning. Along a direction parallel to the specimen surface, the parts were scanned. At the same time, the synchronal encoder acquired position information of scan parts. The experiment system of acquisition imaging shown in Fig. 1.

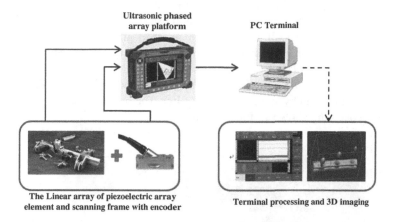

Fig. 1. Gathering image experimental system

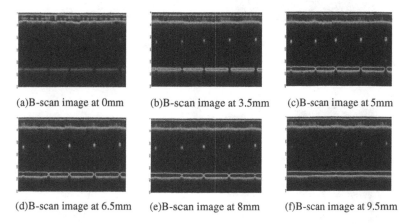

(a)B-scan image at 0mm   (b)B-scan image at 3.5mm   (c)B-scan image at 5mm

(d)B-scan image at 6.5mm   (e)B-scan image at 8mm   (f)B-scan image at 9.5mm

Fig. 2. B-scan images at different position

System communicated and transmitted data through the UDP protocol between Ultrasonic phased array platform and PC terminal. When PC terminal open protocol for transferring data, the scanners was started. According scanning precision setting, ARM transferred the data to PC, which was collected by each of the piezoelectric transducer located in each position, and then data imaging was achieved in the PC terminal. Figure 2 is the B-scan images of the five rows of short-horizontal holes with 10 mm depth and different pore sizes in the scanning position of 0 mm, 3. 5 mm, 5 mm, 6. 5 mm, 8 mm and 9. 5 mm.

3   Digital Image Processing

Affected by the inherent characteristics of the ultrasonic testing, imaging process will form speckle noise [8] and clutter clusters. Since the thermal noise and phase effects noise [9] generated by the amplifier circuit, the 2D slice images have issues such as low contrast and blurred edges. It increases the average gray value of the local area, resulting in a difficulty of image interpretation [10, 11]. In order to improve the image contrast, edge feature and SNR, better the segmentation and reconstruction of defect features, 2D slice images need to be preprocessed. We conducted a preliminary filtering process to collection data by video smoothing filter. The results of unfiltered processing and filtered processing shown in Figs. 3 and 4.

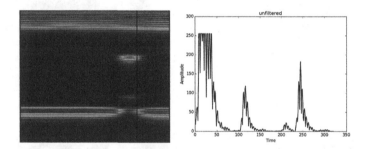

Fig. 3. Atlas and waveforms with unfiltered process

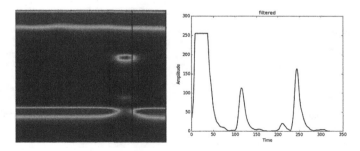

Fig. 4. Atlas and waveforms with filtered process

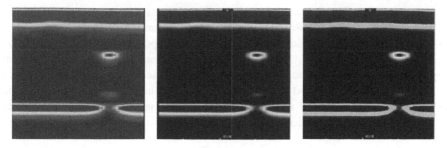

Fig. 5. Image processing and threshold extraction

Due to industrial detection defects mostly are holes and crevice, and acoustic beam reflect by the defect interface without considering the acoustic beam attenuation in the defect, so gray change during which need not be considered. Using thresholding technology [12] and setting a threshold range, clearest form of 2D defect was obtained, and the results shown in Fig. 5.

4   3D Reconstruction

Now common three-dimensional reconstruction techniques are generally divided into two categories. One is that intermediate geometric grid is constructed in 3D space, and then surface rendering is completed by the computer [13]. Another method is that 3D space is divided into different voxels, and then volume rendering of 3D image is obtained. Based on surface rendering and volume rendering technology, combined their respective advantages and needs of industrial detection, we used a new hybrid rendering method based on region growing technology to achieve 3D reconstruction.

The principle of region growing is that pixels of similar properties are composed to form regions. By choosing a seed pixel as a growing point, the similarity (generally average gray value) with pixels of surrounding area in the threshold range is compared. If the results are consistent, then they will be connected as a region. The growth plane extends into the space which can achieve visualized segmentation.

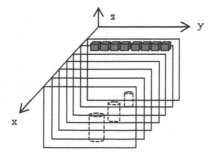

Fig. 6. Gathering image map

The entire scanning area was placed in (x,y,z) Cartesian coordinates. In this, each position value f(x,y,z) was corresponding to voxels of current position. The use of 3D region growing technique, which select a voxel as a seed point $f(x_i,y_j,z_k)$, then look for a calculation point $f(x_{(i\pm1)},y_{(j\pm1)},z_{(k\pm1)})$ of adjacent position in the threshold range and reorganize entity image with the image adjacent pixels of the threshold as shown in Fig. 6.

This reconstruction method not only was able to retain complete information inside the object, but abandoned the unnecessary information, reduced the amount of calculation, and improved calculation speed. So it more suitable for practical applications.

5 Experiment

Experiment relied Phascan Phased Array platform secondary development module. The instrument had 32 independent receiver channels and 128 independent transmit channels, could set freely 256 focal laws of a maximum number. The probe was fixed to the scan frame with encoder. The stepping sampling precision of encoder was set to 0. 1 mm.

The center frequency of the piezoelectric element array, array element length, array element width, array element spacing and array element number were respectively 5 MHz, 10 mm, 0.5 mm, 0.6 mm and 64. Each excitation number of array elements, step and sampling frequency were respectively 17, 1 element and 40 MHz. In the experiment, the test blocks with different diameters vertical hole defects were selected, water and oil were used as couplant.

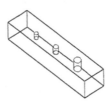

Fig. 7. 3D diagram of test block

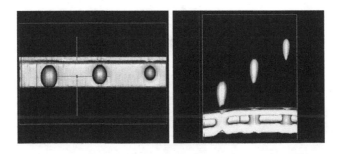

Fig. 8. 3D imaging of vertical hole and its top view

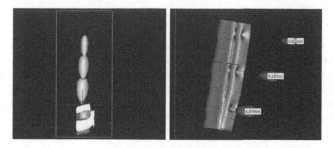

Fig. 9. Front and side view of the 3D imaging of vertical hole

Table 1. Measurement results

Holes/MM	Measurement diameter/MM	Absolute error/MM	Relative error (%)
$\varphi 3$	3.02	0.02	0.7
$\varphi 4$	4.18	0.18	4.3
$\varphi 5$	5.07	0.07	1.4

Test block was plexiglass material, thickness 20 mm. In the test block, three group vertical holes were machined by 30 mm equal intervals, which diameters were $\varphi 3$ mm, $\varphi 4$ mm, $\varphi 5$ mm, and depths were 5 mm, 10 mm, 15 mm. Defect forms of different depths and vertical diameter were simulated, as shown in Fig. 7.

Figures 8 and 9 were 3D imaging images and orthographic views of test block. The measurement results were shown in Table 1.

6 Conclusion

By the precision scanning with encoder, the spatial information of 3D image is accurate. Measurement information of length, trend and diameter can be directly obtained on the 3D image, and the precision can meet the needs of detection. Imaging results can directly reflect information of defect reserved in the material, which contain defect shape, size and direction. It establishes a good foundation for developing of 3D imaging detection experiments based on ultrasonic phased array, that about a variety of materials and defect types. Limited to the material characteristics effect in the industrial ultrasonic nondestructive testing, acoustic waves cannot penetrate the defect to reflect the information of lower part of defect, because of the total reflection on the upper surface of defect. Subsequent research may focus on to scan in multi-dimensional and integrate the scan data to obtain more integrative defect information and form a more intuitive 3D defect image.

Acknowledgements. This work is financially supported by National Natural Science Foundation of China (Nos. 51265044 and 11374134), by Postgraduate Innovation Foundation of Nanchang Hangkong University (No. YC2015043).

References

1. Jing, R.K., Li, J.Z., Zhou, H.L.: Research progress of ultrasonic NDT technology. Foreign Electron. Measur. Technol. **7**, 28–30 (2012)
2. Li, Y.: Industrial application of real-time three dimensional imaging of phased array ultrasound. Nondestr. Test. **5**, 10–13 (2011)
3. Jin, S.J., Yang, X.L., Chen, S.L.: Development and application of ultrasonic phased array inspection technology. J. Electron. Measur. Instrum. **9**, 925–934 (2014)
4. Zhao, Q.: The research of 3D reconstruction algorithm of medical ultrasound images. Shandong University of Science and Technology (2010)
5. Dalichow, M., Dennis, M., Kroening, M., et al.: Advances in 3-D ultrasonic imaging for quantitative flaw. In: Proceedings of the 9th International Conference on NDE in Relation to Structural Integrity for Nucle. 2012-212.8.206.21
6. Yang, P., Guo, J.T., Shi, K.R.: Progress in 3D imaging by 2D phased array. Nondestr. Test. **4**, 177–180 (2007)
7. Shi, K.R., Guo, Y.M.: Phased Array Ultrasonic Imaging and Testing. Higher Education Press, Beijing (2010). pp. 352–353
8. Zhang, Q., Li, B.: Formation principle and model of ultrasonic speckle noise. Electron. Technol. Softw. Eng. **14**, 118–120 (2014)
9. Valentino, D.J., et al.: Volume rendering of 3D medical ultrasound data using direct feature mapping. IEEE Trans. Med. Imaging **3**, 517–525 (1994)
10. Li, X.L.: Numerical three reconstruction of cast defects and effect of cast defects on service performance of casting. Nanchang University (2012)
11. Sun, C.L., Gang, T., Wang, C.X.: Three dimensional imaging based on ultrasonic linear phased array probe. In: 2014 IEEE Far East Forum on Nondestructive Evaluation/Testing, pp. 98–101 (2014)
12. Brandt, T., Paul, M.: Classification Methods for Remotely Sensed Data. CRC Press, Boca Raton (2009). pp. 37–38
13. Chen, T.: Three-dimension visualization of medical volumetric data-set based on VTK. J. Clin. Med. Eng. **5**, 767–769 (2011)

A Survey on Deep Neural Networks for Human Action Recognition based on Skeleton Information

Hongyu Wang[(✉)]

Department of Applied Mathematics,
Northwestern Polytechnical University, Xi'an, China
whyer@mail.nwpu.edu.cn

Abstract. Human action recognition has been a significant topic in the field of computer vision. As deep learning develops, the application of deep neural network in related research is gradually more prevalent. This paper provides a survey of deep neural networks for human action recognition based on skeleton information. The detailed description about each method is explained and several related main datasets are briefly introduced in this paper, all papers are published ranging from 2013 to 2015, which provides an overview of the progress in this area.

Keywords: Action recognition · Deep learning · Skeleton information

1 Introduction

Human action recognition has been drawing more and more attention in computer vision, which is partially due to the development of vision sensors. The information from vision sensors is so abundant and comprehensive that it is qualified to be analyzed for human action recognition [1]. Motion capture data (Mocap) is considered as a main kind of source data. As the low-cost and high-mobility sensors (such as Kinect) appear, more and more researchers have shifted their attention to human action recognition based on the human skeleton information [2] and the related researches become significantly worthy.

Nowadays, the study on Human Action Recognition in general involved four aspects: gestures, actions, interactions and group activities [3]. Gesture refers to a static state which is about a certain movement of body such as standing or bowing. Action generally consists of several sequential gestures such as running, waving. Interaction indicates a correlative action involving two persons or a person and an object such as brawling. Group activity always involves many persons such as meeting. The overview process of handling these tasks consists of feature extraction, action representation learning and classification [4].

In the recent years, many researches has published to find many ways to solve the problems on human action recognition and many methods are proposed on skeleton data [5]. Moreover, deep neural networks are especially and widely employed. There are several frequent and classical deep neural networks such as DBN (Deep Belief

F. Xhafa et al. (eds.), *Recent Developments in Intelligent Systems and Interactive Applications*,
Advances in Intelligent Systems and Computing 541, DOI 10.1007/978-3-319-49568-2_47

Network) [6], RNN (Recurrent neural Network) [7], Denoising Autoencoder [8]. As known, DBN could be viewed as a probabilistic generative model. The component of DBN is RBM (Restricted Boltzmann Machines) which is a stochastic neural network to learn probability distribution from data. Unlike feedforward neural network, RNN works on extract temporal feature in data effectively because the output of each neuron in RNN is not only as input for next neuron but also acts on itself. Denoising Autoencoder is intended to simulate animal's vision and is designed in order to cope with unlabeled data. It performs feature extraction in form of unsupervised learning.

This paper presents the state-of-the-art methods about human action recognition skeleton information in recent years. All involved papers are published from the year 2013 to 2015. Most of them are from CVPR, ECCV, etc. Through the review of these papers, a general framework of the methods is shown in Fig. 1. A detailed discussion on relevant methods is provided in following sections.

The rest of this paper are organized as follows. In Sect. 2, the datasets employed in experiments are reviewed briefly. Sections 3 and 4 describe the methods on single networks and hybrid networks in detail respectivly. The conclusion and future works are presented in Sect. 5.

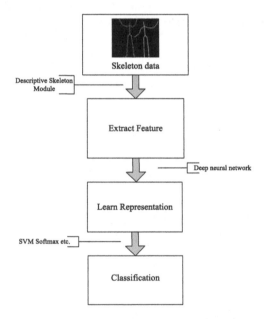

Fig. 1. A general framework of the methods

2 Datasets

In this section, it provides the description of datasets that are regarded as a benchmark in the current research (as shown in Table 1). Some datasets are classic and acknowledged such as MHAD. In addition, there are also some new datasets and several datasets for specific problems.

Table 1. A comparison of datasets

	Action types	Resolution	# of sequences	Format
HDM05	70	N/A	1500	C3D
MSR Action3D	20	320 × 240	567	ASCII
MHAD	11	640 × 480	660	PGM/ASCII
CMU Mocap	109	320 × 240	2605	ASF

HDM05 Database. HDM05 [9] contains more than three hours of recorded and well-documented motion capture data in the C3D as well as in the ASF/AMC data format. Specifically, HDM05 is comprised of more than 70 types of action executed by 10 to 50 actors (in Fig. 2). In this database, most of the sequences have been performed several times by all five actors according to the guidelines fixed in a script. The script is divided into five parts and each part is subdivided into several scenes.

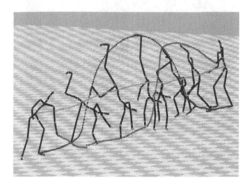

Fig. 2. Example from HDM05 dataset

Berkeley Multimodal Human Action Database (MHAD). This database [10] includes 11 classes of actions performed by 12 youths. And all the persons performed 5 repetitions of each action, yielding about 660 action sequences which correspond to about 82 min of recording time. The set of actions comprises of the following: (1) actions with movement in both upper and lower extremities, e.g., jumping in place, jumping jacks, throwing, etc., (2) actions with high dynamics in upper extremities,

Fig. 3. One class of action in MHAD dataset

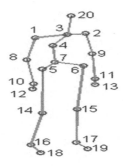

Fig. 4. Skeleton model in MSR Action3D

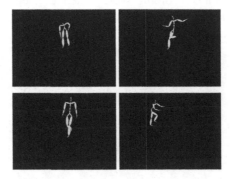

Fig. 5. Four classes of actions in CMU Mocap

e.g., waving hands, clapping hands, etc. and (3) actions with high dynamics in lower extremities, e.g., sit down, stand up. A sample is shown in Fig. 3.

MSR-Action3D Dataset. MSR-Action3D dataset [11] contains twenty actions: high arm wave, horizontal arm wave, hammer, hand catch, forward punch, high throw, draw x, draw tick, draw circle, hand clap, two hand wave, side-boxing, bend, forward kick, side kick, jogging, tennis swing, tennis serve, golf swing, pick up & throw. As shown in Fig. 4. A skeleton has 20 joint positions. The resolution is 320 × 240 in this dataset.

CMU Motion Capture Database. The CMU Motion Capture Dataset [12] is collected by Carnegie Mellon University. It is a free database for all use and contains various actions ranging from locomotion to sports and pantomime. There are 109 types of action and more than 100 subjects in it. The data is stored as ASF format and the resolution is 320 × 240. Some samples are showed in Fig. 5.

3 Methods Based on Single Network

The structure, algorithm, model of all reference articles will be presented in the lines below.

The method that is proposed by D. Wu and L. Shao in this paper is called Deep Dynamic Neural Networks (DDNN) [13]. It is a semi-supervised hierarchical dynamic framework taking both skeleton and depth images as input but this survey only focuses on its skeleton module. This paper only considers 11 upper body joints based on the assumption that recognition tasks are just relevant to upper body. According to 3D coordinates of joints, posture features and motion features are defined. These two features are concentrated and are inputted into a deep belief networks to extract high level features. Taking into consideration that the skeletal features are continuous, the Gaussian RBM is used to model the energy term in the above DBN [14]. Then the emission probabilities of the hidden states are outputted after the high level features are extracted through the DBN. Unlike traditional hidden markov model (HMM), the method productively adds an ergodic states to the HMM in order to perform both action segmentation and recognition. Afterwards the representations can be learned from both skeletal data and depth images. The framework in this paper is a data-driven approach, which brings about more discriminative information. The HMM is extended by introducing an ergodic states so that it can be capable to segment and to recognize action sequences simultaneously.

J. Wang et al. has proposed a novel model called actionlet ensemble [15] for action recognition using depth images and skeletal data. There are two types of skeletal features in this paper. The 3D joint positions are employed to characterize the motion of the body and the local occupancy pattern is to describe the interaction between the human subject and the objects. Inspired by the Spatial Pyramid approach in [16], a descriptive representation called Fourier temporal pyramid is designed which is used to represent the temporal patterns of actions. Based on the above features, the actionlet ensemble approach is proposed to deal with the errors of the skeleton tracking and characterize the intraclass variations. An actionlet is defined as a conjunctive structure of features for a subset of the joints. For increasing the number of the discriminative actionlets, a data mining technique is designed to select them. Afterwards an SVM is trained on each selected action let as an action let classifier and an actionlet ensemble is established by combining these classifiers linearly. With a joint feature map defined on data and labels, an actionlet ensemble could be learned by applying multiple kernel learning approach. This method is insensitive to noise as well as translation and is capable of handle view changes. Moreover, human actions with subtle differences can be discriminated.

For modeling the long-term contextual information of temporal dynamics of human skeleton, Y. Du et al. has established an end-to-end hierarchical recurrent neural network (RNN) [17] whose basic module is a bidirectional recurrent neural network (BRNN) as subnet. Considering human physical structure, the human skeleton is divided into five parts according to arm, trunk and leg, which is used for input. As an innovative improvement, this paper replaces the nonlinear units with LSTMs [18] in order to vanishing gradient and error blowing up problems. As for architecture, the framework in this paper is composed of nine layers. Besides input layer, there are three sets of BRNNs and fusion layer arranged alternately followed by a BRNN with LSTM, a fully-connected layer and a softmax layer. With the skeletal data inputted, the features could be extracted and fused through the hierarchical BRNN architecture. The fully-connected layer and a softmax layer perform classification using the learned representation from the former layers. This is the first paper to apply hierarchical RNN for skeletal based action recognition. The method can capture the spatiotemporal features of action sequences without complex preprocessing.

4 Methods Based on Hybrid Networks

The model proposed by Z. Yu and M. Lee is a hybrid of multiple timescale recurrent neural network (MTRNN) [19] and deep learning neural network (DN) for recognizing walking, running and swinging. MTRNN is adopted for performing dynamic action recognition while DN is capable of static posture recognition. The two major components of MTRNN are slow context layer and fast context layer, which are modeled by a special type of RNN. Utilizing self-organizing map, MTRNN can accept vision signals to predict without supervision. The prediction of MTRNN rely on a prior knowledge of human action so the initial state needs to be modified when the action change. In order to handle the issue, a DN is used to choose the proper initial state, The DN receive the visual information of the current time step from Kinect and the same dimensional data from MTRNN in the adjacent time step. This process has the capability to capture the dynamic features and correct initial states of the current action sequences. This hybrid method gives a compensation for MTRNN. The combination of MTRNN and DN can extract static and dynamic features simultaneously.

K. Cho and X. Chen proposed a novel method [20] to archive human action recognition from skeleton data by introducing deep neural networks. The method in this paper is based on joint distribution model of feature in each frame, which consists of the relative positions of joints (PO), temporal difference (TD) and the normalized trajectory of the motion (NT). For utilizing more information in layers, this paper proposed a hybrid multi-layer perceptron containing an MLP [21] and a deep autoencoder [22]. The MLP and the deep autoencoder are trained with the same set of parameters, which is used to classify and reconstruct respectively. By introducing a hyperparameter λ ranging from 0 to 1, the supervised learning and the unsupervised learning, namely MLP and deep autoencoder are combined. Given a frame, the hybrid MLP can model the posterior probability distribution of classes. With the assumption that the class of each frame only depends on the features of the frame, the classifier is just trained for frame-level classification. This method combines supervised learning and unsupervised learning, which perform the reconstitution of features and classification simultaneously. The deep autoencoder visualizing the features, more distinctive information can be extracted and it makes it possible to study what deep neural networks learned.

In order to understand human intension by analyzing actions effectively, Z. Yu and M. Lee proposed a novel method [23] in 2015 which combine multiple timescale recurrent neural networks (MTRNN) and stacked denoising auto-encoder (SDA). In this paper, supervised MTRNN, an extension of MTRNN has been applied. With the context layers modeled by CTRNN introduced, MTRNN is capable for dynamic action classification. In addition, SDA aims to predict human intention by analyzing the distance between the human's hand and the objects. For catching the scale-invariant features of the object, speeded up robust features (SURF) [1] is employed to find proper matches between image and the object and the output of SURF is used as the input of SDA. When supervised MTRNN and SDA work cooperatively, SDA needs to be trained before supervised MTRNN because the output of the code layer in SDA is fed into the slow context layer in supervised MTRNN. By this way, skeletal data is considered as compensation for dynamic characteristics. The method takes both the action signals and the information of the objects into account. With the unsupervised learning and the supervised learning complementary to each other, it is able to recognize the human intention more accurately.

5 Conclusion

As the application of human action recognition becomes more and more widespread, it has not only been an active area but also a challenging task in computer vision. In recent years, a trend of the research on human action recognition is popular, which address the issue by establishing deep neural networks as deep learning neural networks can learn hierarchical nonlinear function relation in order to model the vision system.

Most of conventional methods construct handcrafted features for recognition relying on domain knowledge, which is time-consuming and inefficient. As one of the most significant type of data for recognizing human actions, skeleton data shows the effectiveness. To tackle with the deficiencies of the handcrafted feature, the methods with deep learning architecture are driven by data, which are capable to extract features from original data automatically.

References

1. Poppe, R.: A survey on vision-based human action recognition. Image Vis. Comput. **28**(6), 976–990 (2010)
2. Wei, Z.Q., Wu, J.A., Wang, X.: Research on applied technology in human action recognition based on skeleton information. Adv. Mater. Res. **859**, 498–502 (2013)
3. Saad, A., Mubarak, S.: Human action recognition in videos using kinematic features and multiple instance learning. IEEE Trans. Pattern Anal. Mach. Intell. **32**(2), 288–303 (2010)
4. Tanaya, G., Rabab Kreidieh, W.: Learning sparse representations for human action recognition. IEEE Trans. Pattern Anal. Mach. Intell. **34**(8), 1576–1588 (2012)
5. Presti, L.L., Cascia, M.L.: 3D skeleton-based human action classification: a survey. Pattern Recogn. **53**, 130–147 (2015)
6. Salama, M.A., Ella Hassanien, A., Fahmy, A.A.: Deep belief network for clustering and classification of a continuous data. In: The IEEE International Symposium on Signal Processing and Information Technology, pp. 473–477. IEEE Computer Society (2010)
7. Gers, F.A., Schraudolph, N.N., Schmidhuber, J., et al.: Learning precise timing with LSTM recurrent networks. J. Mach. Learn. Res. **3**(1), 115–143 (2003)
8. Memisevic, R.: Gradient-based learning of higher-order image features. In: IEEE International Conference on Computer Vision, pp. 1591–1598. IEEE (2011)
9. Müller, M., Röder, T., Clausen, M., et al.: Documentation mocap database HDM05. Computer Graphics Technical Reports (2007)

10. Ofli, F., Chaudhry, R., Kurillo, G., Vidal, R., Bajcsy, R.: Berkeley MHAD: a comprehensive multimodal human action database. In: 2013 IEEE Workshop on Applications of Computer Vision (WACV), Tampa, FL, pp. 53–60 (2013)
11. Li, W., Zhang, Z., Liu, Z.: Action recognition based on a bag of 3D points, pp. 9–14 (2010)
12. Fernando, D.L.T., Hodgins, J., Bargteil, A., et al.: Guide to the Carnegie Mellon University Multimodal Activity (CMUMMAC) Database. Carnegie Mellon University (2009)
13. Wu, D., Shao, L.: Deep dynamic neural networks for gesture segmentation and recognition. In: Agapito, L., Bronstein, Michael M., Rother, C. (eds.) ECCV 2014. LNCS, vol. 8925, pp. 552–571. Springer, Heidelberg (2015). doi:10.1007/978-3-319-16178-5_39
14. Wu, D., Shao, L.: Leveraging hierarchical parametric networks for skeletal joints based action segmentation and recognition. In: IEEE Conference on Computer Vision and Pattern Recognition (CVPR), pp. 724–731. IEEE (2014)
15. Wang, J., Liu, Z., Wu, Y., Yuan, J.: Learning actionlet ensemble for 3D human action recognition. IEEE Trans. Pattern Anal. Mach. Intell. **36**(5), 914–927 (2014)
16. Lazebnik, S., Schmid, C., Ponce, J.: Beyond bags of features: spatial pyramid matching for recognizing natural scene categories. CVPR **2**, 2169–2178 (2006)
17. Du, Y., Wang, W., Wang, L.: Hierarchical recurrent neural network for skeleton based action recognition. In: Computer Vision and Pattern Recognition. IEEE (2015)
18. Baccouche, M., Mamalet, F., Wolf, C., Garcia, C., Baskurt, A.: Sequential deep learning for human action recognition. In: Salah, A.A., Lepri, B. (eds.) HBU 2011. LNCS, vol. 7065, pp. 29–39. Springer, Heidelberg (2011). doi:10.1007/978-3-642-25446-8_4
19. Yu, Z., Lee, M.: Continuous motion recognition using multiple time constant recurrent neural network with a deep network model. In: Yin, H., Tang, K., Gao, Y., Klawonn, F., Lee, M., Weise, T., Li, B., Yao, X. (eds.) IDEAL 2013. LNCS, vol. 8206, pp. 118–125. Springer, Heidelberg (2013). doi:10.1007/978-3-642-41278-3_15
20. Cho, K., Chen, X.: Classifying and visualizing motion capture sequences using deep neural networks. In: International Conference on Computer Vision Theory and Applications, pp. 122–130 (2013)
21. Haykin, S.S.: Neural Networks and Learning Machines. China Machine Press, China (2009)
22. Hinton, G.E., Salakhutdinov, R.R.: Reducing the dimensionality of data with neural networks. Science **313**(5786), 504–507 (2015)
23. Yu, Z., Kim, S., Mallipeddi, R., et al.: Human intention understanding based on object affordance and action classification. In: International Joint Conference on Neural Networks. IEEE (2015)

Picture Reconstruction Methods Based on Multilayer Perceptrons

Yiheng Hu$^{(\boxtimes)}$, Zhihua Hu, and Jin Chen

College of Engineering, Shanghai Polytechnic University, Shanghai, China
yhhu@sspu.edu.cn

Abstract. Multilayer Perceptrons (MLPs) with Back Propagation (BP) training algorithm have been successfully utilized for solving a wide variety of real world engineering problems, such as pattern classification, character recognition, function approximation, clustering and forecasting. In this paper, a MLP classifier is built by using BP algorithm for picture reconstruction. According to the engineering consideration of the effectiveness and efficiency, the optimal parameters of neural networks are carefully selected. Moreover, the contribution of data size for reconstruction accuracy and time consuming are checked. The number of epochs for training is also an important fact for time consuming which has strong relationship with the avoid overtraining algorithm.

Keywords: Multilayer Perceptrons · Backpropagation · Effectiveness · Efficiency

1 Introduction

The multilayer perceptron (MLP) is a feed-forward, supervised learning network which consists of an input layer, one or more hidden layers and an output layer [1]. Over the past years, MLPs has been successfully utilized for solving diverse problem with a popular incremental algorithm known as Back-Propagation (BP) training algorithm [2]. Recently, MLPs with BP algorithm are still widely applied to solve a variety of real world problems, such as pattern classification, character recognition, function approximation, clustering and forecasting [3, 4].

In this paper, a two-coloured picture shown in Fig. 1, is used as an example for classification using MLP. The interior of modified rectangular boundary is pictured in black and assigned to class A and the exterior of the boundary is pictured in white and assigned is class B. Apparently the classes A and B are not linearly separable, so the MLP with BP algorithm can be used for this project. To ensure the network can be well trained, the dataset comprises the parts, the training and testing datasets. In the training process all the weights are adapted in an efficient way. And then the testing dataset determine the accuracy of the testing procedure.

From the whole experimental process, the results including the learning curve of the training process and the tabulation of the computed weights will be obtained for analysis. According to the comparison and analysis of the results, the neural network will be properly tuned to achieve a perfect compromise between effectiveness and efficiency.

© Springer International Publishing AG 2017
F. Xhafa et al. (eds.), *Recent Developments in Intelligent Systems and Interactive Applications*,
Advances in Intelligent Systems and Computing 541, DOI 10.1007/978-3-319-49568-2_48

Fig. 1. Cover of two-color

2 Multilayer Perceptron

A multilayer perceptron (MLP) is a kind of feed-forward network, consisting of a number of neurons which are connected by weighted links. The neurons are organized in several layers, namely an input layer, one or more hidden layers, and an output layer. The input layer receives an external activation vector and passes it via weighted connections to the neurons in the hidden layers. They compute the activations and pass them to the neurons in output layers. In the MLP network both output neurons and hidden neurons are computational neurons. Figure 2 shows the architectural graph of a MLP with two hidden layers. The first hidden layer is fed from the input layer and its outputs are fed into the next hidden layer and separate weights are applied to the sum going into each layer [1].

There are two basic types of signals in the network: function signal and error signal. A function signal comes in at the input layer and flows forward to the output layer. An error signal is produced at the neuron of output layer and propagates backward though the network. Figure 3 illustrates the flow directions of function signal and error signal.

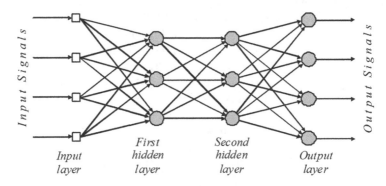

Fig. 2. Neural network architecture

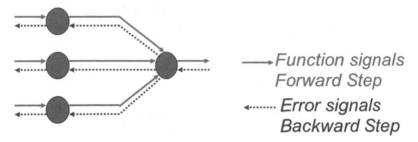

Fig. 3. Forward and backward step

According to Neural Networks and Learning Machines [1, 6], the MLP has three distinctive characteristics. Firstly, the mode of each neuron in the network includes a nonlinear activation. Secondly, the network includes one or more hidden layers, which enables the network to learn complex task. Thirdly, the network exhibits a high degree of connectivity, which is determined by the synapses of the network.

3 Picture Reconstruction

3.1 Sampling Strategy

In order to be more efficiency on learning, the feature of the object is analyzed. In this case study, the pixels around the edges and corners are much more important than other pixels. There are some ways to sample pixels around the features. Basically there are 3 ways for edge feature sampling: uniform sampling, linear distance sampling and exposure distance sampling [5]. Apparently, if the pixels being parallel with edge features is sampled, it could be more efficient, as illustrated in Fig. 4.

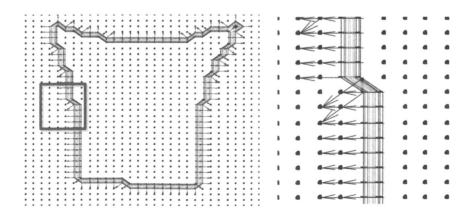

Fig. 4. The sampling should be along the direction of the edge

To implement the edge sampling, we have introduced the image technique called mathematical morphology, which contains two operators: erosion and dilation operation, as shown in Fig. 5.

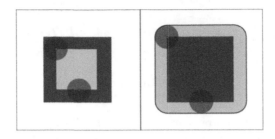

Fig. 5. Erosion and dilation operation

The erosion and dilation method can help us sample the pixels parallel with edge features. For different strategy of sampling method, we just need to choose the different values of the diameters of the disk and apply the morphology operation.

As for corner feature, we just use a 9*9 windows to capture the details of corner. The results of three sampling strategies are shown in Fig. 6.

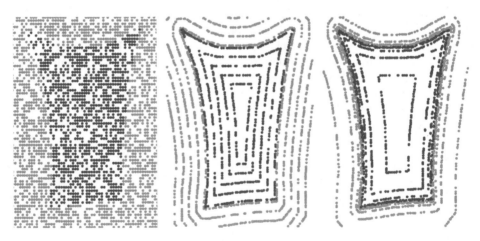

Fig. 6. Different ways of sampling (a) uniform sampling (b) linear distance sampling perpendicular to edge feature (c) exposure distance sampling perpendicular to edge feature

3.2   Experimental Work

To investigate the best way to resolve the conflict between effectiveness of the trained network and the efficiency of the computational procedure, experiments are discussed to compare the performance of the network.

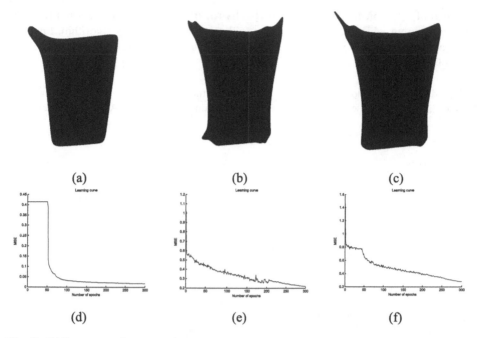

Fig. 7. Different sampling strategies results. (a), (b) and (c) are reconstruction results and (d), (e) and (f) are learning curves.

In this experiment, best sampling strategy for most effectiveness and efficiency is investigated. The neural network includes 3 layers, 1 input layers, one output layers, one hidden layer with 21 neurons. Different strategies are investigated for the original size of picture is 600*700 = 420000. The experiment results are shown in Fig. 7.

In this experiment, according to the learning curve, it can be seen that at around 250 epochs, the BP algorithm has been convergent. The convergent time is nearly the same. Reconstruction accuracy (the error rate between the reconstructed picture and original picture) shows that exposure distance sampling could have a better result.

Another thing attention should be paid that although the MSE reached quite low (about 0.02) in uniform sampling strategy, the reconstruction accuracy is the worst among the three strategies. This can be evidence that the algorithm can only be judged by test data. The MSE can only judge the convergence of the algorithm.

4   Conclusion

In this paper, methods of picture reconstruction are investigated by using neural network based on Multilayer Perceptrons. In order to achieve the highest accuracy rate, optimal parameters of neural networks are carefully selected.

According to the experimental work, there is a trade-off between effectiveness and efficiency. As a matter of fact, the data size contributes most on the reconstruction accuracy and time consuming. Engineer should carefully choose the best data size for the need of the project. The number of epochs for training is also an essential fact for time consuming which has strong relation to the avoid overtraining algorithm. Finally, the reconstruction of the original picture is achieved with the accuracy rate of nearly 99.70 % in about 90 s.

Acknowledgment. I should express my gratitude for funding support of Control Theory and Control Engineering Discipline XXXPY1609 and the funding number is A20NH1609B21-92.

References

1. Haykin, S.: Neural Networks and Learning Machines. McMaster University, Hamilton (1993)
2. Tuli, R.: Character recognition in neural networks using back propagation method. In: IACC IEEE 3rd International, pp. 593–599 (2013)
3. Ghosh, A.: Hybrid optimized back propagation learning algorithm for multi-layer perceptron. Int. J. Comput. Appl. **57**(0975–8887), 1–6 (2012)
4. Joy, C.: Comparing the performance of backpropagation algorithm and genetic algorithms in pattern recognition problems. Int. J. Comput. Inf. Syst. **2**(5), 7–12 (2011)
5. Harris, C., Stephens, M.J.: A combined corner and edge detector. In: Alvey Vision Conference, pp. 147–152 (1988)
6. Prechelt, L.: Early stopping — but when? In: Montavon, G., Orr, G.B., Müller, K.-R. (eds.) Neural Networks: Tricks of the Trade. LNCS, vol. 7700, pp. 53–67. Springer, Heidelberg (2012). doi:10.1007/978-3-642-35289-8_5

Scattered Workpiece Identification Using Binocular Vision System

Shengli Li[1(✉)], Yunfeng Gao[1], and Chao Li[2]

[1] State Key Laboratory of Robotics and System, Harbin Institute of Technology, 92 West Dazhi Street, Nan Gang District, Harbin 150001, China
lishengli623@gmail.com
[2] Wuhu HIT Robot Technology Research Institute Co., Ltd., Electronic Industrial Park, JiuJiang District, Wuhu 241007, China

Abstract. In this paper the authors presents a vision system that has developed to aid bin picking tasks. The intention of the proposed system is to detect and identify randomly piled workpieces in the bin. The strategy described here is based on two steps: detecting workpieces and identifying candidates. The geometric feature was applied to obtain candidates and multi-feature fusion based on SVM was employed to find pickable workpieces. Experimental results demonstrate the good performance of the proposed approach and the system is robust against noise and illumination.

Keywords: Geometric feature · Feature fusion · SVM

1 Introduction

The automated robot-aided manufacturing is developing rapidly in recent years. One of the key task is identifying automatically and locating workpieces, thus robots can pick them up accurately. In general, workpieces are transported in containers or on conveyors. When they are in containers, objects are randomly oriented and clustered in an unstructured environment. Therefore identifying and precisely locating the piled object are essential, and make the automated manufacturing more flexible.

Robotic random bin-picking technologies have been studied for many years. With the commercialization of 3D sensor, such as Kinect, laser range finders, many Bin-picking systems currently utilize 3D data to locate workpieces. The research of Domae [1] shows that grasping position can be generated without the 3D model of object, only the gripper model is applied. Ghita [2] introduced a vision system with a grid projector, which identify piled parts also by using CAD models. Though these sensors directly give the objects 3D pose, there are still some problems, the required 3D model sometimes cannot be provided in real applications and most sensors are mounted on fixed frames, which limits the sensor range.

Stereo vision applied for random bin-picking has also been widely investigated, based on the features, such as texture, shape feature. Rahardja [3] proposed a vision system to identify and locate the stacked parts in the bin. This system identifies the target by landmark features. To identify the pick-able workpiece, Jong-Kyu Oh [4] has presented a geometric feature matching method instead of the stereo matching. But

© Springer International Publishing AG 2017
F. Xhafa et al. (eds.), *Recent Developments in Intelligent Systems and Interactive Applications*, Advances in Intelligent Systems and Computing 541, DOI 10.1007/978-3-319-49568-2_49

before image processing, this system need to register the required features of objects in advance. Hybrid method using multi-features from acquired images to recognize and locate objects has been proposed in [5]. Nevertheless there also exist limits: a complex scene will prevent getting the true pose estimation and the candidate object sometimes may be false consequence, which causes picking failed.

To solve the aforementioned problems, we present a robust vision system for random bin-picking tasks in this work. The binocular structure is mounted on the robot end effector, makes a large sensor range for it. An extended Geometric Feature extraction algorithm is presented to detect the workpiece, which can also be applied on different objects. We then employ a voting strategy to locate initial estimated objects. To confirm the final pickable target, we extract different features to represent the object, including HOG, color histogram, LBP etc. Then we use SVM classifier to identify the target.

The rest of this paper is organized as follows: Sect. 2 introduces the overview of the proposed vision system. Section 3 gives a detail description of Geometric Feature extraction and Sect. 4 presents hybrid feature representation and classifier training algorithm. Comparative experimental results are drawn in Sect. 5.

2   Overview of Vision System

The whole architecture of the vision system is described in this section. The presented system consists of two cameras, a ring-shaped LED light and other image processing modules (Fig. 1).

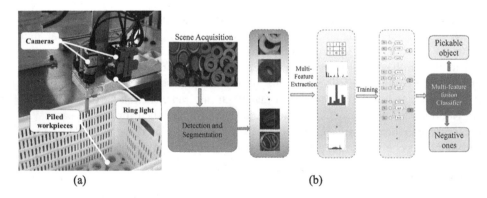

Fig. 1. Overview of our vision system for random bin-picking.

The first module is to detect geometric feature. As workpieces are piled together, a modified feature detection is utilized to obtain their edges. Then edges are employed to detected candidate workpieces. Then future extraction is applied to get multi features and then future fusion is utilized to achieve higher accuracy. For the next step, the SVM classifier is trained to classify the positive object from negative ones and the training process can be conducted offline to improve efficiency. The final step is to employ the trained classifier to classify detected workpieces.

3 Geometric Feature Extraction

In our vision system, geometric features extracted from the input image are used to detect objects. To extract features, we adopt an edge detection algorithm. Conventional edge detection algorithm, such as Canny and Laplacian, sometimes fail to obtain complete edges and usually generate negative edges. Thus we use a state-of-the-art edge detector [6] instead, which is efficient and robust. This algorithm can detect a series of edgelets, including line segments, short elliptical arcs. And edges of objects are composed of these edgelets. The approach first deals with the higher gradient magnitude to find edgelets, then pixels in these edgelets with similar gradient directions grow to one candidate region. NFA (Number of False Alarms) is utilized to eliminate detected negative regions.

In order to reduce the negative results due to the noise, we employ the scaled image instead. Since Gaussian filter has suppression to low gradient magnitudes, we use the median blur instead. By employing the modified algorithm, we can obtain a serials of edgelets {E1 E2 ... En}, where En represents a continuous candidate edge. Edgelets mean the object edges, shown in Fig. 2 in blue. Then we employ voting strategy to get coarse detected objects. Since objects are stacked randomly, we can see some negative results, which reducing the grasping efficiency. To solve this problem, we adopt an identification strategy to distinguish pickable objects from the piled ones.

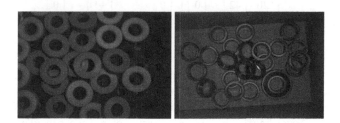

Fig. 2. Edgelets detection results. (Color figure online)

4 Multi-feature Fusion and Classifier Training

Various features have been proposed and utilized recently to identify objects, such as surf, gradient histogram, corner features, and ORB etc. Descriptive ability of single feature is limited and cannot correctly identify objects. Moreover, sparse features like ORB and surf are not able to be detected in some cases. Therefore, we use composite feature to improve stability.

HOG (Histogram of Oriented Gradient) and LBP (Local binary patterns) are applied to acquire higher identification accuracy. HOG is a descriptor of appearance and shape, can identify different shapes among objects. And LBP can perfectly descript the texture feature of object.

HOG descriptor [7] is obtained by computing and combining the local gradient orientation histogram. The local appearance and shape of parts can be described by the gradient magnitude and orientation. Specifically the gradient value can be calculated with two filter kernel, $[-1, 0, 1]$ and $[-1, 0, 1]^T$.

For each cell, a normalized orientation-based histogram is compiled, to effectively reduce the illumination effect. Several cells compose a descriptor block. And The HOG descriptor is then the concatenated vector of all these histograms.

LBP [8] is used to fine describe the local textural feature of objects, and robust to illumination changes. Its value indicates the relationship between one pixel and its neighbors, and can be calculated as

$$LBP(x_c, y_c) = \sum_{p=0}^{p-1} 2^p s(i_p - i_c) \qquad (1)$$

Since the LBP value is computed according to the value of central pixel and its neighbors, it tends to be sensitive to rotation and noise. Thus we modify the descriptor in circle regions. And to avoid the influence of noise, we set a threshold to get the binary code. For each pixel shown in Fig. 3, we adopt two LBP codes.

60	53	64		1	1	1			0	0	0
36	39^{+4}_{-4}	48	$\Rightarrow$	0		1	*and*		0		0
28	20	42		0	0	0			-1	-1	0

Fig. 3. Illustration of the modified descriptor.

SVM has superiority in the small sample size problem and generalization capabilities. We use SVM as the classifier in this paper. This algorithm can construct an optimal hyper plane in high-dimensional space between two categories using the given training samples. According to the obtained hyper plane, we can identify candidate objects to several classes (Fig. 4).

Since we have got several features, a multi-feature fusion algorithm is presented. The classifier with fusion features is designed as follows: we first construct the SVM classifier on each feature separately, then automatically learn the weight coefficient of each feature according to the given training samples, such as H_weight and L_weight. Finally an adaptive classifier is developed according to the weight coefficient. The output of the classifier is defined as $\max\{p_1, p_2, \ldots, p_n\}$. Where p_n is the probability that belongs to each class, and can be calculated as

$$p_n = H\_weight * H\_p_n + L\_weight * L\_p_n + O\_weight * O\_p_n + \cdots \qquad (2)$$

Where H_pn and L_pn are probability.

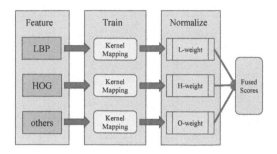

Fig. 4. Overview of the multi-feature fusion strategy.

5   Identification Result

In order to evaluate the proposed system in this paper, we have implemented some experiments. The vision system is composed of two Basler CMOS cameras mounted on an EFORT ER-20C industrial robot, a ring-shaped LED light and a standard PC (3.4 GHz CoreI7 with 4G RAM). All the image processing and calibration code were implemented in C++.

The extraction of geometric features is used to extract the candidate object in images. Since extracted objects are mixed with positive and negative ones. We need to distinguish them in training samples. Figure 5 shows some candidates.

Fig. 5. Some candidates in training data.

Since candidates are detected, multiple features are then extracted to train SVM classifier. To confirm the effectiveness of multi-feature fusion we proposed, individual feature is adopted in experiments. In this paper, each type of workpiece has about 500 training samples. We made 15 trails for each classifier training. 683 of 1073 samples were used to train the classifier, others used for testing. Figure 6 shows the experiment results of different features.

We can observe that, the accuracy of using LBP or HOG is low, and appropriate feature fusion does improve the accuracy compare to the single feature. But overmuch features will reduce the stability. Therefore, the selected features in our method is adequate to identify the object and the identification accuracy rate is about 98.2 % (Table 1).

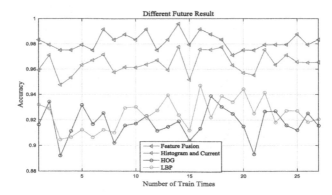

Fig. 6. Accuracy rate of different features.

Table 1. The average accuracy of each trail

Different futures	HOG	LBP	HOG + LBP	Histogram and fused feature
Accuracy	91.75 %	92.45 %	98.19 %	96.46 %

6   Conclusion and Future Work

This paper presents a robust, accurate vision system for random bin-picking, which can identify pickable objects in a bin. We have introduced an effective feature extraction strategy based on image gradient, which can obtain candidate targets. We subsequently proposed an identification method, enables us to get positive objects based on multi-features fusion and support vector machine. The experimental results verify the stability and accuracy of the vision system, with an average of success rates of 98.2 % on different workpieces.

In future, we will improve the generality of this system, thus workpieces with complex shapes can be easily identified.

References

1. Domae, Y., Okuda, H., Taguchi, Y., Sumi, K., Hirai, T.: Fast graspability evaluation on single depth maps for bin picking with general grippers. In: Proceedings of the 2014 IEEE International Conference on Robotics and Automation, pp. 1197–2004 (2014)
2. Ghita, O., Whelan, P.F.: A bin picking system based on depth from defocus. Mach. Vis. Appl. **13**(4), 234–244 (2003)
3. Rahardja, K., Kosaka, A.: Vision-based bin-picking: recognition and localization of multiple complex objects using simple visual cues. In: IEEE/RSJ International Conference on Intelligent Robots and Systems, pp. 1448–1457 (1996)
4. Oh, J.K., Lee, S.H., Lee, C.H.: Stereo vision based automation for a bin-picking solution. Int. J. Control Autom. Syst. **10**(2), 362–373 (2012)

5. Lin, H I., Chen. Y.Y., Chen, Y.Y.: Robot vision to recognize both object and rotation for robot pick-and-place operation. In: 2015 International Conference on Advanced Robotics and Intelligent Systems (ARIS), pp. 1–6. IEEE (2015)
6. Grompone von Gioi, R., Jakubowicz, J., Morel, J.-M., Randall, G.: LSD: a fast line segment detector with a false detection control. IEEE Trans. Pattern Anal. Mach. Intell. **32**(4), 722–732 (2010)
7. Dalal, N., Triggs, B.: Histograms of oriented gradients for human detection. In: Proceedings of the IEEE Conference on Computer Vision and Pattern Recognition (2005)
8. Ojala, T., Pietikäinen, M., Mäenpää, T.: Multiresolution gray-scale and rotation invariant texture classification with local binary patterns. IEEE Trans. Pattern Anal. Mach. Intell. **24**, 971–987 (2002)

Investigation on the Sugar Content Distribution of Grape Using Refraction and Vision Processing Model

Jun Luo[1(✉)], Qiaohua Wang[2], Yihua Tang[2], Ruifang Zhai[1],
Hui Peng[1], Shanmei Liu[1], Liang Wu[3], and Yuhua Zong[3]

[1] College of Informatics, Huazhong Agricultural University,
Wuhan 430070, China
luojun81@163.com
[2] College of Engineering, Huazhong Agricultural University,
Wuhan 430070, China
[3] North Xinjiang Fruit & Vegetable Industry Development Company Limited,
Bole 833400, China

Abstract. According to the refractive distinctiveness of the soluble solids, the refraction index of light will increase in a certain proportion in accordance with the increase of the concentration of the soluble solid, which provides a feasible method for measuring the variation of the sugar content. The critical angle had been applied to express the concentration variation of soluble solids in normal refraction system; thus the final result of soluble solids in every grape will be defined by Brix degree, which represents the sugar content of grape. In the experiments, the standard reference wavelength of refraction is 589.3 nm, the circumstance temperature will be set at 20°C, and then the Brix degree of many red grapes will be investigated in the above refraction system. This work focus on the upper, the middle and the bottom positions of each bunch red grape, many randomly selected grapes were used to test by refraction-based optical method. Through the experiments, the sugar content results show that the sugar content of the upper part of the grape fruits is relative higher than that of the bottom fruits.

Keywords: Grape · Sugar content · Brix degree · Refraction method

1 Introduction

For decades, grape production has been considered as a green and sustainable development industry [1–3], relying on the feasible grape cultivation techniques. Usually, as the arrival of the grape harvest season, the daily production will increase so fast that the grape enterprises have to face two major problems [4–6]. One is grape quality non-destructive testing problem in postharvest stage; the other is grape preservation problem in warehousing and logistics parts. It should be noted that the grape preservation techniques have been successfully applied in wide range of every storage parts but grape quality non-destructive testing still in difficulties. So far we did not find fast and suitable methods to measure the quality of whole string fresh grapes. Usually

© Springer International Publishing AG 2017
F. Xhafa et al. (eds.), *Recent Developments in Intelligent Systems and Interactive Applications*,
Advances in Intelligent Systems and Computing 541, DOI 10.1007/978-3-319-49568-2_50

soluble solids were used to measure the taste quality for analyzing the entire bunch of grapes, and the index of soluble solids can be determined by standard Brix degree. According to the refractive characteristics of soluble solids, the refraction index of light will increase in a certain proportion in accordance with the increase of the concentration of the soluble solid, which provide a feasible method for measuring the variation of the sugar content. According to the model of the refraction system, the critical angle can determine the concentration variation of soluble solids in the refraction system; thus the final result of soluble solids in every grape will be defined by Brix degree, which represents the sugar content of grape. In the experiments, the standard refraction wavelength is 589.3 nm, the circumstance temperature will be set at 20°C. This work focus on the upper, middle and bottom positions of each bunch red grape, many randomly selected grape fruits was used to test by refraction-based optical method. Through the experiments, the sugar content results show that the sugar content of the upper part of the grape fruits is relatively higher than that of the bottom fruits.

2 Refraction-Based Method

2.1 The Principle of Refraction

The refraction phenomenon will occur when the light come from one medium into another object. As shown in Fig. 1, in the refraction-based optical system, the angle of incidence and the angle of refraction can be defined by θ_i and θ_r respectively.

Usually, the refraction ratio n can be defined by Eq. (1).

$$n = \frac{\sin \theta_i}{\sin \theta_r} \tag{1}$$

Let us focus on the incidence process in the optical refraction model, the angle of refraction will be greater than the angle of incidence, when the light incident from the denser medium into the sparse medium. It can be expected that, the angle of refraction will be equal to 90° when the incident angle is increased to a certain angle, and then the refracted light will not be occurred, although the angle of incidence is increased, this phenomenon can be called total reflection, also known as total internal reflection. The

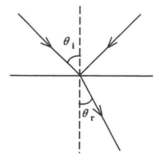

Fig. 1. The optical principle of refraction

critical angle is such an incident angle which satisfies the angle of refraction equals to 90°, and can be used to calculate the characteristics of the particular medium.

2.2 The Principle of Brix Degree

Usually, the Brix degree is the sucrose content of 100 g sucrose solution, which has been defined by international committee of uniform method sugar analysis. It should be noted that the Brix degree of different grape type will be different with each other, and the Brix degree of particular grape in different growth stage will be different with each other as well. It should be noted that the refraction n is corresponding to concentration of the soluble solids in grape juice, and the Brix degree has an obvious relationship with the concentration of the soluble solids, thus the Brix degree will be high correlation with the refraction n defined by Eq. (1).

Table 1. The relationship between the Brix degree and the concentration of the sucrose solution

Brix degree	Concentration of the sucrose solution
10	1.038143
11	1.042288
12	1.046462
13	1.050665
14	1.054990
15	1.059165
16	1.063460
17	1.068779
18	1.072147
19	1.076537
20	1.080959
21	1.085414
22	1.089900
23	1.094420

Generally, the Brix degrees of grapes are located in the range of 10 to 23, thus the relationship between the Brix degree and the concentration of the sucrose solution can be seen in Table 1. The sugar content increases along with the concentration, therefore, the sugar content is positive related to concentration of grape juice.

3 Vision Model of Grape

According to the basic growth characteristics of the entire bunch grapes, as can be seen in Fig. 2, it is easy for us to see that the length of the whole string grape from the upper to the bottom which is very long, thus the different positions of every grape, such as the upper part, the middle upper part, the middle part, the middle and lower part and the

The upper

The middle

The bottom

Fig. 2. The relative position of the entire bunch of grape

bottom part, may cause differentiated sugar content. In accordance with the basic physiology growth principle of grape tree, the upper part of the whole string grape is near to the root can absorb more natural nutrients from the root easily, besides the bottom part is lack of sunlight will lower the fruit sugar content slightly.

Therefore the position relationships of the entire bunch grape have highly associated with the sugar content distribution of grape. In this work, red grape samples will be used to investigate the sugar content distribution.

4   Experiments and Results

Usually soluble solids was used to measure the taste quality for analyzing the entire bunch of grapes, and the index of soluble solids can be determined by Brix degree. According to the refractive characteristics of soluble solids, the refraction index of light will increase in a certain proportion in accordance with the increase of the concentration of the soluble solid, which provide a feasible method for measuring the variation of the sugar content. Therefore the critical angle can be applied to express the concentration variation of soluble solids in the refraction system; thus the final result of soluble solids in every grape will be defined by Brix degree, which represents the sugar content of grape. In the experiments, the standard refraction wavelength is 589.3 nm, the circumstance temperature will be set at 20°C, and then the Brix degree of red grapes will be investigated using refraction-based method.

Let us focus on the sugar content distribution of the first bunch of red grape in Fig. 3, the sugar content results show that the randomly selected four grape fruits in the upper part were 19, 19.2, 19.2 and 18.2, respectively, the sugar content of four grape fruits in the middle upper part were 18, 18.6, 18 and 17.5, respectively, the sugar content of four grape fruits in the middle part were 18.2, 18,17.7 and 17.5, respectively, the sugar content of four grape fruits in the middle and lower part were 17.5, 17.7, 17.9 and 17.2, respectively, the sugar content of four grape fruits in the bottom part were 17, 16.8, 17.2 and 16.8, respectively. Besides, the average values of the upper part, the middle upper part, the middle part, the middle and lower part and the bottom part were 18.9, 18.025, 17.85, 17.575 and 16.95. It is obvious that the average sugar content of

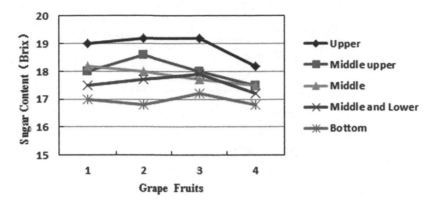

Fig. 3. The sugar content distribution of the first bunch of red grape

the upper part of the entire bunch grape is relatively ∼ 1.95 higher than that of the bottom grape fruits.

From Fig. 4, the sugar content results show that the randomly selected four grape fruits in the upper part were 16, 14.9, 15 and 14.7, respectively, the sugar content of four grape fruits in the middle upper part were 14.8, 15.1, 15 and 14.2, respectively, the sugar content of four grape fruits in the middle part were 14.3, 13.8, 14.6 and 14.2, respectively, the sugar content of four grape fruits in the middle and lower part were 13.2, 13.6, 13.8 and 13.6, respectively, the sugar content of four grape fruits in the bottom part were 13, 13.2, 12.8 and 13, respectively. Besides, the average values of the upper part, the middle upper part, the middle part, the middle and lower part, and the bottom part were 14.9, 14.775, 14.225, 13.55 and 13. It is obvious that the average sugar content of the upper part of the entire bunch grape is relatively ∼ 1.9 higher than that of the bottom grape fruits.

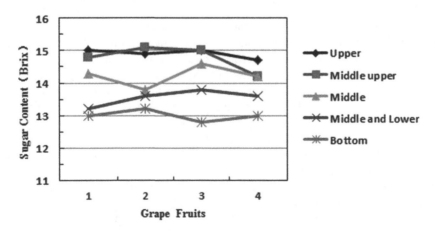

Fig. 4. The sugar content distribution of the second bunch of red grape

5   Conclusions

The length of the entire bunch of grapes is 30 cm or so, according to the basic growth principles of grape trees, the bottom parts of grape fruits are farther than the upper, which will absorb relatively less nutrients from the roots naturally, thus the distance of entire bunch of grape is so long that it cannot be ignored in fact. Therefore the result shows that the sugar content distribution of the whole string of grapes, which is related to the positions of the entire bunch grape. The grape sugar content testing is very important for production quality control of the grape industrial chain, which is an inevitable demand of industrial development. It is expected that the grape sugar content analysis technology will greatly promote the standardization and cultivation of grape, thereby improve the competitiveness of grapes in the world.

Acknowledgements. This research was supported by the Fundamental Research Funds for the Central Universities (grant nos. 2662015QC028 and 2662015PY066), and the National Natural Science Foundation of China (grant nos. 61176052 and 61432007).

References

1. Zalik, K.R.: An efficient E-means clustering algorithm. Pattern Recogn. Lett. **29**(9), 1385–1391 (2008)
2. Leeuwen, C.V., Friant, P., Chone, X., Tregoat, X., Koundouras, S., Dubourdieu, D.: The influence of climate, soil and cultivar on terroir. Am. J. Ennlogy Viticulture **55**(3), 207–217 (2004)
3. Castellarin, S.D., Gaspero, G.D., Marconi, R., Nonis, A., Peterlunger, E., Paillard, S., Adam-Blondon, A.-F., Testolin, R.: Coulor variation in red grapevines: genomic organization, expression of flavonoid 3'-hydroxylase, flavonoid 3', 5'-hydroxylase genes and related metabolite profiling of red cyaniding-/blue delphinidin-based anthocyanins in berry shin. BMC Genom. **7**, 12 (2006)
4. Jeong, S.T., Goto-Yamamoto, N., Hashizume, K., Esaka, M.: Expression of the flavonoid 3'-hydroxylase and flavonoid 3', 5'-hydroxylase genes and flavonoid composition in grape. Plant Sci. **170**(1), 61–69 (2006)
5. Ageorges, A., Fernandez, L., Vialet, S., Merdinoglu, D., Terrier, N.: Four specific isogenes of the anthocyanin metabolic pathway are systematically co-expressed with the red colour of grape berries. Plant Sci. **170**(2), 372–383 (2006)
6. Failla, O., Mariani, L., Brancadoro, L., Minelli, R., Scienza, A.: Spatial distribution of solar radiation and its effects on vine phenology and grape ripening in an alpine environment. Am. J. Enology Viticulture **55**(2), 128–138 (2004)

An Effective Motion Deblurring Method with Sound Extendability

Wei Ji, Jin Fang, and Yong Zhao[(✉)]

School of Electronic and Computer Engineering,
Peking University, Beijing, China
{jiwei991,fangjin}@pku.edu.cn, yongzhao@pkusz.edu.cn

Abstract. Motion Deblurring is a foundational and challenging problem in the community of image and signal processing, has been hotly discussed in recent years. This paper proposes a novel approach to restore motion blurred images using a piecewise function that concatenates a hyper-Laplacian and a Gaussian to approximate the image logarithmic gradient. Besides visual comparisons, quantitative comparisons conducted on a variety of images validate that our proposed method could obtain clearer and finer performance, while preserve the image details well. In addition to natural images, the method possesses sound extendability to other deblurring tasks, such as artificial images and document images.

Keywords: Motion deblurring · Hyper-Laplacian · Gaussian · Sound extendability

1 Introduction and Related Work

Motion blur is the result of a relative movement between camera's perspective and the real scene (e.g., due to camera vibration) during image exposure period. Many photographs are corrupted by motion blur that can significantly degrade image quality, especially under dim light conditions where more time of exposure is needed. A degraded blurry picture with unavoidable information drop-out not only bring a poor visual impression but also lead to undesirable consequences in subsequent procedures.

Since the path of the relative motion can be arbitrary, recovering a latent clear image from input blurry image becomes a much challenging task, particularly single image deblurring. The blur process is generally modeled as a linear convolution of a latent image l with the blur kernel k:

$$b = l \otimes k + n \tag{1}$$

Here l stands for desired sharp image; b is on behalf of a blurry image, that is, degraded image; k represents blur kernel, also known as point spread function (or PSF); n is additive noise usually assumed as following Gaussian distribution; $\otimes$ denotes convolution operator.

If the blur kernel is given as a prior, the task is to estimate the unblurred latent image. This problem is reduced to non-blind deconvolution. In many cases, however,

© Springer International Publishing AG 2017
F. Xhafa et al. (eds.), *Recent Developments in Intelligent Systems and Interactive Applications*,
Advances in Intelligent Systems and Computing 541, DOI 10.1007/978-3-319-49568-2_51

the blur kernel is also not known and estimating blur kernel is essential. The deblurring problem called blind deconvolution is even more ill- posed. One can regard non-blind deconvolution as one inevitable step in blind deconvolution during the course of PSF estimation or after PSF has been computed.

In the past decade, single image deblurring has drawn wide attention. Joshi et al. [1] gives the observation that local color statistics derived from the image can be used for deblurring. Cho et al. [2] presents a technique of fast deblurring, which utilizes sharp step edges as constraint through shock filter. Xiang et al. [3] introduces a supervised learning algorithm, which is developed on a conceptual frame of matrix regression and gradient evolution. Li et al. [4] estimates the sparse coefficient, sharp image and blur kernel alternately based on the sparse prior of dictionary pair.

This paper presents an outstanding method to motion deblurring of a single image, based on a new piecewise function that concatenates a hyper-Laplacian and a Gaussian. This function is chosen in virtue of its excellent approximation to empirical distribution. Compared with state-of-the-art approaches, the outcome of experiments demonstrate that the proposed approach could obtain better deblurring performance, while preserve the image details well. In this paper, we also show its ability to extend to other deblurring tasks, such as artificial images and document images.

2 Proposed Method

Having the forward motion blur model, the latent image l could be recovered from b through solving an inverse problem. Our objective function is given by

$$\min_i \|l \otimes k - b\|^2 + \lambda \psi(l) \tag{2}$$

which is composed of a data term E_{data} (corresponding to likelihood in probability) and a regularization term E_{data}. E_{data} measures the difference between convolved image and blur observation, and is written as

$$\|l \otimes k - b\|^2 \tag{3}$$

E_{prior} is denoted as the function $\psi(l)$, where λ is regularization weight.

2.1 Statistical Priors on Images

Now that the image deblurring is ill-posed, prior knowledge and extra information are usually needed for constraining solutions. A common method is to take advantage of the statistical priors on natural images. Recent research about statistics on natural images has demonstrated that images in real scenes, among which there is wide variation of their absolute color distributions, meet heavy tailed distributions [5] in their gradient values. This is consistent with the intuition that images usually have large-scale regions of stable intensity gradient occasionally interrupted by considerable variation on the edges and occlusion borders. For instance, Fig. 1 gives a natural image

Fig. 1. A natural image and heavy tailed distributions on its gradients

and its gradient-histogram. This gradient distribution indicates that the image mainly consists of small and zero gradients, but a minority of gradients with great magnitudes.

In [6], a hyper-Laplacian function is utilized to approximate gradient-histogram, and the fitting result shows that a hyper-Laplacian with exponent α = 2/3 is a better model of image gradients than a Laplacian or a Gaussian. In this paper, we collect 1000 natural images and their average logarithmic image gradient distribution histogram (blue) is plotted in Fig. 2.

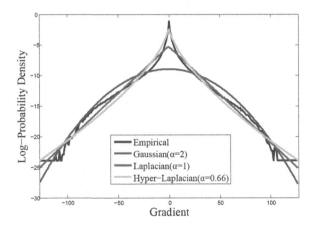

Fig. 2. The curve of empirical gradient-histogram (blue). Laplacian fit (green), hyper-Laplacian fit (purple) and Gaussian fit (red).

As shown in Fig. 2, we find that a natural prior $\psi(l)$ for latent image is actually a piecewise function that concatenates a hyper-Laplacian and a Gaussian, plotted as the concatenating curve. The expression is

$$E_{prior} = \psi(l) = \begin{cases} a\|\nabla l\|^{\alpha}, & |\nabla l| \leq \kappa \\ b\|\nabla l\|^2 + c, & |\nabla l| > \kappa \end{cases} \qquad (4)$$

where ∇l represents a partial derivative for l. κ is the value on which the Gaussian and hyper-Laplacian functions are concatenated. a, b, and c are three parameters. From Fig. 2, we notice that hyper-Laplacian fits empirical distribution closely, when $|\nabla l| \leq 60$. However, when $|\nabla l| > 60$, Gaussian fits empirical distribution more closely than hyper-Laplacian or Laplacian. In this paper, we called the piecewise function as Gaussian-hyper Laplacian function.

Since the density of natural image gradients can be well modeled by the Gaussian-hyper Laplacian function, we collect 1000 artificial images and 1000 document images, then we plot their average logarithmic image gradient distribution histogram separately to reveal the extendability of our method.

From Fig. 3, we can see that the density of artificial image and document image gradients obey the piecewise distribution too. They can be well modeled by the Gaussian-hyper Laplacian function through adjusting the value of κ.

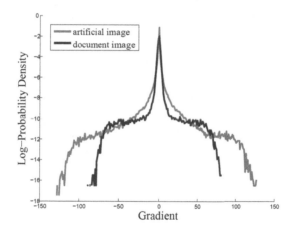

Fig. 3. Gradient distributions on artificial image and document image

2.2   Optimization

When the blur kernel is unknown, we need to enforce the blur kernel constraint. A quadratic form is utilized to constrain the blur kernel term. Then the objective function can be expressed as

$$\min_{l,k} \left(\|l \otimes k - b\|^2 + \lambda \psi(l) + \in \|k\|_2^2 \right) \qquad (5)$$

We adopt an iterative optimization method to solve the problem. First of all, blur kernel is initialized from an input image. Then one can regard the problem as non-blind deconvolution. We utilize two masks to help approximate the gradient distribution with the two piece-wise functions separately. The problem is transformed to

$$\min_l \|l \otimes k - b\|^2 + \lambda_1 \|\nabla l\|^\alpha \circ m_1 + \lambda_2 \|\nabla l\|^2 \circ m_2 \tag{6}$$

Where m_1, m_2 are two masks; λ_1, λ_2 are two parameters; $\circ$ represents the element-wise multiplication operator. An effective scheme to solve non-blind deconvolution is variable splitting. In this work, we use a set of auxiliary variables $\psi = (\psi_x, \psi_y)$ for $\nabla l = (\partial_x l, \partial_y l)$, adding extra condition $\psi \approx \nabla l$. Equation (6) is accordingly updated to

$$\min_{l,\psi} \left(\|l \otimes k - b\|^2 + \lambda_1 \|\psi_x\|^\alpha \circ m_1 + \lambda_1 \|\psi_y\|^\alpha \circ m_1 + \gamma \|\psi_x - \partial_x l\|^2 + \right.$$
$$\left. \gamma \|\psi_y - \partial_y l\|^2 + \lambda_2 \|\psi_x\|^2 \circ m_2 + \lambda_2 \|\psi_y\|^2 \circ m_2 \right) \tag{7}$$

where γ is a weight. When its value is infinitely large, $\psi_x = \partial_x l$, $\psi_y = \partial_y l$,. Now it's possible to iterate between optimizing ψ and l.

Updating ψ: With an estimated l from the previous pass, Eq. (7) is simplified to

$$\min_\psi \left(\lambda_1 \|\psi_x\|^\alpha \circ m_1 + \lambda_1 \|\psi_x\|^\alpha \circ m_1 + \gamma \|\psi_x - \partial_x l\|^2 + \gamma \|\psi_y - \partial_y l\|^2 + \right.$$
$$\left. \lambda_2 \|\psi_x\|^2 \circ m_2 + \lambda_2 \|\psi_y\|^2 \circ m_2 \right) \tag{8}$$

By a few algebraic operations to decompose ψ to all elements $\psi_{i,v}$. $\psi_{i,v}$ is expressed as

$$\psi_{i,v} = \arg\min_{\psi_{i,v}} \left(\lambda_1 |\psi_{i,v}|^2 \circ m_1 + \gamma (\psi_{i,v} - \partial_v l_i)^2 + \lambda_2 |\psi_{i,v}|^2 \circ m_2 \right) \tag{9}$$

To solve the Eq. (9), a lookup table can be constructed offline, from which optimal results can be obtained efficiently.

Updating l: With ψ fixed above, l can be updated. Equation (7) is simplified to

$$\min_l \left(\|l \otimes k - b\|^2 + \gamma \|\psi_x - \partial_x l\|^2 + \gamma \|\psi_y - \partial_y l\|^2 \right) \tag{10}$$

Since the major computation is on convolution, frequency domain operation using Fourier transforms is applied. Equation (10) is updated to

$$\min_{F_i} \left(\|F(l)F(k) - F(b)\|^2 + \gamma \|F(\psi_x) - F(\partial_x)F(l)\|_2^2 + \gamma \|F(\psi_y) - F(\partial_y)F(l)\|_2^2 \right) \tag{11}$$

Accordingly, we compute the optimal l by

$$l = F^{-1}\left(\frac{\overline{F(k)}F(b) + \gamma\overline{F(\partial_x)}F(\psi_x) + \gamma\overline{F(\partial_y)}F(\psi_y)}{\overline{F(k)}F(k) + \gamma\overline{F(\partial_x)}F(\partial_x) + \gamma\overline{F(\partial_y)}F(\partial_y)}\right) \qquad (12)$$

Using above two steps we alternatively update ψ and l until convergence. Afterwards, we adopt an iterative optimization method to solve the blur kernel k. Equation (5) is simplified to

$$\min_{l,k}\left(\|l \otimes k - b\|^2 + \in \|k\|_2^2\right) \qquad (13)$$

It can be solved by FFT (Fast Fourier Transform).

3   Experiments and Results

We apply our method to diverse natural images, artificial images and document images, for the sake of demonstrating the performance.

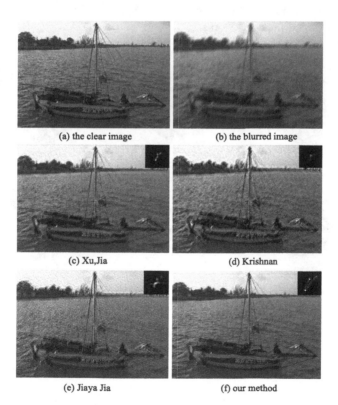

(a) the clear image   (b) the blurred image

(c) Xu,Jia   (d) Krishnan

(e) Jiaya Jia   (f) our method

Fig. 4. The outcome of experiments on natural image

Figure 4 exhibits one example of degraded natural image which is deblurred through the proposed algorithm, and some comparisons with Xu and Jia's algorithm [7], Krishnan's algorithm [8] and Jiaya Jia's algorithm [10]. Our method presents clearer visual effect, less ringing, and sharper image details (e.g., the words on sailing boat). Certain blur and ringing artifacts are still included in the outcomes of [8, 10]. The result of [7] is not favorable with some noise.

Figure 5 displays a degraded artificial image deblurred by our algorithm as an example and the comparisons with other methods. Our approach produces an excellent restored result with less visual ringing and finer details. Certain blur and ringing are still included in the outcomes of Xu and Jia's algorithm [7] and Krishnan's algorithm [8]. The result of Jiaya Jia's algorithm [10] exhibits unpleasant ringing artifacts.

Figure 6 demonstrates one example from degraded document images which is restored through the proposed method, and some comparisons with Xu and Jia's algorithm [7], Zhong's algorithm [9], Krishnan's algorithm [8], Jiaya Jia's algorithm [10], and Jinshan Pan's algorithm [11]. Our method breeds an enjoyable deblurred effect with less noise and ringing. [9, 11] yield lower-quality images. [8] makes the image more blurred. [10] contains many ringing artifacts. [7] also contains a few ringing artifacts at the edge.

In addition we utilize PSNR (peak signal-to-noise ratio) as the quantitative metrics. Average PSNR is computed on the natural images, artificial images and document images and compare among different methods. We give the PSNR in Fig. 7, where our method outperformed previous works.

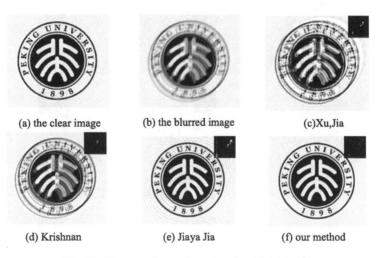

(a) the clear image   (b) the blurred image   (c)Xu,Jia

(d) Krishnan   (e) Jiaya Jia   (f) our method

Fig. 5. The experimental results of artificial image

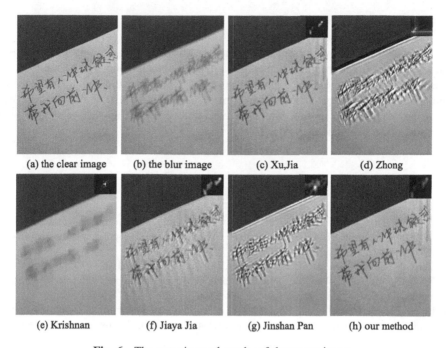

(a) the clear image   (b) the blur image   (c) Xu,Jia   (d) Zhong

(e) Krishnan   (f) Jiaya Jia   (g) Jinshan Pan   (h) our method

Fig. 6. The experimental results of document image

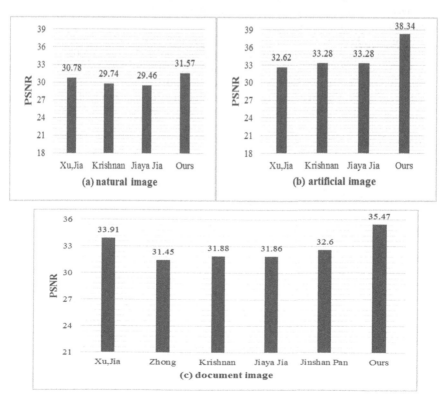

Fig. 7. The average PSNR on the natural, artificial and document images

4 Conclusions

In this paper, the authors present a novel approach to restore motion blurred images using piecewise function that concatenates a hyper-Laplacian and a Gaussian to fit the logarithmic gradient. Contrast experiments on all sorts of natural images and applications of artificial images and document images deblurring illustrate that our approach which can remove the motion blur effectively and preserve the image details well, outperforms the previous methods.

References

1. Joshi, N., Zitnick, C. L., Szeliski, R., Kriegman, D.J.: Image deblurring and denoising using color priors. In: IEEE Computer Society Conference on Computer Vision and Pattern Recognition, pp. 1550–1557 (2009)
2. Cho, S., Lee, S.: Fast motion deblurring. ACM Trans. Graph. **28**, 89–97 (2009)
3. Xiang, S., Meng, G., Wang, Y., Pan, C., Zhang, C.: Image deblurring with matrix regression and gradient evolution. Pattern Recogn. **45**, 2164–2179 (2012)
4. Li, H., Zhang, Y., Zhang, H., Zhu, Y., Sun, J.: Blind image deblurring based on sparse prior of dictionary pair. In: 21th International Conference on Pattern Recognition, pp. 3054–3057 (2012)
5. Levin, A., Weiss, Y., Durand, F., Freeman, W.T.: Understanding and evaluating blind deconvolution algorithms. In: IEEE Computer Society Conference on Computer Vision and Pattern Recognition, pp. 1964–1971 (2009)
6. Krishnan, D., Fergus, R.: Fast image deconvolution using hyper-laplacian priors. In: Advances in Neural Information Processing Systems, pp. 1033–1041 (2009)
7. Xu, L., Jia, J.Y.: Two-phase kernel estimation for robust motion Deblurring. In: European Conference on Computer Vision, vol. 4, pp. 81–84 (2010)
8. Krishnan, D., Tay, T., Fergus, R.: Blind deconvolution using a normalized sparsity measure. In: Conference on Computer Vision & Pattern Recognition, vol. 42, pp. 233–240 (2011)
9. Zhong, L., Cho, S., Metaxas, D., Paris, S.: Handling noise in single image Deblurring using directional filters. In: Proceedings of the IEEE Conference on Computer Vision and Pattern Recognition, vol. 9, pp. 612–619 (2013)
10. Shan, Q., Jia, J.Y., Agarwala, A.: High-quality motion deblurring from a single image. ACM Trans. Graph. **27**, 15–19 (2008)
11. Pan, J., Hu, Z., Su, Z., Yang, M. H.: Deblurring text images via L0-Regularized intensity and gradient prior. In: IEEE Conference on Computer Vision and Pattern Recognition, pp. 2901–2908 (2014)

Automatic Image Segmentation of Grape Based on Computer Vision

Jun Luo[1(✉)], Yue Wang[1], Qiaohua Wang[2], Ruifang Zhai[1], Hui Peng[1],
Liang Wu[3], and Yuhua Zong[3]

[1] College of Informatics, Huazhong Agricultural University,
Wuhan 430070, China
luojun81@163.com
[2] College of Engineering, Huazhong Agricultural University,
Wuhan 430070, China
[3] North Xinjiang Fruit and Vegetable Industry Development Company Limited,
Bole 833400, China

Abstract. Currently, some rapid, accurate and non-destructive detection methods based on computer vision techniques have been widely used in quality inspection of variety of fruits' and grading aspects. However it should be noted that the shapes of the whole string grapes are obviously different with each other, for example, some grapes' shape is circularity, the other shape may be ellipse, which will influence the contour extraction accuracy of non-destructive detection. This work develops a method for real-time image detection and segmentation of grape, using HSV color model and vision technology to extract the color features of grape skin and background objects, and then segment the effective area of grape. Through the experiments, the image segmentation results of the different background-based whole string grapes show that the peripheral contour can be detected by HSV color model, which is relatively robust in different imaging condition.

Keywords: Grape · Image segmentation · Color model · Computer vision

1 Introduction

The total production of fresh grapes in China has significantly grown for decades, but the export production of fresh grapes does not increase obviously. According to the taste demands of consumers, the quality of the grape is the most important factor to be considered. Therefore the non-destructive detection method is such a technique for automatic grading of the whole string fresh grapes after the picking process of grapes [1–3]. Currently, some rapid, accurate and non-destructive detection methods based on computer vision techniques have been widely used in quality inspection of fruits. and grading aspects. However, it should be noted that the shapes of the whole string grapes are obviously different with each other, for example, some grapes' shape is circularity; the other shape may be ellipse, which will influence the contour accuracy of non-destructive detection [4–7]. This work develops a method for real-time image detection and segmentation of grape, using HSV color model and vision technology to

F. Xhafa et al. (eds.), *Recent Developments in Intelligent Systems and Interactive Applications*,
Advances in Intelligent Systems and Computing 541, DOI 10.1007/978-3-319-49568-2_52

extract the color features of grape skin and background objects and then segment the effective area of grape.

2   The Color Model of Grape

2.1   The Color Components of HSV Model

Usually the color phenotypes are used to express the external totally vision characteristics of grape. However RGB color model is limited to simple background, it is expected to develop an adaptive model to solve complex background in portable imaging environment. Therefore hue, saturation and value parameters were chosen in the following experiments.

The component of hue is abbreviated as H, which can be defined by Eq. (1).

$$H = \begin{cases} \frac{1}{6}\frac{G-B}{MAX-MIN}, & R = MAX \\ \frac{1}{6}\left(2 + \frac{B-R}{MAX-MIN}\right), & G = MAX \\ \frac{1}{6}\left(4 + \frac{R-G}{MAX-MIN}\right), & B = MAX \end{cases} \tag{1}$$

The color component of saturation is abbreviated as S, which can be defined by Eq. (2).

$$S = \frac{MAX - MIN}{MAX} \tag{2}$$

The color component of value is abbreviated as V, which can be defined by Eq. (3).

$$V = \frac{MAX}{255} \tag{3}$$

As demonstration in the above, it should be noted that the symbols MAX and MIN are defined by Eqs. (4) and (5), respectively.

$$MAX = \max(R, G, B) \tag{4}$$

$$MIN = \min(R, G, B) \tag{5}$$

2.2   The HSV Characteristics of Grape

As shown in Table 1, several regions of interest (ROIs) have been chosen to investigate the mean values of HSV characteristic parameters, hue, saturation and value. According to de color features distribution of background, the white background, the light color background and the deep color background were used as sample ROIs.

From Table 1, the light intensities of the backgrounds will decrease as the vision effect changing from white to deep color, besides, the variation of grapes' light intensities have a high corresponding with different typed grape, meanwhile, the hue

Table 1. The mean values of characteristics

ROI	Hue	Saturation	Value
White background	0	0	1
White background	0	0	1
White background	0	0	1
Light color background	0.11	0.16	0.98
Light color background	0.18	0.27	0.93
Light color background	0.17	0.12	0.97
Deep color background	0.64	0.33	0.32
Deep color background	0.58	0.57	0.44
Deep color background	0.61	0.54	0.48
Cyan-Blue grape	0.51	0.16	0.57
Cyan-Blue grape	0.59	0.21	0.48
Cyan-Blue grape	0.57	0.22	0.51
Red grape	0.94	0.37	0.16
Red grape	0.88	0.31	0.19
Red grape	0.85	0.20	0.20
Black grape	0.71	0.09	0.64
Black grape	0.61	0.15	0.63
Black grape	0.67	0.16	0.58

and the saturation components vary in a certain proportion as well. According to the above HSV characteristics, it is expected that the H, S and V components can be used to extract different ROIs, which segment the effective grape regions for further processing.

3   Computer Vision-Based Image Acquisition

The computer vision-based image acquisition system is shown in Fig. 1, the CCD is a generic USB2.0 web camera named ASUS winvideo was used as an image acquisition setup in the experiments, which works in 640×480 pixels with 30 fps operation

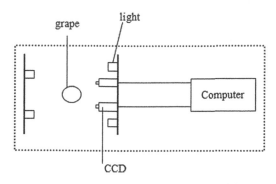

Fig. 1. Computer vision-based image acquisition system in the experiments

settings, the light is a annulus LED light for enhancing the luminance in the imaging condition and the entire bunch grape will be placed on a stationary position, such as white or deep color background platform, the white background maybe bring in a shadow and the deep color background maybe reflect light in a certain degree. However it is expected that the developed image segmentation method can be suitable for different imaging condition and different typed grapes.

4   Experiments and Results

4.1   Color Model-Based Image Segmentation Method

The variations of the ROIs' vision characteristics have a high corresponding with the hue, the saturation and the value components. According to the above HSV characteristics, it is expected that the H, S and V components can be used to extract different ROIs. From Fig. 2, a method based on color model will be used to achieve the image segmentation in the experiments.

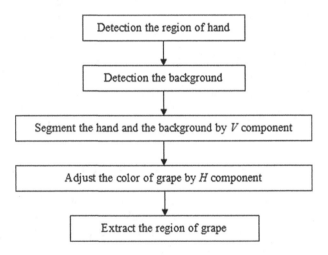

Fig. 2. The steps of image segmentation using HSV color model.

4.2   Image Segmentation of Red Grape

From Fig. 3(a), the ROIs can be divided into three different areas, including red grape, very deep color background and one hand of a person. In accordance with the basic principle of human's vision, it is obvious for us to distinguish these three ROIs.

Let us focus on the HSV characteristic model, the light intensities of the backgrounds will decrease as the vision effect changing from hand to deep color, and the variation of grapes' light intensities are between the hand and the deep color background, Therefore, the H, S and V components can be used to extract different ROIs, which can segment the effective grape regions according to Fig. 2.

Fig. 3. (a) Three ROIs-based red grape image in the experiments, (b) image segmentation result of (a) (Color figure online)

In Fig. 3(b), the result shows that the method basically extracts the red grape region, the very deep color background was set to zero, and then converted to pure black background, and meanwhile, the hand of one person was removed basically in the result. Therefore, the method demonstrated above based on color model is feasible for red grape.

4.3   Image Segmentation of Black Grape

From Fig. 4(a), the ROIs can be divided into three different areas, including black grape, light color background and one hand of a person. In accordance with the basic principle of human's vision, it is obvious for us to distinguish these three ROIs.

As shown in Fig. 4(b), the result completely extracts the black grape region, the light color background was set to zero, and then converted to pure black background, and meanwhile, the hand of one person was removed completely in the result.

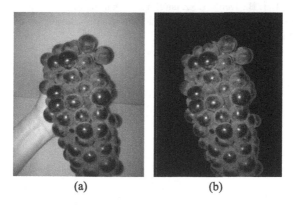

Fig. 4. (a) Three ROIs-based black grape image in the experiments, (b) the extraction result of black grape region (Color figure online)

5   Conclusions

This work develops a method for real-time image detection and segmentation of grape by using HSV color model and vision technology to extract the color features of grape skin and background objects and then segment the effective area of grape. In the experiments, the image segmentation results of the different background-based whole string grapes show that the peripheral contour can be detected by HSV color model, it is expected that the *H*, *S* and *V* components can be used to extract different ROIs, which segment is the effective grape regions for further processing, which is relatively robust in different imaging condition.

Acknowledgements. This research was supported by the Fundamental Research Funds for the Central Universities (grant nos. 2662015QC028 and 2662015PY066), and the National Natural Science Foundation of China (grant nos. 61176052 and 61432007).

References

1. Wang, W., Paliwal, J.: Separation and identification of touching kernels and dockage components in digital images. Can. Biosyst. Eng. **48**, 1–7 (2006)
2. Zalik, K.R.: An efficient E-means clustering algorithm. Pattern Recogn. Lett. **29**(9), 1385–1391 (2008)
3. Ageorges, A., Fernandez, L., Vialet, S., Merdinoglu, D., Terrier, N.: Four specific isogenes of the anthocyanin metabolic pathway are systematically co-expressed with the red colour of grape berries. Plant Sci. **170**(2), 372–383 (2006)
4. Feng, Q.C., Cheng, W., Zhou, J.J., Wang, X.: Design of structured-light vision system for tomato harvesting robot. Int. J. Agric. Biol. Eng. **7**(2), 19–26 (2014)
5. Wang, X., Gao, Z., Xiao, H., Wang, Y., Bai, J.: Enhanced mass transfer of osmotic dehydration and changes in microstructure of pickled salted egg under pulsed pressure. J. Food Eng. **117**(1), 141–150 (2013)
6. Da, W.S., Tadhg, B.: Inspection and grading of agricultural and food products by computer vision systems-a review. Comput. Electron. Agric. **36**(2), 193–213 (2002)
7. Lü, Q., Cai, J.R., Liu, B., Deng, L., Zhang, Y.J.: Identification of fruit and branch in natural scenes for citrus harvesting robot using machine vision and support vector machine. Int. J. Agric. Biol. Eng. **7**(2), 115–121 (2014)

Automatic Color Detection of Grape Based on Vision Computing Method

Yue Wang[1], Jun Luo[1(✉)], Qiaohua Wang[2], Ruifang Zhai[1], Hui Peng[1], Liang Wu[3], and Yuhua Zong[3]

[1] College of Informatics, Huazhong Agricultural University,
Wuhan 430070, China
luojun81@163.com
[2] College of Engineering, Huazhong Agricultural University,
Wuhan 430070, China
[3] North Xinjiang Fruit & Vegetable Industry Development Company Limited,
Bole 833400, China

Abstract. The vision characteristics-based on image detection techniques are functional method, which are appropriate for many industrial and agricultural applications. The color detection method is such a technique for non-destructive grading of the grape after the harvest process, which have been widely used in a variety of fruits' quality inspection and grading aspects based on vision computing method. However, the shapes of the entire bunch of grapes are obviously different with each other, which will influence the accuracy of non-destructive detection. This work develops a method for image detection and color grading of red and black grape samples, which uses vision computing technology to extract the effective color features of red and black grape samples, the experimental results show that the effective region extraction and color detection algorithms are feasible for color detection of grape.

Keywords: Grape · Image processing · Color detection · Vision computing

1 Introduction

Currently the vision characteristics-based image detection techniques are functional method, which are suitable for many industrial and agricultural applications. The color detection method is such a technique for non-destructive grading of the grape after the harvest process [1–3], which have been widely used in a variety of fruits' quality inspection and grading aspects based on vision computing method. However, the shapes of the entire bunch of grapes are obviously different with each other, which will influence the accuracy of non-destructive detection [4, 5]. This work develops a method for image detection and grading method of grape, using vision computing technology to extract the effective color features of grapes' peel and background objects.

© Springer International Publishing AG 2017
F. Xhafa et al. (eds.), *Recent Developments in Intelligent Systems and Interactive Applications*,
Advances in Intelligent Systems and Computing 541, DOI 10.1007/978-3-319-49568-2_53

2 The Vision Computing Model of Grape

2.1 The Overall Implementation Process

The overall implementation process of vision computing model are shown in Fig. 1, the total system can be divided into the following steps. For the first, it is essential for us to set up a vision computing platform and then use standard industrial camera to acquire the images of grapes, besides, a series of image processing algorithms based on HSV color mode will be used to extract the effective region of grape, finally, a particular dada evaluation algorithm will be applied to investigate the color characteristics immediately.

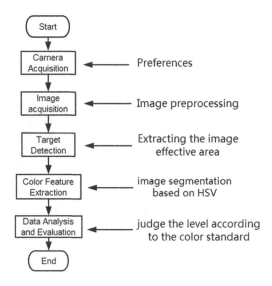

Fig. 1. The process steps of vision computing model in the experiments

2.2 The HSV-Based Vision Model

Usually, the color phenotypes are used to express the external totally vision characteristics of grape. However RGB color model is limited to simple background; it is expected to develop an adaptive model to solve complex background in portable imaging environment. Therefore hue, saturation and value parameters [6] are chosen in the following experiments.

2.3 Image Acquisition Setup of Grapes

In the image acquisition process, the camera typed JAI AD080GE was used as an image acquisition setup in the experiments, which works in 1024 × 768 pixels with 15fps operation settings, the light is a LED white light in the imaging condition. It should be noted that the camera is a GigE version, which is different from other USB-typed cameras according to the hardware support package.

3   Experiments and Results

3.1   Effective Region Extraction and Color Detection of Red Grape

As shown in Fig. 2, three regions of interest (ROIs), such as red-purple region, cyan-blue region and deep color background, can be clearly distinguished in the origin image of red grape. It is important for us to extract the effective regions of grape, usually, the extraction conditions of red-purple grape region and cyan-blue grape region have a high corresponding with the HSV color model.

Fig. 2. The origin image of red grape (Color figure online)

In the experiments, the extraction conditions of cyan-blue grapes can be defined by Eq. (1).

$$H \geq \frac{1}{6} \; and \; H < \frac{1}{2} \tag{1}$$

According to the analysis of the origin red grape image, the extraction conditions of red-purple grapes can be defined by Eq. (2).

$$H \geq 0.5 \; or \; H \leq 0.167 \; and \; V > 0.135 \; and \; S > 0 \tag{2}$$

It should be noted that the components of H, S, and V in the Eqs. (1) and (2), are hue, saturation and value in the HSV color model.

From Fig. 3, the red-purple grape region and cyan-blue grape region have been segmented clearly.

After extraction the effective regions of red-purple grape and cyan-blue grape, a particular detection criteria should be applied to evaluate the color grade of grape.

The overall proportion of color distribution can be expressed by Eq. (3).

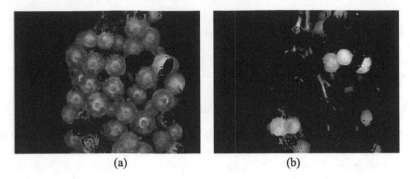

(a)   (b)

Fig. 3. (a) Red-purple region of red grape, (b) cyan-blue region of red grape (Color figure online)

$$\eta = \frac{the\ area\ of\ red - purple\ region}{the\ area\ of\ cyan - green\ region} \qquad (3)$$

The detection criteria for evaluating the color grade of grape can be defined by Eq. (4).

$$Grade = \begin{cases} 1, & if \quad \eta > 0.97 \\ 2, & if \quad 0.93 \leq \eta \leq 0.97 \\ 3, & if \quad \eta < 0.93 \end{cases} \qquad (4)$$

In Eq. (4), the value 1, 2, 3 are the best, the better and low-quality color grade, respectively.

Finally, the grade of the origin red grape given in Fig. 3 is grade 3, which means that the color grade of entire bunch red grape is low quality.

3.2   Effective Region Extraction and Color Detection of Black Grape

As shown in Fig. 4, three regions of interest (ROIs), such as high saturation region, low saturation region and deep color background can be clearly distinguished in the origin image of red grape. It is important for us to extract the effective regions of grape, usually; the extraction conditions of high saturation grape region and low saturation grape region have a high corresponding with the HSV color model.

In the experiments, the extraction conditions of low saturation grapes can be defined by Eq. (5).

$$H \geq 0.5\ and\ H < 0.833 \qquad (5)$$

The extraction conditions of high saturation grapes can be defined by Eq. (6).

Fig. 4. The origin image of black grape

$$H \geq 0.5 \ or \ H \leq 0.833 \ and \ V > 0.115 \ and \ S > 0.5 \tag{6}$$

As can be seen in Fig. 5, the high saturation grape region and low saturation grape region have been segmented clearly.

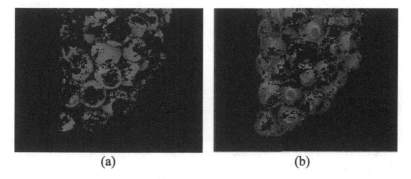

(a)   (b)

Fig. 5. (a) High saturation region of black grape, (b) low saturation region of black grape

After extraction the effective regions of high saturation grape and low saturation grape, a detection criteria should be applied to evaluate the color grade of grape.

The overall proportion of color distribution can be expressed by Eq. (7).

$$\eta = \frac{the \ area \ of \ high \ saturation \ region}{the \ area \ of \ low \ saturation \ region} \tag{7}$$

The detection criteria for evaluating the color grade of grape can be defined by Eq. (8).

$$Grade = \begin{cases} 1, & if \quad \eta > 0.5 \\ 2, & if \quad 0.3 \leq \eta \leq 0.5 \\ 3, & if \quad \eta < 0.3 \end{cases} \tag{8}$$

In Eq. (8), the value 1, 2, 3 are the best, the better and low-quality color grade, respectively.

Finally, the grade of the origin black grape given in Fig. 4 is grade 1, which means that the color grade of entire bunch black grape is high quality.

4 Conclusions

The vision characteristics-based image color detection method is such a technique for non-destructive grading of the grape after the harvesting process, which have been widely used in a variety of fruits' quality inspection and grading aspects based on vision computing method. This work develops a method for image detection and color grading of red and black grape samples, which uses vision computing technology to extract the effective color features of red and black grape samples, the experimental results show that the effective region extraction and color detection algorithms are feasible for color detection of grape. It is expected that the non-destructive color detection based on vision computing method will be applied in the agricultural engineering widely.

Acknowledgements. This research was supported by the Fundamental Research Funds for the Central Universities (grant nos. 2662015QC028 and 2662015PY066), and the National Natural Science Foundation of China (grant nos. 61176052 and 61432007).

References

1. Ageorges, A., Fernandez, L., Vialet, S., Merdinoglu, D., Terrier, N.: Four specific isogenes of the anthocyanin metabolic pathway are systematically co-expressed with the red colour of grape berries. Plant Sci. **170**(2), 372–383 (2006)
2. Failla, O., Mariani, L., Brancadoro, L., Minelli, R., Scienza, A., Murada, G., Mancini, S.: Spatial distribution of solar radiation and its effects on vine phenology and grape ripening in an alpine environment. Am. J. Enology Viticulture **55**(2), 128–138 (2004)
3. Leemans, V., Destain, M.-F., Magein, H.: On-line fruit grading according to their external quality using machine vision. Biosyst. Eng. **83**(4), 397–404 (2002)
4. Zhai, H., Du, Y.P., et al.: On the development situation of Chinese grape industry. J. Fruit Sci. **24**(6), 820–825 (2007)
5. Zhao, Y.C., Xu, L.M.: Automated strawberry grading system based on image processing. Comput. Electron. Agric. **71**(1), S32–S39 (2010)
6. Xu, K., Lu, X., Wang, Q., Ma, M.: Online automatic grading of salted eggs based on machine vision. Int. J. Agric. Biol. Eng. **8**(1), 35–41 (2015)

Design Study on the Digital Correlator Using for Radio Holography

Linfen Xu[1,2(✉)] and Yingxi Zuo[1]

[1] Purple Mountain Observatory/Key Laboratory of Radio Astronomy,
Chinese Academy of Sciences, Nanjing 210008, China
{lfxu, yxzuo}@pmo.ac.cn
[2] University of Chinese Academy of Sciences, Beijing 100049, China

Abstract. In this paper the author presents a digital correlation design for the radio holography receiver of DATE5 (the 5-m Dome A Terahertz Explorer). Signal digitization is made at the IF outputs of the receiver and correlation is implemented in time domain. A 4-component algorithm for complex correlation is proposed. With this algorithm, the systematic error is much lower and the SNR is expected to have a 3 dB improvement comparing with the previous 3-component algorithm.

Keywords: Digital correlator · Algorithm · DATE5 · Radio holography

1 Introduction

One terahertz telescope called the 5-m Dome. A Terahertz Explorer (DATE5) has been proposed to operate at Dome A, Antarctica, to exploit one of the best observing conditions at terahertz bands on the earth [1]. Error budget for the main reflector panels setting is only 5 μm rms, requiring higher reflector measurement accuracy (~ 3 μm rms). Near-field radio holography is chosen as the method of on-site measurement of the reflector for DATE5 panel adjustment [2]. Theoretically, radio holography is based on the integral transform relationship between the radiation pattern and the aperture field distribution of a reflector antenna [3]. From the measured complex radiation pattern, one can derive the aperture field distribution with its phase term directly related to the reflector surface deviations. The obtained surface accuracy is mainly determined by the signal-to-noise ratio (SNR) of the measured radiation pattern [3]. Block diagram of the dual-channel holography receiver system for DATE5 is shown in Fig. 1. The receiver frontends operate at 3 mm waveband (e.g. 106 GHz). A beacon signal is simultaneously received by both receivers and is converted to intermediate frequency (IF) signals of 30 MHz. During measurement the AUT (antenna under test) scans around the beacon to obtain its 2-D pattern. The digital correlator is a key component of the holography receiver system to measure the complex radiation pattern of the main antenna.

A digital correlator for radio holography can be implemented in different ways. Wang et al. described a so-called FX implementation [4], in which the signals from both receiver channels are first decomposed into spectral components using FFT, and

F. Xhafa et al. (eds.), *Recent Developments in Intelligent Systems and Interactive Applications*,
Advances in Intelligent Systems and Computing 541, DOI 10.1007/978-3-319-49568-2_54

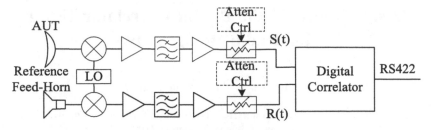

Fig. 1. Block diagram of the dual-channel holography receiver system

then the correlation is obtained by multiplying the corresponding spectral component of the two receiver channels. Because in this case only the signal located frequency bin is used, the computation efforts for other frequency components may be wasted. Perfetto et al. adopted a time domain implementation, which involves analog I/Q (in-phase and quad-phase) demodulation and baseband digital correlation using DSP [5]. This paper presents a digital correlator design for the radio holography receiver of DATE5, using a time-domain implementation and digitizing signals at the IF outputs. The structure and algorithm are proposed. Simulations for systematic and random errors are also performed.

2   Scheme and Algorithm of the Digital Correlator

One simple block diagram of the digital correlator is shown in Fig. 2. The dual-channel signals are defined as $S = A_s\cos(w_{IF}t + \theta_1)$ and $R = A_r\cos(w_{IF}t + \theta_2)$. They are synchronously sampled and quantized with the ADCs at a sampling rate of 100 MHz. In the DDC module, by mixing with signals from the NCO (numerically controlled oscillator, 29.980 MHz), the digitized signals are then decomposed into two pair of baseband I/Q components. Considering the small frequency drift of the LO (local oscillator) in Fig. 1 and the beacon, the baseband signals choose a frequency of 20 kHz rather than zero. Low pass filters (LPF) must be followed to eliminate the sum frequency components. Decimation may be made to decrease the computation rate. Since decimation is equivalent to resampling, however, the choice of cutoff frequency of the LPFs must satisfy the Nyquist sampling criterion. With the I/Q components one can achieve the correlation of the dual-receiver signals.

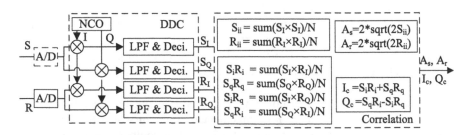

Fig. 2. Data flow of the digital correlator

2.1   Correlation Algorithm

The complex correlation is usually obtained by using only three components of the I/Q outputs [5–7], i.e. S_I, S_Q and R_I, referred to as 3-component algorithm. The formulas are shown in Eqs. 1 and 2. It is obvious that the integration of the double-frequency term will leave a systematic error if the integration time is not an integer number of the signal cycles. The error will exhibit a damped ripple as the integration time increases.

$$I_c = \frac{1}{N}\sum_{n=1}^{N}(S_I \cdot R_I) = \frac{1}{8}A_sA_r\cos(\Delta\theta) + \frac{1}{8}A_sA_r\frac{1}{N}\sum_{n=1}^{N}\cos(4\pi ft_n + \theta_1 + \theta_2) \tag{1}$$

$$Q_c = \frac{1}{N}\sum_{n=1}^{N}(S_Q \cdot R_I) = \frac{1}{8}A_sA_r\sin(\Delta\theta) + \frac{1}{8}A_sA_r\frac{1}{N}\sum_{n=1}^{N}\cos(4\pi ft_n + \theta_1 + \theta_2) \tag{2}$$

In order to avoid such systematic error, a 4-component algorithm is presented, which is based on 4 components of the I/Q outputs. The computing process is shown in the correlation module of Fig. 2, where N is the sampling size and can be seen as the integration time (selectable from 10 to 1000 ms). Supposing that the DDC outputs are without any I/Q imbalance, the correlation-output I_c and Q_c can be are finally derived as shown in Eqs. 3 and 4 respectively. It is obvious that the complex correlation, $C = I_c + jQ_c$, will not have any systematic error in this case. Moreover, the correlation amplitude has been doubled, compared with the former algorithm.

$$I_c = \frac{1}{4}A_sA_r\cos(\Delta\theta) \tag{3}$$

$$Q_c = \frac{1}{4}A_sA_r\sin(\Delta\theta) \tag{4}$$

2.2   Dynamic Range Requirement of the ADC

The dynamic range of an ADC is usually chosen based on the SNR at its input. In engineering the requirement of SNR for a measurement is always related to a noise bandwidth or integration time. Since digital signal processing which includes filtering or integration will significantly improve the signal-to-noise ratio (SNR) by suppressing the noise, the ADC's dynamic range is usually not necessary to meet the final SNR requirement for the measurement system.

The SNR improvement of the digital correlator is found by computing $SNR_{gain} = SNR_{out} - SNR_{in}$ in dB [8]. Assuming that the integration time is 50 ms, thus the equivalent noise bandwidth (ENBW) will be $1/(2 \times 50 \text{ ms}) = 10$ Hz [10], so the improvement becomes:

$$SNR_{out} - SNR_{in} = 10 \lg \left(\frac{noise\ power\ input\ to\ ADC}{noise\ power\ through\ signal\ integration} \right)$$

$$= 10 \lg \left(\frac{Nyquist\ bandwidth}{2 \times ENBW} \right) = 10 \lg \left(\frac{50\ MHz}{2 \times 10\ Hz} \right) = 64\ dB \qquad (5)$$

The dynamic range of the ADCs should be larger than the system SNR requirement minus the SNR gain of the digital correlator. Thus for a receiver channel if we would like to obtain a peak SNR of 85 dB at integration time of 50 ms [2], dynamic range of the ADC sampling at 100 MHz should be greater than 21 dB. It should be noticed that ADC's bit-width will also cause I/Q imbalance errors [9], so ADCs of 8-bit or more are preferred.

3   Error Analysis

3.1   Systematic Effect

The systematic error variation with integration time is simulated for both algorithms and the results are shown in Fig. 3. When the simulation is made at short intervals of integration time, as shown in Fig. 3(a), one can observe clearly the damped ripple with 3-component algorithm, but no ripple with 4-component algorithm. Peaks of the amplitude-error ripple with the 3-component algorithm appear when the integration includes an extra 1/8 (3/8, 5/8 or 7/8) period of the 20 kHz baseband signal. The results are in agreement with Eq. 1 through Eq. 4.

Usually the baseband signal frequency is not recognized exactly because of the frequency drifts of the beacon and LO. Thus it is hardly possible to make integration of integer number of cycles. For this reason we have to use the worst case (i.e. the ripple peaks) as an evaluation of the systematic error for 3-component algorithm. In Fig. 3(b) the integration time is chosen at each of which the error of the 3-component algorithm reaches a local peak. As the integration time increases, the error curves for both algorithms decrease and then get flat, and the error of 4-component algorithm is smaller and decreases faster.

3.2   Effect of Noise

Phase noise of the beacon and LO will not affect the correlation results, because it is common in both receiver channels and will be eliminated by the correlation which makes a phase difference. The noise in the two receivers, however, are independent, and must be taken into consideration.

Receiver noise comes mainly from the mixer and the first-stage amplifier at the frontend, and is usually treated as white noise. Assuming the input SNRs to the ADCs are 20 dB and −20 dB respectively, we have made a simulation to check the output SNR of the correlator with the integration time ranging from 0.1 ms to 100 ms. At each integration time the correlating calculation repeats 100 times, and at each repeat the independent random noises are added at the inputs. By this means we obtain the mean

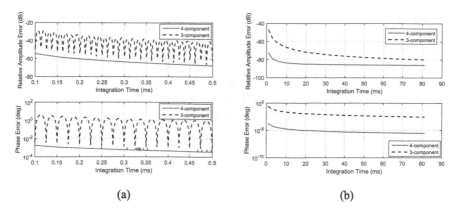

Fig. 3. Relative amplitude error and phase error versus integration time

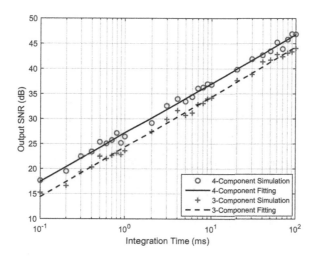

Fig. 4. SNR of the digital correlatior output versus integration time

value and the rms value (which treated as the error or noise amplitude) and consequently the SNR of correlation output. Simulation results are shown in Fig. 4. One can see that the SNR with the 4-component algorithm is about 3 dB greater than that with the 3-component. Using the 4-component algorithm the output SNR is approximately 44 dB at 50 ms integration time. A 64 dB SNR improvement has been obtained compared with the weaker input SNR, being in agreement with Eq. 5. It can be noticed that the SNR of the digital correlator output mainly depends on the weaker signal, in our case [10].

4 Conclusion

We have presented a digital correlator design using for radio holography of the DATE5 antenna. Signal digitization is made at the IF outputs of the receiver and correlation is implemented in time domain. A 4-component algorithm for complex correlation is proposed. With this algorithm, the systematic error is much lower and the SNR is expected to have a 3 dB improvement comparing with the 3-component algorithm.

Acknowledgments. This work is partly supported by the National Natural Science Foundation of China (Grant Nos. 11190014 and 11373073), and by the Operation, Maintenance and Upgrading Fund for Astronomical Telescopes and Facility Instruments, budgeted from the Ministry of Finance of China (MOF) and administrated by the Chinese Academy of Sciences (CAS).

References

1. Yang, J., Zuo, Y.X., Lou, Z., Cheng, J.Q., et al.: Conceptual design studies of the 5 meter terahertz antenna for Dome A, Antarctica. Res. Astron. Astrophys. **13**(12), 1493–1508 (2013)
2. Zuo, Y., Lou, Z., Yang, J., Cheng, J.: Design study on near-field radio holography of the 5-meter Dome A terahertz explorer. In: ISAP 2013, pp. 25–28 (2013)
3. Baars, J.W.M., Lucas, R., Mangum, J.G., Lopez-Perez, J.A.: Near-field radio holography of large reflector antennas. IEEE Antennas Propag. Mag. **49**(5), 24–41 (2007)
4. Wang, J., Fan, Q., Li, B.: The implementation of a correlator for microwave holographic measurement. Astron. Res. Technol. **6**(4), 280–291 (2009)
5. Perfetto, A., Addario, L.D'.: Holography receiver design, October 2010. https://www.cv.nrao.edu/~demerson/holocdr/larry/rxDesign.pdf
6. An, H.: Research of near-field holography experiment for reflector antenna. Qinghai Normal University, pp. 42–46, Xining (2015)
7. Serra, G., Busonera, G., Pisanu, T., et al.: Microwave holography system for the Sardinia radio telescope. In: Proceedings of the SPIE 8444, Ground-based and Airborne Telescopes IV, 84445W, September 2012
8. Karras, T.J.: Equivalent noise bandwidth analysis from transfer functions, NASA Technical report, November 1965
9. Somann, J.P., Kim, Y.C.: Characterization of in-phase/quad-phase digital downconversion via special sampling scheme. In: ICECS 2006, pp. 224–227 (2006)
10. Addario, L.R.D'.: Holography antenna measurements: further technical considerations, 12m Telescope Memo No. 202, NARO, Charlottesville, November 1982

Edge-Adaptive Structure Tensor Nonlocal Kernel Regression for Removing Cloud

Guohong Liang[1,2(✉)], Ying Li[1,2], and Junqing Feng[1,2]

[1] School of Computer Science, Northwestern Polytechnical University,
Xi'an 710129, China
liangguohong321@163.com
[2] School of Science, Air Force Engineering University, Xi'an 710051, China

Abstract. This paper the authors applies structure tensor matrix as a tool in order to survey the anisotropic structure of remote sensing image and combines nonlocal kernel regression methods for removing cloud tasks. The method utilizes both local structural regularity and the nonlocal self-similarity properties in remote sensing images. The nonlocal self-similarity takes advantages of observation that image patches incline to repeat themselves in remote sensing images. The non-local prior avails of the redundancy of similar patches remote sensing images, while the local prior consider that a target pixel can be computed by a weighted average of its neighbors. Experimental results indicate that our new algorithm better than both the steering kernel regression in persevering edge and improve the visual quality.

Keywords: Structure tensor · Kernel regression · Nonlocal Self-similarity · Cloud

1 Introduction

Remote sensing images have been used widely in agriculture crop monitoring, Land-cover classification, landscape ecological change detection and so on. However, owing to the effect of atmosphere condition, cloud cover is one of the most noise elements in remote sensing image. Land surface information is contained by region covered by thin cloud [1], Removal cloud is a very critical step when there is one image by cloud.

Removing cloud from remote sensing image is a traditional subject, so many different kinds of approaches have been advanced, such as Tasseled Cap Transformation (K-T Transformation) filtering, homomorphic filtering method, image fusion method, histogram matching method, methods based on machine learning multispectral image, statistical methods based on Bayes and so on. Tseng et al. putted forward an algorithm for generating cloud-free mosaics SPOT images [2].

© Springer International Publishing AG 2017
F. Xhafa et al. (eds.), *Recent Developments in Intelligent Systems and Interactive Applications*,
Advances in Intelligent Systems and Computing 541, DOI 10.1007/978-3-319-49568-2_55

2   Related Work

2.1   Adaptive Structure Tensor Kernel Function [3]

This section gives conception of the structure tensor and proposes an edge-adaptive structure tensor matrix to kernel regression.

Let u be a image, its structure tensor is a field of symmetric matrix, which includes in each element information on orientation and intensity of the surrounding structure of u:

$$J_\rho(\nabla u_\sigma) = k_\rho * (\nabla u_\sigma \otimes \nabla u_\sigma) \triangleq \begin{pmatrix} J_{11} & J_{12} \\ J_{21} & J_{22} \end{pmatrix} \tag{1}$$

The tensor product $\nabla u_\sigma \otimes \nabla u_\sigma$ which is to enhance the geometric structure such as a thin line, edge and corner is a symmetric and positive semi-define matrix, v_1 and v_2 are orthogonal eigenvectors of the matrix $J_\rho(\nabla u_\sigma)$.

$$v_1 = \begin{pmatrix} 2J_{12} \\ J_{22} - J_{11} + \sqrt{(J_{22} - J_{11})^2 + 4J_{12}^2} \end{pmatrix}, v_2 = v_1^\perp \tag{2}$$

The homologous eigenvalues are presented by:

$$l_{1,2} = \frac{1}{2}\left(J_{11} + J_{22} \pm \sqrt{(J_{22} - J_{11})^2 + 4J_{12}^2}\right) \tag{3}$$

The eigenvalues convey shape information, $l_1 \approx l_2 \approx 0$ characterize isotropic structure, while $l_1 \gg l_2 \approx 0$ give edges, and corner yield $l_1 \geq l_2 \gg 0$. Definition of the coherence measure is:

$$c = \frac{(l_1 - l_2)^2}{(l_1 + l_2)^2} \tag{4}$$

In this paper, we plan a matrix S which drives the anisotropic interpolation in kernel regression. if the eigenvalue l_1 is large, in order to reduce the diffusivity v_1 perpendicular to edges, so S is defined as:

$$S = \gamma[v_1 \ v_2]\begin{pmatrix} \lambda_1 & 0 \\ 0 & \lambda_2 \end{pmatrix}\begin{bmatrix} v_1^T \\ v_2^T \end{bmatrix} \tag{5}$$

Corresponding Eigenvalues λ_1, λ_2 are given by:

$$\lambda_1 = \begin{cases} 1 & c < \bar{c} \\ c_1 e^{-c} & else \end{cases}, \lambda_2 = 1 \tag{6}$$

where c_1 is a constant and $\bar{c}$ is a threshold that controls the coherence.

Adaptive structure tensor matrix is defined as:

$$H_i^{ST} = h\mu_i S_i^{-\frac{1}{2}} \tag{7}$$

2.2   Kernel Regression [4]

The kernel regression framework defines its data model in 2-D as

$$y_i = r(x_i) + \varepsilon_i, \quad i = 1, 2, \cdots, P \tag{8}$$

The local expansion of the regression function is provided by

$$
\begin{aligned}
r(x_i) &= r(x) + \{\nabla r(x)\}^T (x_i - x) + \frac{1}{2}(x_i - x)^T \{Hr(x)\}^T (x_i - x) + \cdots \\
&= r(x) + \{\nabla r(x)\}^T (x_i - x) + \frac{1}{2}\mathrm{vec}^T \{Hr(x)\}^T \mathrm{vec}\{(x_i - x)(x_i - x)^T\} + \cdots
\end{aligned}
\tag{9}
$$

Specifically, we can depend on the local expansion of the function using the Taylor series assuming that the image is locally smooth to some order,

$$r(x_i) = a_0 + a_1^T (x_i - x) + a_2^T tril\{(x_i - x)(x_i - x)^T\} + \cdots \tag{10}$$

3   Proposed Method and Algorithm

Adaptive structure tensor kernel function is given by:

$$K_{H_i^{ST}}(x_i - x) = \frac{\sqrt{\det(S_i)}}{2\pi h^2 \mu_i^2} exp\left\{ -\frac{(x_i - x)^T S_i (x_i - x)}{2h^2 \mu_i^2} \right\} \tag{11}$$

Thus the model of kernel regression can be denotes as:

$$\min_{\{\alpha_n\}} \sum_{i=1}^{P} \left[y_i - \alpha_0 - \alpha_1^T (x_i - x) - \alpha_2^T \mathrm{vec}\{(x_i - x)(x_i - x)^T\} - \cdots \right]^2 K_{H_i^{ST}}(x_i - x) \tag{12}$$

Equation (12) as a weighted least-squares problem can be reformulated into a matrix form:

$$\hat{a}_i = \arg \min_{a_i} E(a)$$

$$= \arg \min_{a_i} \| y - \Phi a_i \|_{W_{H_{x_i}}}^2 \tag{13}$$

$$= \arg \min_{a_i} (y - \Phi a_i) W_{H_{x_i}} (y - \Phi a_i)$$

where $W_{H_{x_i}} = diag(H_{x_i})$, and Φ is defined by

$$\Phi = \begin{bmatrix} 1 & (x_i - x_j)^T & vech^T\{(x_i - x_j)(x_i - x_j)^T\} \\ \vdots & \vdots & \vdots \end{bmatrix} \tag{14}$$

we differentiate E(a) with regard to a

$$\frac{\partial E(a)}{\partial a} = 2\Phi^T W_{H_{x_i}} (\Phi a - y)$$

and the solution is

$$\hat{a}_i = (\Phi^T K_{H_i^{ST}} \Phi)^{-1} \Phi^T W_{H_{x_i}} y \tag{15}$$

Equation (9) can be as follows [7]:

$$\hat{z}(x_i) = \arg \min_{z_i} E(z_i)$$

$$= \arg \min_{z_i} \sum_{j \in P(x_i)} [y_j - z(x_i)]^2 w_{ij}$$

$$= \arg \min_{z_i} \| y - 1 z(x_i) \|_{W_{x_i}}^2 \tag{16}$$

$$= \arg \min_{z_i} (y - 1 z(x_i)) W_{x_i} (y - 1 z(x_i))$$

solution to (10), we differentiate $E(z_i)$ with regard to z_i

$$\frac{\partial E(z_i)}{\partial z_i} = 2 \times 1^T W_{x_i} (1 z(x_i) - y)$$

In order to solve for z_i. Setting it to be zero, we get

$$z(x_i) = [1^T W_{x_i} 1]^{-1} 1^T W_{x_i} y \tag{17}$$

the proposed NL-KR model is created as

$$\hat{a} = \underset{a}{arg\,min}\, E(a)$$

$$= \underset{a}{arg\,min}\, \overbrace{\frac{1}{2} w_{ii} \| R_{x_i} y - \Phi a \|^2_{W_{H_{x_i}}}}^{local} + \overbrace{\frac{1}{2} \sum_{j \in P(x_i)\setminus\{i\}} w_{ij} \| R_{x_j} y - \Phi a \|^2_{W_{H_{x_i}}}}^{nonlocal}$$

$$= \underset{a}{arg\,min}\, \frac{1}{2} \sum_{j \in P(x_i)} w_{ij} \| R_{x_j} y - \Phi a \|^2_{W_{H_{x_i}}}$$

$$= \underset{a}{arg\,min}\, \frac{1}{2} \sum_{j \in P(x_i)} \| R_{x_j} y - \Phi a \|^2_{W_{H_{x_i}}}$$

(18)

$$w_{ij} = exp\left(-\frac{\| R_{x_i} y - R_{x_j} y \|^2_{w_G}}{2\sigma^2} \right) \tag{19}$$

we differentiate the above function in (13) with respect to a to get the regression coefficients

$$\frac{\partial E(a)}{\partial a} = \Phi^T \sum_{j \in P(x_i)} w_{ij} W_{H_{x_j}} (\Phi a - R_{x_j} y)$$

Setting it to be zero, we can have the estimation for a as

$$\hat{a} = [\Phi^T (\sum_{j \in P(x_i)} w_{ij} W_{H_{x_j}})\Phi]^{-1} \Phi^T \sum_{j \in P(x_i)} w_{ij} W_{H_{x_j}} R_{x_j} y \tag{20}$$

thence, $\hat{r}(x_i) = \hat{a}_0 = e_1^T \hat{a}$.

Algorithm   Edge-adaptive Structure Tensor Nonlocal Kernel Regression (ESTNKR)

Input: Noisy image y, and parameters N.

Return: Denoised image $\hat{x} = x^N$.

Initialize: the present denoising estimation $x^0 = y$,estimate the image gradient ∇y ;

For every pixel j ,do

For each pixel location x_i on the image grid, do

Construct the similar patch index set $P(x_i)$ with current denoising estimation x^{j-1}

Compute the structure kernel function H_{x_i} use (7) ;

Construct the spatial weight matrix $\tilde{W}_{x_j}$ using estimated H_{x_j} for all $j \in P(x_i)$;

Establish the regression coefficients with (22) and update the present estimation of x^t at x_i with $x^t(x_i) = e_1^T \hat{a} = \hat{a}(1)$;

Update the image gradient ∇y at x_i as $\nabla y_{x_i} = [\hat{a}(2), \hat{a}(3)]^T$

4   Experimental Result

Some experiments have been carried out to reveal the validity of cloud removing approach mentioned in this paper [8]. The experiments result is illustrated as Fig. 1.

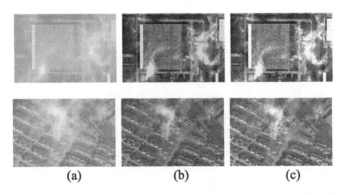

(a)   (b)   (c)

Fig. 1. Original images (a) and Advanced MSR (b) and New method in this paper (c)

References

1. Gabarda, S., Cristobal, G.: Cloud covering denoising through image fusion. Sci. Direct. **25**, 523–530 (2007)
2. Takeda, H., Farsiu, S., Milanfar, P.: Kernel regression for image processing and reconstruction. IEEE Trans. Image Process. **16**(2), 349–366 (2007)
3. Takeda, H., Farsiu, S., Milanfar, P.: Robust kernel regression for restoration and reconstruction of images from sparse noisy data. In: Proceedings of the International Conference on Image Processing (ICIP), Atlanta, GA, October 2006
4. Xingfang, J., Xiang, L., Wei, M.: Research of new method of removal thin cloud and fog of the remote sensing images. In: SOPO 2010. IEEE (2010)
5. Kong, J., Hu, G., Liang, D.: Thin cloud removing approach of color remote sensing image based on support vector machine. In: 2010 Asia-Pacific Conference on Wearable Computing Systems (2010)
6. Irani, M., Peleg, S.: Motion analysis for image enhancement: resolution, occlusion and transparency. J. Vis. Commun. Image Represent. **4**(4), 324–335 (1993)
7. Mahmoudi, M., Sapiro, G.: Fast image and video denoising via nonlocal means of similar neighborhoods. IEEE Signal Process. Lett. **12**(12), 839–842 (2005)
8. Coupe, P., et al.: An optimized blockwise nonlocal means denoising filter for 3-D magnetic resonance images. IEEE Trans. Med. Imaging **27**(4), 425–441 (2008)
9. Kervrann, C., Boulanger, J., Coupé, P.: bayesian non-local means filter, image redundancy and adaptive dictionaries for noise removal. In: Sgallari, F., Murli, A., Paragios, N. (eds.) SSVM 2007. LNCS, vol. 4485, pp. 520–532. Springer, Heidelberg (2007). doi:10.1007/978-3-540-72823-8_45
10. Buades, A., Coll, B.: A non-local algorithm for image denoising. In: Proceedings of IEEE CVPR, pp. 60–65 (2005)

Mobile Communication
and Wireless Network

The Research on Bluetooth Technologies for Non-invasive FHR Monitoring

Rongyue Zhang, Mingrui Chen, and Wei Wu$^{(\boxtimes)}$

College of Information Science and Technology,
Hainan University, Haikou, China
340448683@qq.com, mrchen@hainu.edu.cn,
wuwei@idsse.ac.cn

Abstract. The instrument of Fetal Heart Rate (FHR) monitoring based on the technique of ultrasonic Doppler is harmful to the baby. In order to overcome the limitations a new method is proposed in this paper which is a non-invasive FHR monitoring method based on low-power Bluetooth technology. and is combined with the most widely used smartphone and low-power Bluetooth technology in the paper. In the first step the sensor SCP1000-D01 picked up mixed signal from maternal's abdomen. In the next course of action the FHR signal is separated from the mixed signal by pre-amplification processing circuit, signal processing circuit, A/D conversion and optical coupling circuit. Results show that the signal collected by the sensor is processed by Blind source separation algorithm based on two order statistics to extract real-time clear FHR signal. The extracted signal is transmitted to the smartphone through the low-power Bluetooth technology and reflected in the phone.

Keywords: Sensor · Low power consumption · Blind source separation · Algorithm · FHR

1 Introduction

In a landmark policy initiative the nation adopted a comprehensive "two children" policy on 1st January, 2016. Such a decision in turn will substantially increase the population in short run. It is expected statistically that the number of the target population having two children will figure around 90 million, especially in big cities. However, implementation of the policy has led to significant increase in elderly pregnant women and the tension of the medical resources. In this paper the monitoring method proposed is not only to ease the tension of the medical resources but also more importantly to increase the awareness on maternal and child health. Continuous monitoring on the psychological status of pregnant women and frequent contact between pregnant women and doctors has confirmed to reduce the incidence of prematurity and miscarriage [1]. Due to the inconvenience of fetal heart monitor, it needs to get the main physiological parameters in a non-invasive fetal heart rate (FHR) monitoring methods, acquired the relevant information by analyzing FHR parameters to determine the health of the fetus. The normal fetal heart rate should be $120 \sim 160$ times per minute, not only applying traditional stethoscope but also using

F. Xhafa et al. (eds.), *Recent Developments in Intelligent Systems and Interactive Applications*,
Advances in Intelligent Systems and Computing 541, DOI 10.1007/978-3-319-49568-2_56

just amplification function of the ultrasonic Doppler instrument is difficult to accurately measure [2]. However, the ultrasonic is not suitable for long-term monitoring. Once the frequency of the ultrasonic exceeds 2 Hz would be dangerous to the fetus and ultrasonic detection method is very sensitive to body's movement. Compared with the QRS complex analysis, ultrasonic technology is not accurate [3].

At present, it is received to pick up FECG signals by two means: one is the scalp electrode detection method and the other is the abdominal ECG detection method [4]. In terms of the two methods compared with each other, abdominal FHR monitoring is the preferred method of non-invasive FHR monitoring because of almost no efficacy on foetus and gravida. It could be infected to pregnant women and fetal by the scalp electrode detection method. The three-electrode structure is used at the method of abdominal ECG detection, and the electrode is fixed to the location of pregnant women's belly and chest, which extracts precisely to the maternal and fetal ECG signals. The extracted signals also need to design the hardware circuit for filtering, noise removal and design software to get the final FECG. At present, the method is widely accepted for simplicity, long-term monitoring and non-invasive.

It is proposed to be improved non-invasive FECG extraction algorithm on the basis of the combination of MEMS pressure sensor technology, Bluetooth wireless technology and smart phones. It is likely to be non-invasive monitoring FHR and accurately reflect the fetal ECG signal on the phone at any time when implementing the ambulatory monitoring of FHR.

2   Structure of the System Introduction

Using the design of low-power domestic FHR monitoring system in the paper as Fig. 1. The smartphone is looked up on as monitoring equipment of Bluetooth intelligent in the system and it is not only used widely in the crowd but also an economically viable option. Therefore, it easily makes the phone and the FHR monitoring module connected, received and transmitted data via Bluetooth technology as long as the preparation of the corresponding mobile phone software. Furthermore, the data also be sent by Smartphone to the hospital and care for doctors to diagnose the disease in pregnant women.

2.1   The Sensor SCP1000-D01

Since this research systems need monitoring the clear data through a series of processes, transmission of data via Bluetooth to the phone APP and computers and display

Fig. 1.   Structure of the system

of data, selecting the sensor is very important. The system chose a digital pressure sensor SCP1000-D01 which has high accuracy, high resolution and low power consumption. The sensor is based on 3D-MEMS technology absolute pressure sensor, achieved accurately sub-meter level of the resolution and 1 m accuracy under normal operation, so that it is suitable for advanced medical applications. Furthermore, the sensor SCP1000-D01 is based on 3D-MEMS technology which is a three-dimensional structure by processing silicon and its package contacts for easy installation and assembly. The sensor made with this technique has excellent precision, small size, low power consumption and it is likely to measure the acceleration of three mutually perpendicular directions. Sealing structure of the sensor SCP1000-D01 makes particles or chemicals not able to enter the sensor thus ensuring its reliability. The built-in A/D converter circuit in the sensor SCP1000-D01 reduces design system's weight and circuit complexity, which complies with the requirements of system.

2.2 FHR Acquisition and Processing Module

Above all, the original heterogeneous FHR collected by acceleration sensor is translated into the analog voltage signal. Then by means of three steps it achieves the differential amplifier through the preamplifier circuit. It removes the noise of FHR signal by the signal processing circuit which has the filtering and a series of intermediate signal processing to convert the electrical signal to the digital signal through the A/D conversion and transmits it the Bluetooth module by an optical isolator device, during this conversion and transmission process signal processing circuit and A/D converter and photoelectric coupler are shown in Figs. 2 and 3:

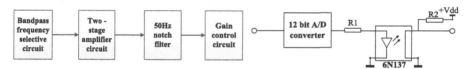

Fig. 2. Intermediate signal processing circuit

Fig. 3. A/D converter and optocoupler

2.3 Wireless Module

The wireless Bluetooth module used in the study is a low-power (BLE4.0) and single-mode chip CC2541. The single-mode chip will combine a variety of power of electronic devices, for example, RF transceiver, 8051 MCU, programmable flash memory, 8 KB RAM and so on. The chip CC2541 is designed for Bluetooth low power consumption as well as Power-optimized chip with private 2.4 Hz applications. The chips operating frequency of 32 MHz and 12 bits analog to digital converter with 8 channel resolution and programmable can be low overall BOM cost to build the strong network node. The chip is compliant with the Bluetooth 4.0 protocol stacks, perfectly compatible with Apple and android smartphone, which satisfies the design requirements of the system.

Table 1. CC2541 low power parameters

Project	Parameter
Operating mode RX(receive mode)	≥ 17.9 mA
Operating mode TX (transmission mode 0dBm)	18.2 mA
Power mode 1(suspend to 4 μs)	270 μs
Power mode 2(open sleep timer)	1 μs
Power mode 3(external interrupt)	0.5 μs
Range of wide power supply voltage	2 V–3.6 V

Bluetooth (BLE) 4.0 which combines traditional Bluetooth technology, high-speed technology and low energy technology, compared with the older version of the Bluetooth technology, has reduced the power consumption by 90 %. It is targeted at wireless solutions whose cost and power requirements are high and widely used in many fields of home entertainment, health care, security and so on. And it supports the transporting of ultra short packet which ranges in groups from 8 bits to 27 bits within 2Mbps data transmission rate, and the range of maximum transmission can be more than 100 meters (depending on the application, different distances). The chip has a higher safety factor for using AES-128 CCM encryption algorithm to encrypt and authenticate data packets; Table 1 is low-power parameter of the chip CC2541:

3 Processing Algorithm of FHR Signal

Collection of the mother and fetus mixed ECG signal from the mother's abdomen has the following characteristics:

(1) Maternal and fetal ECG mixed signals would be ignored for its short-time transmission, therefore the mixed signal observed can be regarded as a linear instantaneous mixture model with a delay of 0 [5].
(2) Maternal and fetal ECG statistically independent of each other [6].
(3) The pseudo-cyclical. Due to the above three typical features, this article uses the blind source separation algorithm based on second-order statistics which takes advantage of characteristics of sequence structure of two order statistics of sample data and source signal to achieve blind source separation signal to extract the fetal ECG, and R-wave is an important basis for determining the timing FECG structure.

Steps of periodic element analysis algorithms to extract the FECG: abdominal observed signal x(t) pre-processed and pre-whitening processed to obtain a signal z (t) = Tx(t), and it is evaluate z(t) for correlation matrix Rz(0). Moreover, R wave is estimated by dyadic wavelet transform and modulus maximum, once more it find the optimal time delay $\tilde{\tau}_t$ which is used to calculate the time-delay of dispersion matrix $Rz(\tilde{\tau}_t)$. The separative matrices W is calculated by generalized eigenvalue decomposition of matrix pair $(Rz(0), Rz(\tilde{\tau}_t))$, y(t) = $W^T x(t)$ is a combination of estimation of fetal ECG signals. Take $y_1(t)_1$ as the estimated value of the fetal ECG signal. Taking

$(y_1(t)_1, y_1(t + \tau t)_1, y_1(t + 2\tau t)_1, y_1(t + 3\tau t)_1)^T$ regard as the input of updating observation signal and return the step of pre-whitening processing to iteration number of n times and purifying the FHR signals. Finally, the estimated value of fetal ECG is $y_1(t)_n$.

The principle of blind source separation of the whole system (Fig. 4):

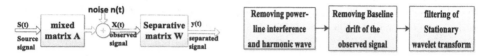

Fig. 4. Principle diagram of the model of blind source separation

Fig. 5. The pre-processing of abdominal observation signal

The algorithm of blind source separation as follows [7]:

$$X_i = \sum_{j=1}^{n} a_{ij} s_j(t) + n_i(t), i = 1, 2, \cdots, m \tag{1}$$

The output of the acceleration sensor is X_i, a_{ij} is a mixing coefficient, S_j is the j-th source signal, $n_i(t)$ is observation noise or interference of the sensor. The signal which is the collection of the FHR data from abdomen of pregnant women transfer to the hardware acquisition system by electromagnetic coupling. The noise or interference of the ECG signal mainly includes power-line interference, baseline drift and EMG interference, the pretreatment process of the abdominal FECG signal is shown in Fig. 5. Elimination of power frequency interference and harmonic as follow:

$$H_{notch}(z) = \frac{bN(z)}{N(\rho^{-1}z)} = b\frac{1 - z^{-D}}{1 - az^{-D}} \tag{2}$$

$$a = \frac{1 - \beta}{1 + \beta}, b = \frac{1}{1 + \beta}, \beta = \tan\left(D\Delta\varpi/4\right), D = \frac{f_s}{f_1}, \Delta\varpi = \frac{2\pi\Delta f}{f_s} \tag{3}$$

In the Formula (3), $f_s = 1000$ Hz represents the sampling frequency of the system and f_1 represents the fundamental frequency of power-line interference.

Median filtering algorithm is used to remove baseline drift:

$$y(i) = x'(i) - x(i) \tag{4}$$

and Pre-whitening algorithm of observation signal (random vector x) [8]:

$$Z(t) = C_0 S(t) \tag{5}$$

In the Eq. (4), $y(i)$ is expressed as the baseline correction of the abdominal ECG signals, the sequence of original signal is $x(t) = \{x(1), \cdots, x(N)\}$, the width of the sort window is "L" ($1 \leq L \leq N$), starting from the original signal sequence at any location i-th, the signal x' in the sort window is obtained, and it is known as $x'(i) = \{x(i), \cdots, x(i+L-1)\}$. Taking the median of x' will be sorted as output and the estimated value of the limit drift signal will be regarded as $x_0(i)$. In the Eq. (5), $S(t)$ as the source signal, $Z(t)$ is pre-whitening signal, the relationship between them is the orthogonal transformation, where $C_0 = TA$ is orthogonal matrix, $T = \Lambda^{-1}Q^T$ is whitening matrix, eigen value decomposition of correlation function matrix Rx of mixed signal vector x is $T = Q\Lambda^2 Q^T$. The matrix Λ^2 is a diagonal matrix, the diagonal elements $\lambda_1^2, \lambda_2^2, \cdots, \lambda_n^2$ is characteristic value of matrix Rx, the column vector of orthogonal matrix "Q" is Eigenvectors Orthonormal which is corresponding to these characteristic values.

Basic processing of R wave detection in FECG [9]: (Fig. 6)

Fig. 6. Basic processing of R wave detection in FECG

It is assumed that the original signal has been $f(t) \in L^2(R)$, the algorithm of j scale "dyadic wavelet transform" for signal is expressed as Formula (6). $\Psi_{a,b}(t) = 2^{-j/2}\Psi[2^{-j}(t-b)]$ is dyadic wavelet sequence which is obtained by translating and scaling the mother wavelet $\Psi(t)$. $a = 2^j$ (j is decomposition level) is the stretch factor, b is the translation factor. d_j set to j scale wavelet detail coefficients of $W_f(a,b)$, and if there is a dot $(t_0, d_j(t_0))$:

$$W_f(a,b) = \langle f, \Psi_{a,b} \rangle = \int f(t)\overline{\Psi_{a,b}(t)}dt < \infty \tag{6}$$

$$|d_j(t)| \leq |d_j(t_0)|, t \in (t_0 - \delta, t_0 + \delta) \tag{7}$$

The modulus maxima dot of the wavelet transform is $(t_0, d_j(t_0))$, and $d_j(t_0)$ is the modulus maxima. The modulus maxima of any point in a neighborhood of t_0 (except t_0) is set to 0, and the set of all the modulus maxima in the time domain $(0, t)$ is called the modulus maximum sequence

$$\text{Amax}: \text{Amax} = \text{Amax}_m + \text{Amax}_f \tag{8}$$

Comparison between R wave measured by this algorithm and R wave measured by Shehada algorithm [10] as Table 2.

Periodic Element Analysis Algorithm principle: looking for periodic structure included in the n-dimensional observation vector x, cyclical measure is defined as:

$$\varepsilon(w, \tau) = \frac{\sum_t |s(t+\tau) - s(t)|^2}{\sum_t |s(t)|^2} \tag{9}$$

$$\varepsilon(w, \tau) = \frac{w^T A_x(\tau) w}{w^T R_x(0) w} = 2\left[1 - \frac{w^T R_x(\tau) w}{w^T R_x(0) w}\right] \tag{10}$$

Take $s(t) = w^T x(t)$ into the Formula (9) obtained Eq. (10), τ is the interesting of cycle length, $w = [w_1, w_2, \ldots, w_n]^T$ is defined as the transformation vector of $s(t) = w^T x(t)$, w and periodic structure $s(t)$ is obtained by minimizing the $\varepsilon(w,\tau)$. The $R_x(0) = E[x(t+\tau)x(t)^T]$ is zero delay correlation matrix of x(t), $R_x(\tau) = E[x(t+\tau)x(t)^T]$ is the delay correlation matrix of x(t), the time delay is τ, $A_x(\tau)$ can be expressed as:

$$A_x(\tau) = E\big[[x(t+\tau) - x(t)][x(t+\tau) - x(t)]^T\big] = 2(R_x(0) - R_x(\tau)) \tag{11}$$

Generalized eigenvalue decomposition of the matrix (A,B) which is composed of "A" and "B" can be expressed as Formula (12). In the Formula (12), $A,B \in R_{N \times N}$, characteristic matrix $w = [w_1, w_2, \ldots, w_n]$ which is composed of feature vector w_i is also Diagonalization of transformation matrix "A" and "B", "I" is the unit matrix, Generalized eigenvalue corresponding to w_i is the diagonal matrix $\lambda_i (\lambda_1 \geq \lambda_2 \geq \cdots \geq \lambda_n)$, according to the properties of the symmetric matrix, the largest generalized eigenvalue λ_i corresponds to the feature vector w_i, the corresponding Rayleigh quotient's maximum value:

$$W^T A W = D, W^T B W = I \tag{12}$$

$$Max(J(w)) = Max\left(\frac{w^T A w}{w^T B w}\right) \tag{13}$$

The improvement of optimal time delay τ_t[8]:

$$\tilde{\tau}_t = \min\{\tau_f | \Phi_f(t+\tau_f) = \Phi_f(t), \text{ and } \min\{\tau_m | \Phi_m(t+\tau_m) = \Phi_m(t)\}, \ |\tau_m - \tau_f| \geq \delta, \delta > 0\} \tag{14}$$

ϕ_f, ϕ_m is the fetal ECG phase and maternal ECG phase respectively, τ_f and τ_m represent the optimal time delay of the delay correlation matrix respectively. The Formula (14) indicates the lower limit ($\geq \delta$) of the sample point delay difference of Fetal ECG delay and Maternal ECG delay. The computer simulation of the FHR signal is obtained by Second-order statistics of Blind Source Separation Algorithm as shown in the following picture (Fig. 7):

4   System Software Design

4.1   Main Program Schematic

The main program schematic noninvasive fetal heart monitoring system based on Bluetooth networks as Fig. 8. The system of main program schematic includes the following components:

Table 2. Comparison of accuracy of R wave

Name	Accuracy of R wave detection
This algorithm	95.3 %
Shehada algorithm	91.2 %

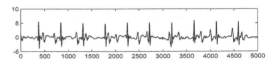

Fig. 7. FHR signal

(1) The Bluetooth module of the system is initialized.
(2) The FHR signal is collected, analysed and calculated.
(3) Initialize mobile application software configuration.
(4) FHR monitoring module and mobile devices to communicate via Bluetooth module, Bluetooth module receives the control command from the mobile phone and to control the operation of the FHR monitoring module, the phone receive and parse the effective data displayed and stored through the Bluetooth module transmission.

4.2   Mobile Terminal

Mobile phone monitoring terminal must support the version of Bluetooth 4.0 or higher and network functions, such as iPhone android 4.3 or later phones etc. The system select the HUAWEI P8 mobile phone, 2G memory, 5 inches of the display, support for Bluetooth 4 and WLAN hotspot, mobile 4G network etc. In addition, android software development tools are the Android 4.3 version. The phone terminal is connected with the FHR monitoring module through the wireless Bluetooth, and the detected data are displayed in real time. The combination of package ID, packet length, data, and the data of checksum bits regards as a transmission format, through the package ID to identify the FHR data. The structure of the system's mobile terminal is shown in Fig. 9.

5   Summary

In this piece of research work the authors describe the design of a household non-invasive FHR monitoring system based on low-power wireless Bluetooth network. Due to the low power consumption of Bluetooth and sensor, the system's battery will be longer. Furthermore, the system deals with FHR signal through Second-order

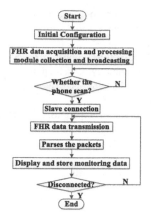

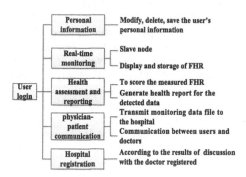

Fig. 8. The main program schematic

Fig. 9. Application structure

statistics of Blind Source Separation Algorithm and it is likely to monitor FHR for a long time in anywhere. Moreover, the measured waveform can be clearly reflected in the phone. Currently remote wireless fetal electronic monitoring is becoming very significant in terms of its wide applicability, Furthermore, development prospects has been an important content in our family and community guardianship in perinatal medicine. If the system is combined with the Internet and the cloud computing to introduce a function which uploads the FHR data to the hospital server, it will be very convenient for the doctors to analyse, backup and manage the FHR data. As a result, it also has great use for full implementation of "two children" policy.

Acknowledgments. The Authors express great sense of gratitude and indebtedness for the invaluable support and contributions from Dr. Chen Mingrui and Mr. Wu. This work has been partially funded by Social development science and technology special in Hainan province (No. 2015SF32).

References

1. Katz, M., Gill, P., Newman, R.: Detection of preterm labor by ambulatory monitoring of uterine activity: a preliminary report. Obstet. Gynecol. **68**(6), 773–778 (1986)
2. Nizhong, F., Guowei, G.: Fetal heart rate monitor designed with a new type of MEMS accelerometer. Sens. World **12**(3), 41–43 (2006)
3. Jinmei, S.: The extraction of fetal ECG signals and the application and development of the remote monitoring system. Pract. Electron. **11**, 219 (2013)
4. Hainan, Y.: The FECG detection algorithm and implementation based on BSS. Harbin University of Science and Technology, Harbin (2011)

5. Sugumar, D., Vanathi, P.T., Mohan, S.: Joint blind source separation algorithms in the separation of non-invasive maternal and fetal ECG. In: Electronics and Communication Systems (ICECS), pp. 1–6 (2014)
6. Yue, T.: Fetal electrocardiogram (FECG) acquisition and extraction algorithm based on blind source separation theory. South China University of Technology, Guangzhou (2012)
7. Kun, C.: Research on fetal electrocardiogram extraction with blind source separation method. South China University of Technology, Guangzhou (2011)
8. Hongyuan, Z.: Theoretical and experimental studies of blind source separation. Shanghai Jiao Tong University, Shanghai (2000)
9. Wenlong, T.: Research on non-invasive fetal ECG extraction. Tianjin University of Technology, Tianjin (2015)
10. Shehada, D., Khandoker, A.H.: Non-invasive extraction of fetal electrocardiogram using fast independent component analysis technique.In: Biomedical Engineering (MECBME), pp. 349–352 (2014)

No-Reference Network Packet Loss Video Quality Assessment Model Based on LS-SVM

Jin Wang[1,2](✉) and Yi bin Hou[1](✉)

[1] School of Software Engineering,
Beijing University of Technology, Beijing, China
805372192@qq.com, wangjin1204@emails.bjut.edu.cn,
ybhou@bjut.edu.cn
[2] Shijiazhuang Tiedao University, Shijiazhuang, China

Abstract. The authors in this paper propose to build a NS2 + MyEvalvid simulation platform for the purpose of qualitative evaluation of a video signal which emerge through the network transmission. The paper also emphasizes on feature extraction using Least squares support vector machine method to establish no-reference network packet loss video quality assessment model based on LS-SVM. The experimental results show that LS-SVM's training speed is fast and the proposed model displays greater level of accuracy than the other models.

Keywords: No-reference · Network packet loss · Least squares support vector machine

1 Introduction

In order to obtain a good Quality of Experience (QoE) [1, 2], we research on the factors affect the video's quality and use them establish no-reference network packet loss video quality assessment model based on LS-SVM [3–5]. In related work, the reference [6] mainly proposed packet loss's scale characteristic. References [7, 8] researches in network traffic and discusses several existing long-range dependence model. Reference [9] using MPEG4 codec and HD video under MyEvalvid platform research on the influence of packet loss on QoE and set up the mapping model of packet loss rate and the Quality of experience in matlab environment. Reference [2] presented an adaptive scheme that is the experience quality driven in order to optimize the content supply and network resource utilization for application of video in wireless networks. Reference [10] considers the packet loss concentration and packet loss rate's impacts on the quality of video, proposed a no-reference network packet loss video quality assessment model. Reference [11] considering different frame type drop and packet loss rate impacts on video quality, put forward a kind of method don't need to decode video which called no-reference video quality assessment model. Reference [12, 13] puts forward support vector machine (support vector machinse, SVM) which has become the current research hotspots in the field of machine learning, it has excellent learning performance. Reference [14] for problems which exists in least squares support vector machine parameters optimization, proposed least squares support vector machine

© Springer International Publishing AG 2017
F. Xhafa et al. (eds.), *Recent Developments in Intelligent Systems and Interactive Applications*,
Advances in Intelligent Systems and Computing 541, DOI 10.1007/978-3-319-49568-2_57

(LS-SVM) parameters self-tuning optimization algorithm which is a cross validation. On the whole, this paper finally establishes no-reference network packet loss video quality assessment model based on LS-SVM.

2 The Least Squares Support Vector Machine Theory

Support Vector Machine's basis is the optimal separating hyperplane, as shown in Fig. 2. LS-SVM regression is also known as the least squares support vector machine (LS-SVR) is often used for multivariate nonlinear regression analysis, nonlinear simulation and forecasting, Least squares support vector machine is running fast, accuracy is high. For nonlinear sample data,

$(x_1, y_1), (x_2, y_2) \ldots (x_i, y_i) \ldots (x_l, y_l)$ $x_i, y_i \in R$. Using least squares support vector machine regression for function estimation, then the optimization problem becomes as in (1).

$$\min \frac{1}{2} \|w\|^2 + \gamma \sum_{i=1}^{l} \varepsilon_i^2 \tag{1}$$

Constraints are as in (2).

$$y_i = w^T \phi(x_i) + b + e_i, i = 1, 2, 3, \ldots l \tag{2}$$

The corresponding Lagrangian form

$$L = \frac{1}{2} \|w\|^2 + \gamma \sum_{i=1}^{l} \varepsilon_i^2 + \sum_{i=1}^{l} \alpha_i \left\{ w^T \varphi(x_i) + b + e_i - y^i \right\} \tag{3}$$

By the conditions of KTT, there are 4 equations as shown as in (4), (5), (6), (7). By

$$\partial L / \partial w = 0, obtain \ w = \sum_{i=1}^{l} \alpha_i \phi(x_i) \tag{4}$$

By

$$\partial L / \partial b = 0, obtain \ \sum_{i=1}^{l} \alpha_i = 0 \tag{5}$$

By

$$\partial L / \partial e_i = 0, obtain \ \alpha_i = \gamma e_i \tag{6}$$

By

$$\partial L/\partial\alpha_i = 0, obtain \ w\phi(x_i) + b + e_i - y_i = 0 \tag{7}$$

In the Eq. (3) using (4) and (7) after stripped e_i and w, to solve the least squares support vector machine's the implementation of the form, as in (8)

$$\begin{bmatrix} 0 & 1^T \\ 1 & \Omega + \gamma^{-1}I \end{bmatrix} \begin{bmatrix} b \\ a \end{bmatrix} = \begin{bmatrix} \frac{0}{y} \\ y \end{bmatrix} \tag{8}$$

Among them, $y = [y_1, y_2, \ldots y_N]$, $I = [1, 1, 1, \ldots 1]$, $\alpha = [\alpha_1, \alpha_2, \ldots \alpha_N]$, I is Matrix for the unit, as in (9).

$$\Omega_{i,j} = K(x_i, y_i) = \varphi(x_i)\varphi(x_j), i, j = 1, 2, \ldots l \tag{9}$$

$K(x_i, y_i)$ is the kernel function of support vector machine (SVM). Solving that is shown as in (10).

$$f(x) = \sum_{i=1}^{l} \alpha_i k(x_i, x) + b \tag{10}$$

3 Feature Extraction

3.1 Quantization Parameter Affecting the Users' Quality of Experience

Where the use of quantitative parameter Q value is bigger the video quality QoE will be worse and worse. At the same time we can also find the compression data quantity which use Q value is more, it need more packets to send [15].

3.2 Different Output Link Speeds Have Impact on the Users' Quality of Experience

N = 5, Hurst parameter is 1.5, link = 10 MB. We can assume the output link speed [18] is bigger, self-similar is smaller, Hurst parameter is smaller, packet loss rate is smaller. The output link speed respectively set to 5 MB, 10 MB, 15 MB, 20 MB, under the condition of these settings the flow rate Figures are as shown below Figs. 1, 2, 3 and 4. By observing the above four linetypes of the figures and the area the linetype and the horizontal axis enclosed we can find that: with the increasing of the output link speed, self-similar [16] decreases, Hurst parameter decreases, packet loss rate is smaller. Thus we can came to the conclusion that the output link speed is bigger, self-similar is smaller, Hurst parameter is smaller, packet loss rate is smaller [17].

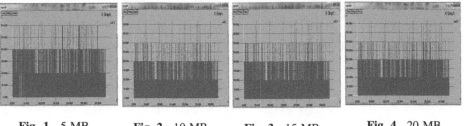

Fig. 1. 5 MB   **Fig. 2.** 10 MB   **Fig. 3.** 15 MB   **Fig. 4.** 20 MB

3.3   Video Content Complexity

In the no-reference video quality evaluation, no need to completely decoding the video. It only requires the decoding to the stream of the Baotou information, and then determine the time complexity of the video, decoded the Baotou and get the information including: Quantization parameter Q, the frame type, code rate, the number of packet loss, the location of the packet loss and each frame on the display's display time. So time complexity model is as follows, among them, σ represent the time complexity of the video: Q is coded quantization parameter; R is the code rate; a and b is parameter for undetermined model. $\sigma = Q(aR + b)$, Time complexity can be used to represent the video content complexity.

3.4   Feature Extraction

Parameters will be divided into three categories [18], the coding parameters, network parameters, content layer parameters. Here first select five representative parameters, encoding parameters QP, network parameters for the packet loss rate and bandwidth that is the output link speed, the output link speed including simplex link speed and duplex link speed, the content layer parameters is video content complexity.

4   Establish LS-SVM's No-Reference Video Quality Assessment Model

4.1   Topology Description and HD Video Options and the Experimental Process

Wired topology structure consists of 4 nodes, between n0 and n1, n2 and n3 are duplex links, link bandwidth is 10 MBPS, delay time is set to 1 ms. Between n1 and n2 is simplex link, the bandwidth is 640 KB, delay time is set to 1 ms. We select HD video sources which mainly have 525 series and 625 series. For src13 video, data points looks less, exploring the parameter values each time how much interval once can not affect obtain the final turning point, and in the case of reduce test times as much as possible to get the packet loss rate's turning points between 0 and 1, mainly using the vernier caliper's main ruler and the principle of the vernier. For src22 video, delay time

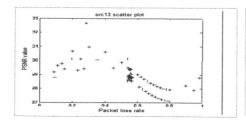

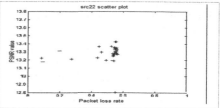

Fig. 5. src13_hrc1_525.yuv   **Fig. 6.** src22_hrc1_525.yuv

setting has no effects on the user's quality of experience and packet loss rate. cLink speed lower limits are 10 MB and 74 KB. Src13 as shown in Fig. 5, src22 as shown in Fig. 6.

For src13 wired network environment use LS-SVM method described below. (1) Enter a value for the X [19]. The input sequence is quantitative parameter, packet loss rate, the simplex output link speed, the duplex output link speed, video's time complexity, and these parameters are equal weights. X = [X1, X2, X3, X4, X5], X1 = [31, 31, 31, 31, 10......]^T, X2 = [0.553 226, 0.550 704, 0.982 976, 0.556 273, 0.549 443,......]^T, X3 = [6.4, 0.64, 0.64, 0.064, 0.064......]^T, X4 = [100, 100, 0.1, 10, 10......]^T, X5 = [19 998 720.1, 19 998 720.1, 19 998 720.1, 19 998 720.1, 6 451 200.03......]^T. (2) Then, enter a value for the Y. Here, Y value represent PSNR value. Y = [14.672 341, 14.682 139, 14.388 424, 14.388 424, 14.388 424......]^T. (3) Setting gam and parameters corresponding kernel function involved. gam = 50; sig2 = 0.5; type = 'function estimation'; LS-SVM requires call two parameters, ls-svm parameters gam and sig2, among them decide adapt error's minimization and smooth degree's regularization parameter is gam, RBF function's parameter is sig2. Type has two type, one is used for classification's classification, one is used for function regression's estimation. (4) Algorithm training, build a model. Use trainlssvm function to achieve establish model. It according to the sample's input and output and the training function's parameters preset, trained the network, get ls-svm's support vector machine and the corresponding threshold value. [alpha, b] = trainlssvm({X, Y, type, gam, sig2, 'RBF_kernel'}). Get, alpha = [−0.016 6, 0.017 5, −1.004 6, −1.083 2, −1.360 2......], b = 0.525 3. Otherwise, use statements [alpha, b] = trainlssvm({X, Y, type, gam, sig2, 'RBF_kernel', 'preprocess'}). Trainlssvm function is LS-SVM, it is one of the important function of toolbox, it is the ls-svm's training function. Kernel function is RBF function use 'RBF_kernel' show. Alpha is support vector, b is threshold value. Data has been normalized use Preprocess show, it can also be indicates that the data is not normalized 'original', when the default is 'preprocess'. Therefor, alpha this array is α_i's value, b = 0.525 3 the threshold value in this formula. For src13 wired network environment such as:

$$(x) = \sum_{i=1}^{l} \alpha_i k(x_i, x) + b \quad \text{so,} \quad f(x) = \sum_{i=1}^{l} \alpha_i k(x_i, x) + 0.5253, \quad \text{among them,} \quad \alpha_i =$$

[−0.016 6, 0.017 5, −1.004 6, −1.083 2, −1.360 2......]. Therefor, for src22 wired use LS-SVM method establish no-reference network packet loss video quality assessment model is the same as src13 wired network environment [20].

Establishment's results as follows: Src22 wired: $f(x) = \sum_{i=1}^{l} \alpha_i k(x_i, x) + b$ so,

$f(x) = \sum_{i=1}^{l} \alpha_i k(x_i, x) + 0.4110$, among them, $\alpha_i = [-34.229\ 9, -42.082\ 5, -43.600\ 8,$
$-7.271\ 9, -7.704\ 3......]$, b = 0.411 0.

4.2   The Analysis of Experimental Results

The evaluation indexes are mainly R-square, RMSE (Root Mean Square Error, RMSE), SROCC (Spearman Rank Order Correlation Coefficient, SROCC), Pearson, and these values are between 0 and 1 [21]. The horizontal header is coefficient of each measures, the vertical header is a variety of methods, for these evaluation indexes, RMSE is the smaller the better, while R-square, SROCC, Pearson are bigger are better. LS-SVM method suitable for small samples training, BP neural network [22] suitable for big samples training, so here is suitable use the LS-SVM method [23, 24]. The experimental results show as Table 1, LS-SVM's performance is better than BP, PSNR and SSIM, and training speed is fast, the model is more accurate.

Table 1. LS-SVM method compared with other classic models

Video	R-square	RMSE	SROCC	Pearson
LS-SVM	0.8889	0.006	0.998	0.9883
BP neural network	0.7999	0.66	0.888	0.45
PSNR	0.588	0.16	0.634	0.71
SSIM	0.666	0.15	0.815	0.83

5   Conclusion

First, study on different output link speeds has a significant impact on the users' quality of experience. Second, set the Mapping model of packet loss rate and the Quality of experience on the influence of packet loss on QoE. Third, establish no-reference network packet loss video quality assessment model based on LS-SVM. Finally experiments show that, the LS-SVM has better generalization ability and the training speed is faster.

Acknowledgments. This work was partially supported by the National Natural Science Foundation of China (No: 60963011, 61162009).

References

1. Kim, H.J., Choi, S.G.: A study on a QoS/QoE correlation model for QoE evaluation on IPTV service. In: 2010 The 12th International Conference on Advanced Communication Technology (ICACT), vol. 2, pp. 1377–1382. IEEE (2010)
2. Khan, A., Sun, L., Jammeh, E., et al.: Quality of experience-driven adaptation scheme for video applications over wireless networks. IET Commun. **4**(11), 1337–1347 (2010)
3. Zhang, F., Steinbach, E., Zhang, P.: MDVQM: a novel multidimensional no-reference video quality metric for video transcoding. J. Vis. Commun. Image Represent. **25**(3), 542–554 (2014)
4. Suykens, J.A.K., De Brabanter, J., Lukas, L., et al.: Weighted least squares support vector machines: robustness and sparse approximation. Neurocomputing **48**(1), 85–105 (2002)
5. Sebald, D.J., Bucklew, J.A.: Support vector machines and the multiple hypothesis test problem. IEEE Trans. Sig. Process. **49**(11), 2865–2872 (2001)
6. Leland, W.E., Taqqu, M.S., Willinger, W., et al.: On the self-similar nature of ethernet traffic (extended version). IEEE/ACM Trans. Networking **2**(1), 1–15 (1994)
7. Karagiannis, T., Molle, M., Faloutsos, M.: Long-range dependence ten years of Internet traffic modeling. IEEE Internet Comput. **8**(5), 57–64 (2004)
8. Willinger, W., Taqqu, M.S., Sherman, R., et al.: Self-similarity through high-variability: statistical analysis of ethernet LAN traffic at the source level. IEEE/ACM Trans. Networking **5**(1), 71–86 (1997)
9. Wang, Z., Wang, J., Xia, Y., Wan, Z., Li, L., Cai, C.: Mapping model of packet loss rate and the Quality of experience on the influence of packet loss on QoE. In: ISITC (2014)
10. Maisonneuve, J., Deschanel, M., Heiles, J., et al.: An overview of IPTV standards development. IEEE Trans. Broadcast. **55**(2), 315–328 (2009)
11. Tao, S., Apostolopoulos, J., Guérin, R.: Real-time monitoring of video quality in IP networks. IEEE/ACM Trans. Networking (TON) **16**(5), 1052–1065 (2008)
12. Chen, J., Ji, G.: Weighted least squares twin support vector machines for pattern classification. In: 2010 The 2nd International Conference on Computer and Automation Engineering (ICCAE), vol. 2, pp. 242–246. IEEE (2010)
13. Chen, J., Ji, G.: Weighted least squares twin support vector machines for pattern classification. In: Proceedings of the 2nd International Conference on Computer and Automation Engineering (2010)
14. Zheng, L., Zhou, H., Wang, C., et al.: Combining support vector regression and ant colony optimization to reduce NOx emissions in coal-fired utility boilers. Energy Fuels **22**(2), 1034–1040 (2008)
15. Wang, Z., Wang, W., Xia, Y., Wan, Z., Wang, J., Li, L., Cai, C.: Visual quality assessment after network transmission incorporating NS2 and Evalvid. Sci. World J. **2014**, article ID 267403, 7 p. (2014). doi:10.1155/2014/267403
16. Willinger, W., Taqqu, M.S., Sherman, R., et al.: Self similarity through high-variability: statistical analysis of ethernet LAN traffic at the source level. IEEEACM Transactions on Networking **5**, 71–86 (1997)
17. Wang, J.: The influencing factors of network packet loss's long-range dependence has impacts on the packet loss rate. Int. J. Multimedia Ubiquitous Eng. **10**(11), 161–172 (2015)
18. Suykens, J.A.K., Vandewalle, J.: Least squares support vector machine classifiers. Neural Process. Lett. **9**(3), 293–300 (1999)
19. Suykens, J.A.K.: LS-SVMlab ToolboxUser's Guide (2005). http://www.esat.kuleuven.ac.be/sista/

20. Seong, W.L., Hee, H.S.: A new recurrent neural network architecture for visual pattern recognition. IEEE Trans. Neural Netw. **8**, 331–340 (1997)
21. Yamagishi, K., Hayashi, T.: Analysis of psycological factors for quality assessment of interctive multimodal service. Electron. Imaging **5666**, 130–138 (2005)
22. Du, H., Guo, C., Liu, Y., et al.: Research on relationship between QoE and QoS based on BP neural network. In: IEEE International Conference on Network Infrastructure and Digital Content, IC-NIDC 2009, pp. 312–315. IEEE (2009)
23. Brown, M., Lewis, H.G., Gunn, S.R.: Linear spectral mixture models and support vector machines for remote sensing. IEEE Trans. Geosci. Remote Sens. **38**(5), 2346–2360 (2000)
24. Zhao, Q., Principe, J.C.: Support vector machines for SAR automatic target recognition. IEEE Trans. Aerosp. Electron. Syst. **37**(2), 643–654 (2001)

Blind Separation of Radar Signals Based on Detection of Time Frequency Single Source Point

Xude Cheng[✉], Fuli Liu, Xuedong Xue, Bing Xu, and Yuan Zheng

Wuhan Mechanical Technology College, Wuhan 430075, Hubei, China
xxdmymail1228@126.com

Abstract. Radar signal sorting is a key part of radar reconnaissance. A blind separation algorithm based on detection of time frequency is proposed in this paper and the problems of underdetermined radar signal sorting can be solved effectively. Firstly, the method is that single source point of each radar source signal was detected. Then; the mixing vector in the corresponding single source point set was estimated by Singular Value Decomposition (SVD). Finally, the mixing matrix simultaneously were estimated by the cluster validation technique based on k-means clustering algorithm, and the radar signals can be got by the mixing matrix and the observed signals. Each time domain waveform of radar source can be sorted based on this method, the time frequency graphs of radar signal can be got and the whole radar signal sorting process is accomplished.

Keywords: Time-frequency transformation · Single source point · Radar signal · Cluster validation · Blind separation

1 Introduction

Radar reconnaissance is a key part of radar countermeasures with which reconnaissance system sorts intercepted mixed signals in the airspace and extracts pulse description words (PDW) of each radar signal source. Finally it determines and positions the threat from each radar radiation source. In literature [1] the author applies blind source separation technology to radar signal sorting by which blind separation is conducted from four-order cumulants matrix in the space so as to realize the process of signal sorting. Simulation experiment also indicates that blind source separation is workable in radar signal sorting with ideal separation effect. In the simulation part of the literature only number of radiation sources is known, sorting outcome of the signals from unknown radiation sources is not mentioned. In literature [2] the author introduces blind separation of radar signals under over-determined condition by the method of fixed-point independent component analysis. Simulation experiment indicates high similarity ratio of separated radar signals in comparison with source signals, with higher SNR, information on clutters is remained, but nothing is mentioned regarding signal sorting outcome under underdetermine condition. In literature [3] the author introduces blind identification algorithm based on k means clustering, by which observed data after normalization is clustered, the clustering outcome is estimation of

© Springer International Publishing AG 2017
F. Xhafa et al. (eds.), *Recent Developments in Intelligent Systems and Interactive Applications*,
Advances in Intelligent Systems and Computing 541, DOI 10.1007/978-3-319-49568-2_58

mixing matrix, which solves the problem of blind decomposition of underdetermined signals. In literature [4], independent component analysis is made with observed signals of equal interval sections, which solves the problem of signal separation under the condition of insufficiently sparse radar signals, but both literatures mentioned above depend much on sparse decomposition statistics of the signals for signal decomposition.

In view of above drawbacks a method based on detection of time-frequency single source point [5, 6] is introduced in this paper, with which radar signal sorting with the number of unknown radiation sources under underdetermined condition is realized without depending upon the condition of sparse signals. The principle of the method based on detection of time-frequency single source point is that, after singular value decomposition of time-frequency single source points of various radar source signals detected, mixing vector of various sets of time-frequency single source points is derived, and finally estimation of mixing matrix is fulfilled by optimized validation method based on k means clustering. With this method, precise estimation is made from mixing matrix, time domain waveform of radar signals is derived from observed matrix and estimated mixing matrix, and signal spectrum is gained from further analysis of time domain waveform of signals, so as to fulfill signal sorting process.

In the process of radar reconnaissance, airborne radar receiving antenna of the reconnaissance aircraft intercepts mixed signals in the airspace, central processing module sorts intercepted mixed signals in the airspace by removing unrelated signals and further process of signals of interest [7]. However, in the process of reconnaissance of radar signals, there're a lot of radars on the ground in most cases, when there're only a few radar receiving antenna on reconnaissance aircrafts. Assume number of radar signals is P, which are radiated to the same airspace at the same time; airborne radar receiving antenna is uniformly arrayed in M order, then airspace signals intercepted by radar alarm receiver can be expressed as follows [8]:

$$x(t) = As(t) + n(t) \tag{1}$$

Where: $x(t) = [x_1(t), x_2(t), \ldots, x_M(t)]^T$ is observed signal; $s(t) = [s_1(t), s_2(t), \ldots, s_P(t)]^T$ is radar emitted signal; $n(t) = [n_1(t), n_2(t), \ldots, n_M(t)]^T$ is antenna noise signal; in the observation model, radar emitted signal and noise signal are mutually dependent; A is mixing matrix, expressed by: $A = [\alpha_1, \alpha_2 \ldots, \alpha_P]^T$. In matrix A, (i, k) element may be expressed by:

$$\alpha(i, k) = b_{ik} e^{-j2\pi f_k \tau_{ik}} \tag{2}$$

Where: b_{ik} means amplitude attenuation of arriving radar signal; τ_{ik} means time delay of arriving radar signal; f_k is signal frequency; spare representation of signals with short time Fourier transformation (STFT) [9–13]. In this paper, LFM signals are sued as target signals, and the signals can be expressed as follows:

$$s(t) = e^{j(\omega_0 t + \omega_1 t^2 + \theta_0)} \tag{3}$$

Where: ω_0 means signal carrier frequency; ω_1 means signal modulation ratio; θ_0 means signal initial phase; assume initial phase = 0 in order to simplify formula derivation; it's derived from STFT to both sides of Formula (1), that:

$$x(t,f) = As(t,f) + n(t,f) \tag{4}$$

Where: t, f spanning plane is whole time-frequency plane, $x(t,f)$, $s(t,f)$ and $n(t,f)$ are STFT outcome of observed matrix, signal matrix and noise matrix, respectively.

Definition 1. if, on the whole time-frequency plane, square of two norms of observed signal $x(t,f)$ is > 0, i.e. $\|x(t,f)\|_2^2 > 0$, point (t,f) is defined as time-frequency support point of observed signal $x(t,f)$; if antenna noise signal is taken into consideration, the criterion of time-frequency support point transforms to $\|x(t,f)\|_2^2 > \xi$, of which ξ is noise threshold.

Definition 2. if, on the whole time-frequency plane, $s_i(t,f) > > s_k(t,f)$ $i \neq k$ appears at random point (t,f), it's considered only signal $s_i(t,f)$ is at point (t,f), so point (t,f) is defined as time-frequency single source point of signal $s_i(t,f)$.

To fulfill signal blind sorting, it's assumes that, each signal has one discrete time-frequency single source point. All radar source signals have multiple time-frequency single source points, thus, if time-frequency single source points of different source signals are detected, estimation of corresponding mixing vectors can be derived; then, form a matrix with estimated mixing vectors, that is mixing matrix. Precisely estimated mixing matrix un-mixes observed signals, and then time domain waveform of each radar source signal is derived for further analysis of time domain waveform.

2 Mixing Matrix Estimation

In fact, above mixing matrix estimation is estimation of various components in mixing matrix. Due to ergodic m values, mixing vector under each m value is estimated; for mixing matrix A, it's actually multiple estimations of various components in A; therefore, mixing matrix estimation can be derived only by clustering analysis of components estimation $\hat{e}_k$. However, given the condition of number of unknown source signals, it's impossible for clustering with traditional k means clustering method in case of lack of priori condition of the number of source signals. In order to prove the validity of the algorithm, 4 LFM signals are selected, $f_s = 5000$ Hz, number of data sampling points N = 10000, starting frequencies of signals are [100, 200, 300, 400], modulation rates are [400, 500, 600, 700], sampling points in one time-width are selected as simulation signals.

Mixing matrix A is:

$$A = \begin{bmatrix} 0.5774+0.0000i & 0.5774+0.0000i & 0.5774+0.0000i & 1.0000+0.0000i \\ 0.2618+0.5146i & -0.5270+0.2359i & 0.3260-0.4765i & -0.9570-0.2901i \\ -0.3400+0.4666i & 0.3846-0.4306i & -0.2092-0.5381i & 0.8317+0.5553i \end{bmatrix} \quad (5)$$

Derived estimation matrix $\hat{A}$ is:

$$\hat{A} = \begin{bmatrix} 0.5795+0.0000i & 0.5784+0.0000i & 0.5804+0.0000i & 0.5774+0.0000i \\ 0.2574+0.5148i & 0.3265-0.4755i & -0.5513-0.1700i & -0.5271+0.2358i \\ -0.3396+0.4664i & -0.2078-0.5382i & 0.4782+0.3186i & 0.3847-0.4304i \end{bmatrix} \quad (6)$$

It's obvious from corresponding mixing matrix estimation that the sorting outcome by this method includes difference in the sequence of signals, but, for radar signal sorting, what is of interest is characteristic information in radar signals, therefore, difference in the sequence of signals has no impact on radar reconnaissance.

After estimation of mixing matrix, recover the signals; according to Formula (6), as long as inverse matrix of estimated mixing matrix is derived, source signal matrix is obtained by multiplying inverse matrix and observation matrix, then:

$$\hat{A}^{-1}x(t) = s(t) \quad (7)$$

Under underdetermined condition, mixing matrix is $M \times P$ dimensions, and $M < P$; thus, for non-square matrix inversion, generalized inverse matrix is used as inverse matrix; for this experiment, estimated number of dimensions of mixing matrix is 3×4; conduct singular value decomposition with estimation matrix, then:

$$[U,S,V] = svd(\hat{A}) \quad (8)$$

In Formula (8), U is 4×4 order unitary matrix; S is 3×4 order diagonal matrix; V is 3×3 order unitary matrix composed of characteristic vectors; then, generalized inverse matrix of matrix $\hat{A}$ is:

$$\hat{A}^+ = VS^+U \quad (9)$$

In Formula (9), "+" means generalized inversion; due to inherent error of Matlab simulation software in deriving inverse matrix, major error exists in deriving inverse matrix of estimated mixing matrix; to solve this problem, reiteration is employed here to reduce and even eliminate such error; time-frequency curve of recovered signals from deriving of generalized inverse matrix by reiteration is as shown in Fig. 1.

Where: $\hat{A}$ is estimation matrix, $\|\ \|_F$ is F norm; then, smaller estimation error indicates higher estimation precision; and bigger estimation error indicates lower estimation precision; in this experiment, sorting effect of this method is measured by estimation error of matrix E_A.

By change of SNR value in this experiment, the same sorting experiment is conducted with the method introduced here under different SNR conditions; in this

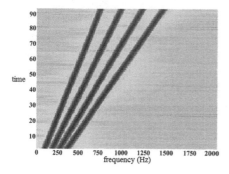

Fig. 1. TF curve of recovered signals

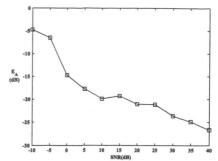

Fig. 2. Impact of SNR on estimation performance

experiment, SNR is increased from −10 dB to 40 dB at a step of 5 dB, and the experiment outcome is as shown in Fig. 2.

Figure 2 indicates that, under the condition of gradually increase of SNR with this method, the value of estimation error of matrix E_A is smaller, in other words, estimation precision is higher; however, within the full range of SNR, E_A value is negative all the time, indicating this method brings ideal effect with radar signal sorting even under the condition of low SNR.

Experiment 2: validates the advantages of this method.

Compare the method introduced here with TIFROM algorithm and the method utilizing traditional k means clustering; under the condition of continuous variation of SNR, compare estimation error of matrix between these methods, so as to weigh the merits and weakness of these methods; in this experiment, SNR is increased from −20 dB to 20 dB at a step of 2 dB; conduct 100 cycles of Monte Carlo risk analysis at each SNR value, and final comparative outcome is as shown in Fig. 3.

Experiment 3: success rate of detection technology based on k means clustering in estimation of number of sort centers.

With the method proposed from detection technology based on k means clustering, number of sort centers is estimated from assumed c_{max} value; this experiment functions to detect success rate of detection technology based on k means clustering; assume c_{max} values are 5, 6 and 7, under the condition of different c_{max} values, number of source signals is 4, i.e. number of sort centers is 4; conduct 100 cycles of Monte Carlo risk analysis under the conditions of SNR variation from −20 dB to 20 dB at a step of 2 dB, precision rate of the number of estimated sort centers with this detection technology under different SNR conditions; and final outcome is as shown in Fig. 4:

Figure 4 indicates that, as long as c_{max} value is bigger than the number of sort centers, number of sort centers can be estimated precisely, and finally realize k means clustering; this experiments has proved the validity of optimized detection technology based on k means clustering, which leads to precise estimation of sort centers; so, as long as set c_{max} value is bigger than number of sort centers, then: $c_{max} > P$ estimation can be achieved.

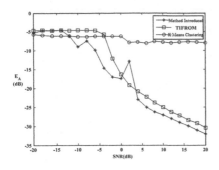

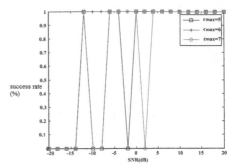

Fig. 3. Comparison of estimation performance of TIFROM

Fig. 4. Impact of different c_{max} on the estimation of number of sort centers

3 Conclusions

In this paper, signal sorting algorithms applied to modern radar reconnaissance are explored and a type of blind separation algorithm based on detection of time frequency single source point is proposed, which is valid in vector estimation of the set of single source points through detection of time-frequency single source points in observed signals, then cluster vector estimation by optimized k means clustering, and finally fulfill estimation of mixing matrix. This method has three main features as follows: it is able to overcome the difficulty of unknown number of source signals; it could effectively solve the problem of blind radar signal sorting under underdetermined condition; and 3; it is adaptive to wide range of SNR. Simulation outcome indicates that this method is able to separate radar signals under underdetermined condition and solve the problem of signal sorting in radar reconnaissance, thus it's valuable in practical application.

References

1. Hong, S., Huangbin, A.: A method to select radar signal based on blind source separation. Mod. Radar **23**(3), 47–50 (2006)
2. Wen-shu, X., Xing-gan, Z., Si-dan, D.: Blind separation of radar signals. J. Nanjing Univ. **42** (1), 38–43 (2006)
3. Li, Y.Q., Cichocki, A., Amari, S.I.: Analysis of sparse representation and blind source separation. Neural Comput. **16**, 1193–1234 (2004)
4. Xiao-jun, C., Hao, C., Bin, T.: Underdetermined blind radar signal separation based on ICA. J. Electron. Inf. Technol. **32**(4), 919–924 (2010)
5. Abrard, F., Deville, Y.: A time-frequency blind signal separation method applicable to underdetermined mixtures of dependent sources. Sig. Proc. **85**, 1389–1403 (2005)
6. Jie, Y., Wen-wen, Y., Hao, Y., Zhen-zhen, G.: Underdetermined blind source separation method based on independent component analysis. J. Vib. Shock **32**(7), 30–33 (2013)
7. Pakhira, M.K., Bandyopadhyay, S., Maulik, U.: Validity index for crisp and fuzzy clusters. Pattern Recogn. **37**, 487–501 (2004)

8. Zongli, R., Liping, L., Guobing, Q., Minggang, L.: Fast fixed-point algorithm based on complex ICA signal model with noise. J. Electron. Inf. Technol. **36**(5), 1094–1099 (2014)
9. Guo-long, L., Shu-min, D.: Study on algorithms of blind separation and DOA estimation about coherent sources. J. Harbin Eng. Univ. **31**(11), 1478–1484 (2010)
10. Bin, G., Woo, W.L., Dlay, S.S.: Single-channel source separation using EMD-subband variable regularized sparse features. IEEE Trans. Audio, Speech, Lang. Process. **19**(4), 961–976 (2011)
11. Klaus, N., Esa, O., Hannu, O.: On the performance indices of ICA and blind source separation. In: 2011 IEEE 12th International Workshop on Signal Processing Advances in Wireless Communications, pp. 461–465 (2011)
12. Ilmonen, P., Nevalainen, J., Oja, H.: Characteristics of multivariate distributions and the invariant coordinate system. Stat. Probab. Lett. **80**, 1844–1853 (2010)
13. Fan, G.U., Hui-gang, W., Hu-xiong, L.I.: A blind speech separation algorithm with strong reverberation. Sig. Proc. **27**(4), 534–540 (2011)

The Mechanisms of Wireless Resource Management of MBSFN

Shuguang Zhang($\boxtimes$), Min Wang, Qiaoyun Sun, and Yu Zhang

Information Department, Beijing City University, Beijing, China
shugzhang@163.com

Abstract. With the rapid development and growth of information society especially of mobile Internet applications. The demand for high bandwidth multimedia is growing fast. The fact that many users in same or adjacent cells receive same data stream at the same time promotes the development of multicast technology with the same frequency in the MBSFN. However the resource allocation in MBSFN is not easy. In this paper, the multi cell cooperative transmission mode of MBSFN and the diversity gain characteristics of UE end are studied. On this basis some resource allocation and scheduling mechanisms are discussed such as Strict Throughput Maximization, Max-Min Fairness, Fair and Equitable Scheme.

Keywords: MBSFN · Resource management · Resource allocation · Resource scheduling

1 Introduction

Due to the rapid development of mobile Internet applications and the popularity of more and more intelligent terminals the demand for high bandwidth multimedia services is increasing exponentially. In mobile cellular system, high efficiency and high quality multicast service is becoming an important research agenda at present and in the future. In order to adapt to the requirements of the development of wireless communication 3GPP proposed multicast/broadcast multimedia service (MBMS) concept in R6 (release 6). During the period of standard formation, 3GPP has introduced the evolution of MBMS standard, which introduced the multicast broadcast single frequency network transmission mode in the transmission mode.

The transmission mode of MBSFN is multi cell cooperative, i.e. a plurality of cells with the same frequency transmit the same multicast service simultaneously in the MBSFN so as to the diversity gain can be obtained more than one cell signal for multicast UEs. It effectively solves the problems of MBMS standards, including wireless resource insufficient usage, the interference of multicast inter cell and the coverage blind spots.

The multi cell cooperative transmission mode of MBSFN and the diversity gain characteristics of UE end making the MBSFN has new characteristics of transmission mode, compared with the traditional transmission mode of P-T-P and P-T-M.

© Springer International Publishing AG 2017
F. Xhafa et al. (eds.), *Recent Developments in Intelligent Systems and Interactive Applications*,
Advances in Intelligent Systems and Computing 541, DOI 10.1007/978-3-319-49568-2_59

2 Wireless Resource Management in MBSFN

Wireless resource management mainly includes two aspects: radio resource allocation and mobility management. However, in the present domestic and foreign research literature, the study of radio resource management for MBSFN is less. The MBSFN multicast service resource allocation generally use simple mode based on the worst UE channel quality resource allocation and there was no more deeper study [1]. In mobility management, the research on UE cell reselection for MBSFN idle state is not involved. Therefore, the resource allocation and the cell reselection characteristics for MBSFN were studied in this paper.

First, although the transmission in MBSFN is different from the traditional multi-cast, it is still a kind mode of multicast transmission. However multicast resource allocation is constrained by the worst channel quality UE, so the UEs with good channel quality cannot make full use of channel resources. As a result, MBSFN will also suffer from this limitation as multicast transmission mode. In MBSFN, all the cells use same resources to send multicast service with same frequency simultaneously.

The worst channel quality UE of a cell will affect the allocation of resources in a single frequency network, if the multicast resource is allocated according to the worst UE. For example, there is a very poor channel quality UE in a power constrained cell in the MBSFN, however other single frequency network cells UE's channel quality is relatively much better. If resources allocating is based on the poor UE's channel quality, the power limited cell need more wireless resources to meet the worst channel quality UE. At the same time, the other cells' UEs also need to use more resources according to the characteristics of unified resource allocation of MBSFN, regardless the fact that the UEs' channel quality of these cells is good or bad. Therefore, all of the UEs' channel resources will be wasted.

3 The Effect of the Worst Channel Users on the Resource Allocation

In MBSFN the worst channel UE will affect the single frequency network resource allocation. As a result, more UEs cannot make full use of the channel resource, which even leads to cells overload, and worsen other UEs' service quality in the cell. In the LTE cellular, a UE is in the idle state if it has access to the network but without connection of radio resource control (RRC) layer to the base station. Due to the absence of RRC layer of unable to connect with the base station uplink interaction, the idle state UE can only receive the paging messages of the base station and cannot receive any service messages. The reselection algorithm is to ensure idle UEs can always reside in the high-quality cells without loss of base station signal [2]. However, according to the 3GPP standard, the idle state UEs will receive MBSFN multicast service in the multicast broadcast single frequency network system.

However, when idle state UEs move from the MBSFN area to the normal cells, which are unicast cells and do not join the single frequency network, the cells will not send the multicast service to the idle state UEs. Therefore, the traditional cell reselection cannot guarantee the service continuity. As a result, the idle state UEs' cell reselection at the edge of the MBSFN not only needs to ensure no loss of base station signal, but also need to ensure multicast service continuity. At the same time, the transmission of multicast traffic in multicast broadcast single frequency network will have diversity gain from multiple cells, while the traditional cell selection considers the signal strength of the cell specific reference signal [3].

Moreover the signal strength is not enough to guarantee the idle state UE multicast channel quality judgment, therefore, cell reselection may be implemented while the multicast channel quality can guarantee multicast service in early, so as that channel switches to waste resources or generate unnecessary handoffs.

To sum up, MBSFN resource management is different from the traditional single cell resource management. As to MBSFN transmission characteristics, the problem of the limit of the MBSFN resource allocation caused by the worst channel quality UE and the problem of service continuity guarantee of the Idle state UE should be solved. In order to make full use of the limited wireless resource and to meet more and more high service requirements, the research on improving the utilization rate of MBSFN wireless resources and guaranteeing MBSFN multicast business continuity strategies of resource management is of great significance.

4 The Wireless Carriers in MBSFN

Multicast broadcast single frequency network (MBSFN) requires transmitting same signal waves from multiple cells at the same time. In this way, the UE receivers will be able to regard a large number of MBSFN cells as a large area. In addition, UE will not be affected by the inter cell interference, and will benefit from multiple MBSFN cell signal superposition. Not only that, such as G-RAKE and other advanced UE receiver technology can solve the multipath spread of the time difference, thereby eliminating the intra-cell interference.

MBSFN is divided into two kinds: MBSFN of the dedicated carrier and MBSFN of the unicast hybrid carrier, here we mainly discuss the design of hybrid carrier MBSFNRS. There are special requirements for the design of MBSFNRS. In a wireless frame of a mixed carrier MBSFN system, usually most of the resources are used for unicast traffic, and only a few sub-frames are used for MBSFN services. The typical scenario is that an isolated MBSFN sub-frame is inserted into continuous unicast sub-frames, which makes the receiving terminal of the MBSFN sub-frame unable to carry out interpolation channel estimation between RSs of the adjacent sub-frames as it receives a unicast signal. So the design of MBSFN RS must be able to support channel estimation within an isolated sub-frame.

5 Resource Scheduling Mechanism for Multicast Broadcasting System

The strong demand for high data rates and providing flexible quality of service (QoS) for large numbers of users has led to the explosive proliferation of mobile communication systems these years. Now more and more applications need transferring information to multiple users, such as traffic geographic information, IPTV, video conferencing and so on. Therefore, when multiple users need the same content in the same or adjacent cell, the multicast transmission mode can be used to allow these users to form an organization and share the allocation of resources. Multicast mode further improves the spectrum efficiency and decreases the transmission power consumption of base stations. Therefore, modern communication must solve the difficult problems of multimedia multicast owing to huge difference of many wireless channels and high mobility of UEs. In order to solve these problems, the multicast technology should be combined with orthogonal frequency division multiple access (OFDMA), resource scheduling and dynamic radio resource distribution (DRA) etc. thus maximizing the spectrum utilization ratio, reducing the transmission power of the base station (BS) and providing users with better quality assurance (QoS) [4].

There is a contradiction due to wireless channel differences in the multicast radio resource management. In order to ensure the reliable receiving for the worst channel quality UEs, sufficiently low data rate is necessary. However, it is better if transmission rate is higher in order to maximize the system resource utilization, so finding a rate tradeoff became an important issue in multicast wireless resource management. In all kinds of related articles, two standards are discussed. The first is the principle of equality fairness, on which each user are expected to have the same rate or resource allocation. The second is the principle of proportional fairness, on which the users' resource distribution is based on their potential ability to receive service traffic. Because fairness and throughput maximization become the main problems in resource distribution, we need a compromise to achieve the desired system performance [5]. In addition to the traditional resource management strategies, cooperative relaying, multi rate transmission and resource allocation while moving are also the content of radio resource management.

5.1 Strict Throughput Maximization

Strict Throughput Maximization is mainly applied where multiple multicast groups exist. Resource allocation among multiple multicast groups must coordinate to achieve system spectrum utilization rate optimization. Strict throughput maximization is a completely overly optimistic approach that it does not apply on resource allocation within the group, because the transmission rate is selected according to the maximum channel gain of the users, which will inevitably cause the scarcity of resources of users within a group. However, strict throughput maximization application in multicast group resource allocation will have great capacity gain, because it will distribute the best time domain, frequency domain and the airspace resources to those who have the potential to make use of system resources to maximize the multicast group. But the capacity gain of the system is obtained under the condition that the partial channel quality is poor, so the idea of this kind of resource allocation is unfair [6].

5.2 Max-Min Fairness

In a multicast system, the maximum and minimum Fairness (Max-Min Fairness) idea attempts to set priority for a user or multicast group with poor channel quality to allow them to receive services at a higher rate. A max-min realization way for the first of all the data flow rate is set to 0, then increases the transmission rate of each data stream at the same ratio until it reaches the capacity limit position. Those whose data flow hasn't been restricted continue to increase transmission rate until all users or multicast group reached the first position or resource allocation completed. Another paper proposed in the single rate multicast transmission, a variety of different rate thresholds can be set based on the average throughput as capacity constraints, so as to achieve the fairness of the multicast group and optimize the throughput.

5.3 Fair and Equitable Scheme

Absolute fairness will result a sharp drop in the overall throughput. While the strict throughput maximization will result in a zero tolerance for the worst throughput multicast group. However, the proportional fair scheme which is an eclectic approach can improve multicast broadcasting system resource scheduling mechanism for high data rate, so as to support a lot of UEs with the flexible QoS. As a result, mobile and wireless communication systems develop and surge explosively in recent years. Many users get the same content in the same or adjacent cell through multicast transmission mode. So multicast mode improves the spectrum efficiency and minimizes the power consumption, while maximizing the use of limited system resources.

Acknowledgement. This work was supported by Beijing Natural Science Foundation (4154072) and by the non-governmental education promotion project of Beijing: Comprehensive practical teaching base for modern communication technology in Beijing City University.

References

1. Zhang, T., Jiang, A., Feng, C.: An adaptive resource allocation mechanism for the MBSFN in LTE. J. Xi'an Electron. Sci. Technol. Univ. **39**(5), 126–131 (2012)
2. Liu, L., Cao, X.: Analysis of cell reselection under idle mode of LTE terminal. TV Technol. **35**(1), 81–84 (2011)
3. Yang, F., Zhou, K.: Research on RSRP measurement based on LTE cell reselection. Electron. Test. (8), 45–50 (2010)
4. Richard, A., Dadlani, A., Kim, K.: Multicast scheduling and resource allocation algorithms for OFDMA-based systems: a survey (2012)

5. Afolabi, R.O., Dadlani, A., Kim, K.: Multicast scheduling and resource allocation algorithms for OFDMA-based systems: a survey. IEEE Commun. Surv. Tutorials **15**(1), 240–254 (2013)
6. Kim, J.Y., Kwon, T., Cho, D.H.: Resource allocation scheme for minimizing power consumption in OFDM multicast systems. IEEE Commun. Lett. **11**(6), 486–488 (2007)

A Novel Design of Sharp MDFT Filter Banks with Low Complexity Based on DPSO-MFO Algorithm

Yang Yan[1(✉)] and Xu Pingping[2]

[1] Department of Electronic and Electrical Engineering, Bengbu University,
Bengbu 233030, China
yangetyan@126.com
[2] National Mobile Communications Research Laboratory, Southeast University,
Nanjing 210096, China

Abstract. A novel filter bank with low complexity is designed for Discrete Fourier Transform (MDFT). Due to high data rates and speed in the multicarrier wireless communication and software defined radio systems it is necessary to obtain low complexity and reduce power consumption. Because of sharp transition, Frequency Response Masking (FRM) technique is applied to design the prototype Finite Impulse Response (FIR) based on Canonic Sign Digit (CSD). The performances of the MDFT filter banks are improved by reducing multipliers of the filter using hybrid Discrete Particle Swarm Optimization-Moth-flame optimization algorithm (DPSO-MFO) which is able to provide excellent performances as a nature-inspired heuristic algorithm.

Keywords: Low complexity · Frequency response masking · Discrete Particle Swarm Optimization · Moth-flame optimization

1 Introduction

In recent years with the rapid development of multicarrier wireless communication the multirate filter banks are studied especially as a key technology in fifth generation communication [1]. The MDFT filter banks have been designed using much method [2–5] because it can remove inherent aliasing cancellation structure. The most important advantage of MDFT filter banks is that it can be got that is the linear phase in both filters called the analysis filter and synthesis filter respectively. Rightly the prototype filter must have linear phase and the perfect reconstruction (PR) of filter banks is required in many applications [6]. In multicarrier wireless communication system to enhance spectral efficiency of the filter banks the sharp transition width filters is required but order of the prototype filter will be increased with high the implementation complexity. Fortunately, this problem can be resolved by frequency response masking (FRM) technique [9] that low same complexity prototype filter can be designed. Thus the MDFT filter banks can be completely derived through the analysis and synthesis filters. The sharp MDFT filter banks with PR are designed via FRM technique. In [2] hybrid search algorithm called HAS-GSA is proved that it is better than other

© Springer International Publishing AG 2017
F. Xhafa et al. (eds.), *Recent Developments in Intelligent Systems and Interactive Applications*,
Advances in Intelligent Systems and Computing 541, DOI 10.1007/978-3-319-49568-2_60

algorithms. In this paper a new hybrid algorithm called DPSO-MFO is proposed for the design of multiplier-free MDFT filter banks with PR based on FRM technology. The number structural adders are reduced, Low power and high speed of operation can be obtained. As we know that the DPSO-MFO algorithm proposed in this paper is not used in the literature so far.

2 Design of Modified DFT Filter Banks with PR

In [6] the structure of the filter banks was modified which lead to all adjacent alias spectra and all odd alias spectra are canceled and there is no real or imaginary part at the sub-bands respectively without a delay of one sampling period. As shown in Fig. 1 the modified non-critically sub-sampled M-channel DFT filter bank with PR is given. The corresponding equation is given as follow:

$$\hat{X}_{DFT}(z) = \frac{2}{M} \sum_{k=0}^{M-1} \sum_{l=0}^{M/2-1} F_k(z) H_k\left(z W_M^{2l}\right) X\left(z W_M^{2l}\right) \tag{1}$$

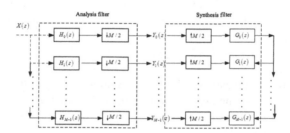

Fig. 1. M-channel filter bank

To meet the PR conditions on the analysis and synthesis filters the low-pass filter $h_0(n)$ used in the analysis filter bank are given as [6] the Type-1 and Type-3 polyphase filters respectively.

$$H_0(z) = \sum_{k=0}^{M-1} z^{-k} G_k\left(z^M\right) \tag{2}$$

Where

$$G_k(z) = \sum_{n=-\infty}^{+\infty} g_k(n) z^{-n} \tag{3}$$

$$g_k(n) = h_0(Mn \pm k), \ k = 0, 1, 2, \dots M - 1 \tag{4}$$

Where M is an integer. For MDFT filter banks, amplitude distortion function is given as [6]

$$T_{dist} = \frac{1}{M} \sum_{0}^{M-1} G_k(z) H_k(z) \tag{5}$$

If the Eq. (4) is satisfied the PR condition MDFT filter banks can be obtained.

3 Design of Prototype Filter Based on FRM

Basic prototype FRM filter architecture is made in [9] as shown Fig. 2.

Where $F_p(z)$ is designed as a shaping filter for the band edge, $F_{com}(z)$ is its complementary filter. N is even order of the linear phase FIR filter $F_{com}(z)$ can be easily obtained from $F_p(z)$ by the following equation.

$$F_{com}(z) = z^{\frac{-(N-1)}{2}} - F_p(z) \tag{6}$$

$F_{mcom}(z)$ and $F_{mp}(z)$ are two masking filters which can eliminate the extra band at the band edge. The FRM filter transfer function $H(z)$ is given as [9].

$$H(z) = F_p(z^L) F_{mp}(z) + F_{com}(z^L) F_{mcom}(z) \tag{7}$$

Based on above specifications after using minimax method the number of order prototype filter's is 2565. The number of multipliers required to implement the filter is 1283. In [2] the length of the sub-filters attains to 221, 85 and 89 respectively. The total number of multipliers is 199.

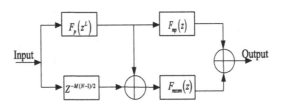

Fig. 2. Basic FRM filter architecture

4 Optimization of Multiplier-Free Sharp MDFT Filter Banks with PR Using DPSO-MFO

4.1 Objective Function Formulation

The implementation complexity of the filter coefficients can be reduced by CSD method and those filter's continuous coefficients can be converted into discrete. As a fractional number g is expressed as follow:

$$g \sum_{j=1}^{P} d_j 2^{R-j} \tag{8}$$

Where P the CSD number length is $d_j = \{1, 0, 1\}$ is a integers and R is a radix point in the range $0 < R < P$. Now a multi objective function. of MDFT filter banks with PR can be formulated under optimal CSD representation. In the prototype filter F_p is defined the passband error and F_s is stopband error. The amplitude distortion error is defined by f_{dist}, where T_{dist} has been defined by Eq. (5).

$$F_p = \max_{0 < \omega < \omega_p} ||H(\omega)| - 1| \tag{9}$$

$$F_s = \max_{\omega_s < \omega < \pi} |H(\omega)| \tag{10}$$

$$F_{dist} = \max_{0 < \omega < \pi} |T_{dist}(\omega) - 1| \tag{11}$$

The final objective function is given as:

$$Minimize\ \delta = \beta_1 F_p + \beta_2 F_s + \beta_3 F_{dist} + \beta_4 p(b_H, B_H) \tag{12}$$

Where $\beta_1, \beta_2, \beta_3, \beta_4$ the weights are in the objective function, b_H is the average number of signed power of two (SPT) terms, B_H is the upper bound.

4.2   Optimization of FRM Filter Using DPSO-MFO Algorithm

A. DPSO algorithm: In [7] that the DPSO automatically searches filter coefficient values through permissible CSD multiplier in the look-up table (LUT). In this way the values of CSD multiplier coefficient is searched though instead of the coefficient values themselves. The optimal values can be obtained. The weights of the objective function for DPSO are obtained as $\beta_1 = 1, \beta_2 = 2, \beta_3 = 1, \beta_4 = 0.1$ by trial and error method.

B. MFO algorithm: MFO algorithm is proposed in [8]. The MFO algorithm the search spaces were explored and exploited with several operators. After applying MFO algorithm to the design for FIR digital filter based on FRM the time of iterations is about 214–500. The optimal FRM filter is designed by the MFO converges to the optimal. The weights of the objective function for DPSO are obtained as $\beta_1 = 1, \beta_2 = 1, \beta_3 = 0.5, \beta_4 = 0.1$ by trial and error method. Note that if the time of iterations is 1–213 there are no feasible solutions.

C. Proposed DPSO-MFO algorithm of filter banks with PR: All of steps for this algorithm are given below:

Step 1   Initialization: A wider search space is provided through choosing proper the initial number of solutions. This number will an integer multiple of selected population size of $PS_{DPSO} + PS_{MFO}$.

Step 2   ***Priority to solution vectors:*** After evaluating the fitness function for each solution vector, the best solution will be obtained its number of $PS_{DPSO} + PS_{MFO}$ should be transferred to the next stage.

Step 3   ***Selecting suitable initial population for DPSO, MFO:*** After step 2 a group random filters are generated the number of PS_{DPSO} can be selected as the suitable initial population of DPSO. At the same time remaining the number of PS_{MFO} from group filters population initialization of MFO algorithm is completed.

Step 4   ***Testing the condition before combining:*** Before combining two algorithms checking previous INV number of iterations. if the mixing is not satisfied the step will go to step 5 otherwise step 7.

Step 5   ***Group and Splitting:*** The group of two algorithms population is built then the population will be spitted into two groups randomly. The size is N_1 and N_2 respectively.

Step 6   ***Updating the population and the best solution:*** Combing DPSO and MFO algorithms populations can be updated. The former best solution of the DPSO-MFO algorithm will be replaced by current.

Step 7   ***The termination:*** Repeat the step 4 and 6 until desired number of iterations can be received. If the algorithm is terminated the best solution is taken as the optimum solution of hybrid DPSO-MFO algorithm.

D. Results and Discussion: All the simulations are run on an Intel core i5 processor operating at 2.4 GHz using MATLAB (R2013b). Magnitude response of Prototype filter based on FRM is shown as Fig. 3. As a result it is know that the PS_{DPSO} is 28, the memory rate is 0.98, pitch adjust rate is 0.01, PS_{MFO} is 28, Gravitational constant is 100, α is 20, INV: 50, the maximum number of iterations is 1000. The prototype filter of FIR is designed based on FRM techniques and optimized by DPSO algorithm. The continuous coefficients of sub-filters are converted into CSD equivalent representation via 14 bit LUT. The LUTs is found to be 15,687, one adder requires 8 LUTs. Hence if number of adders is reduced then the number of LUTs will also be reduced. One multiplier requires 95 LUTs. In order to improve CSD's representation of FRM prototype filter hybrid DPSO-MFO algorithms are applied with more competitive performances than hybrid HAS-GSA in [2] as shown in Fig. 3. The LUTs required

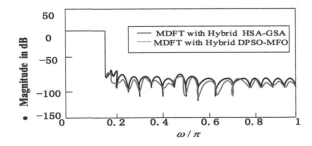

Fig. 3. Magnitude response of prototype filter based on FRM

reduces to 2973. With hybrid DPSO-MFO the number of adders reduces from 306 to 296 as shown in Table 1.

Table 1. Parameters of MDFT filter banks with PR based on FRM.

Method Parameters	Continuous coefficients	CSD rounded coefficients	Hybrid HSA-GSA	Hybrid DPSO-MFO
Amplitude distortion (dB)	0.007986	0.02235	0.00861	0.00912
Passband ripple (dB)	0.007993	0.02214	0.00853	0.00907
Stopband attenuation (dB)	−59.38	−44.35	−57.35	−58.36
Number of multipliers	147	0	0	0
Structural adders	142	142	142	142
Adders of SPT terms		283	306	296
Total adders		425	448	433

5   Conclusion

In this paper, the continuous FRM prototype filter coefficients are reduced by CSD further the performances of the filter bank are improved by hybrid DPSO-MFO that is observed to be better than other algorithms. Hardware complexity is least for multiplier-free MDFT filter bank with PR using multistage FRM method. Low power consumption, low chip area and high speed of operation can be obtained in this design proposed based on hybrid DPSO-MFO algorithm. It is evident from the results that the design will play a very important role in the upcoming applications such as software defined radio, multicarrier wireless communication and portable computing systems.

Acknowledgements. The above work is supported by Nature research fund for key project of Anhui higher education (No: KJ2016A455).

References

1. Kang, X.A.S., Vigo, R.: Simulation analysis of prototype filter bank multicarrier cognitive radio under different performance parameters. Indonesian J. Electr. Eng. Inf. **3**, 157–166 (2015)
2. Sakthivel, V., Elias, E.: Design of low complexity sharp MDFT filter banks with perfect reconstruction using hybrid harmony-gravitational search algorithm. Eng. Sci. Technol. **18**, 648–657 (2015)

3. Fliege, N.: Computational efficiency of modified DFT polyphase filter banks. In: Conference Record of the Twenty-Seventh Asilomar Conference on Signals, Systems and Computers (1993)
4. Fliege, N.J.: Modified DFT polyphase SBC filter banks with almost perfect reconstruction. In: IEEE International Conference on Acoustics, Speech, and Signal Processing, vol, 3, p. III/149 (1994)
5. Elias, E.: Design of multiplier-less sharp transition width MDFT filter banks using modified metaheuristic algorithms. Int. J. Comput. Appl. **88**(2), 1–14 (2014)
6. Karp, T., Fliege, N.J.: Modified DFT filter banks with perfect reconstruction. IEEE Trans. Circ. Syst. II: Analog Digit. Sig. Proc. **46**(11), 1404–1414 (1999)
7. Hashemi, S.A., Nowrouzian, B.: A novel discrete particle swarm optimization for FRM FIR digital filters. J. Comput. **7**(7), 1289–1296 (2012)
8. Mirjalili, S.: Moth-flame optimization algorithm: a novel nature-inspired heuristic paradigm. Knowl.-Based Syst. **89**, 228–249 (2015)
9. Lim, Y.C.: Frequency-response masking approach for the synthesis of sharp linear phase digital filters. IEEE Trans. Circ. Syst. **33**(4), 357–364 (1986)

E-Enabled Systems

Modeling and Evaluating of Decision Support System Based on Cost-Sensitive Multiclass Classification Algorithms

Xiaobo Wu[1(✉)], Hong Sun[1], Zhaohui Wu[1], and Xuna Miao[2]

[1] China Academy of Transportation Sciences, Beijing 100029, China
wuxiaobo1980@163.com
[2] Henan University of Economics and Law, Zhengzhou 450052, Henan, China
miaoxuna1227@126.com

Abstract. Through analyzing the limitations of modeling and evaluating the cost-sensitive multiclass classification algorithms, a series of models based on three classification algorithms are presented. On this basis, expected cost of misclassification as a cost-sensitive metric, which is introduced for evaluating the more cost details of models.

Keywords: Decision support system · Classification algorithms · Cost-sensitive evaluating

1 Introduction

With the rapid improvement of computing storage and Internet technology, the amount of accumulated data is rising at an exponential rate. In this context, big data technology immediately caught great concern form academic institution and commercial organization in recent years. Providing efficient data services based on decision support system, gradually becoming the key way to mine the potential value from big data. Intelligent process of data-information-knowledge-strategy can be achieved by utilizing powerful data management and analysis. As one of the significant research fields in data mining, classification algorithms are applied in many important fields, such as risk assessment, behavior analysis, document retrieval, searching engine category, intrusion detection etc. [1]. Through the analysis of the known class of training set, the classification rules are automatically generated by algorithms, which can be used to predict the class of new data.

The existing decision support system mainly adopts binary classification algorithms, because it is difficult to predict the impact of multiclass behavior accurately. Furthermore, cost details of incorrect classification results usually are not reflected while utilizing the traditional metric just like accuracy [2]. Taking account into the above two deficiency, we investigated the modeling and evaluation based on cost-sensitive multiclass classification algorithms. In this paper, a series of decision support models were carried out based on 3 existing relatively mature classification algorithms such as C4.5 (decision trees classification) [3], RIPPER (Repeated Incremental Pruning to Produce Error Reduction) [4] and PART (Partial decision trees) [5].

© Springer International Publishing AG 2017
F. Xhafa et al. (eds.), *Recent Developments in Intelligent Systems and Interactive Applications*,
Advances in Intelligent Systems and Computing 541, DOI 10.1007/978-3-319-49568-2_61

On this basis, ECM (Expected Cost of Misclassification) [6, 7] as a new metric was introduced for evaluating the performance of cost-sensitive models.

2 Cost-Sensitive Evaluation on Classification Models

2.1 Evaluation Based on Accuracy

The traditional methods are mostly based on the metric, such as accuracy for evaluating the performance of decision support algorithms or models. Accuracy is the most simple, effective and intuitive metric in statistics. For a binary classification problem, 4 different results will occur after a judgment. The two of correct classification results are TP and TN, which indicate true positive and true negative. Symmetrically, the other two of incorrect classification results are FP and FN, which indicate false positive and false negative. The expression of accuracy is as follows.

$$accuracy = \frac{TP + TN}{TP + TN + FP + FN} \tag{1}$$

It is noteworthy that introducing of the TP rate and FP rate has played a certain role on evaluating the predictive capability of binary classification models.

2.2 Analysis of Cost-Sensitive Evaluation

As mentioned before, 2 misclassifications as incorrect classification results may occur in a binary classification problem. Obviously, the costs of misclassification on these 2 situations are quite different in most decision support scenarios. Costs of FN mainly from waste of resources, thus impact is relatively small. In contrast, it can cause serious impact and huge loss directly by FN, especially in some anomaly detection applications. It follows that introducing the cost of misclassification is not only shows different cost details, but also have contribution to evaluate effectiveness of classification models.

Further in multiclass classification problems, cost of misclassification would be more complicated. We introduce an n-dimensional cost matrix in which n denotes the number of classes to reflect the sensitivity accordingly. The principal diagonal elements in matrix represent the cost of the correct classification, and are usually assigned by zero. Other elements of matrix represent the cost of misclassification. The n-dimensional cost matrix is non-symmetric, which is caused by the different cost of misclassification.

2.3 Expected Cost of Misclassification

In order to calculate the cost of classification models in sample set, Expected Cost of Misclassification (ECM), a concept of statistics is introduced. ECM denotes the average cost of every misclassification situation, so that it can be applied in quantitative comparison of classification models on test sets with different sample sizes. The expression is as follows.

$$ECM = \frac{\sum\limits_{i=1}^{m}\sum\limits_{j=1}^{m} Cij \times Nij}{N} \qquad (2)$$

In where C_{ij} denotes cost of misclassification that result is class j when the actual is class i, N_{ij} denotes the number of misclassification instances, and N denotes the size of test set. ECM as a new metric is conducive to evaluate the cost-sensitive classification models in decision support system while the cost issues are not need to be taken into account in previous modeling.

3 Experiments and Performance Evaluation

3.1 Algorithms Modeling

In this paper, experiments of modeling and testing based on 3 classification algorithms (C4.5, RIPPER and PART) were carried out. The two metrics about accuracy and ECM were used to evaluate these models in known and unknown test sets respectively. The numerical results of model performance based on these three kinds of classification algorithms are shown in Figs. 1, 2 and 3.

3.2 Results Analysis

From the simulation results we identify that, models based on three classification algorithms achieve good fitting of train set. It reflects self-learning capability of models on train set are strong. The three algorithms perform exceptional well for determining the recognized behaviors in test set. From the testing results in detecting unknown behaviors, classification algorithms have certain adaptability. Although not as

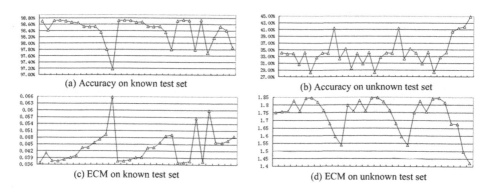

(a) Accuracy on known test set

(b) Accuracy on unknown test set

(c) ECM on known test set

(d) ECM on unknown test set

Fig. 1. Accuracy and ECM of C4.5 models

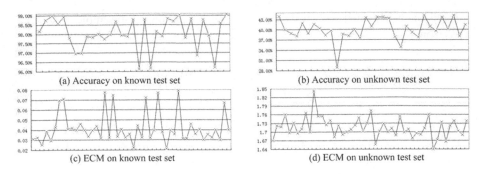

(a) Accuracy on known test set   (b) Accuracy on unknown test set

(c) ECM on known test set   (d) ECM on unknown test set

Fig. 2. Accuracy and ECM of RIPPER models

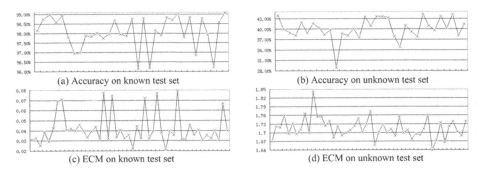

(a) Accuracy on known test set   (b) Accuracy on unknown test set

(c) ECM on known test set   (d) ECM on unknown test set

Fig. 3. Accuracy and ECM of PART models

performance on known behaviors, algorithms are still able to detect the new behavior by some prior knowledge.

Furthermore, the result shows the fact that ECM is directly inversely related to accuracy in general. In some cases, however, models reflect the different ECM while accuracies are almost similar. Especially in extreme cases, the models are with high accuracy but with high ECM as well. Accordingly, the evaluation process is more objective while introducing the metric of ECM.

Horizontal comparison of models based on three classification algorithms on accuracy and ECM are shown in Fig. 4. From the line chart of accuracy we know that performance of RIPPER is best, followed by PART and C4.5. Similarly, we can observe that the performances of three classification algorithms maintain the same comparative relationship while introducing the metric of ECM. Decision trees-based algorithm (C4.5) as the earliest development and relatively mature classification algorithm doesn't show the best predictive capability. In addition, the size of rule set generated by the decision tree is relatively large, which will result in a large resource overhead. Rule-based algorithm (RIPPER) expresses a certain advantage whether in evaluating of accuracy or ECM, particularly the size of rule set is smaller. The shortcoming is that higher time complexity leads to longer modeling time, and therefore it is not conducive to update the rule set by adding new training data. PART which

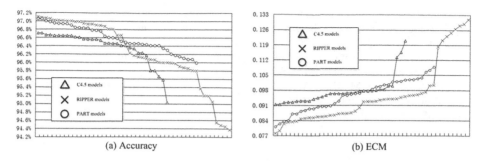

(a) Accuracy   (b) ECM

Fig. 4. Comparisons of 3 kinds of multiclass classification algorithm models

is the combination of decision tree-based algorithm and rule-based algorithm exhibits high stability. The evaluating of accuracy and ECM is at an intermediate level. Meanwhile the size of rule set and the time complexity is comparatively low.

4 Conclusions

Classification algorithms have been widely used in decision support system because of the high adaptive and predictive capability. In this paper, we initially analyzed the limitations of modeling and evaluated the cost-sensitive multiclass classification algorithms by utilizing accuracy as the only metric. A series of models based on three classification algorithms were presented in the next place. Eventually, more cost details of misclassification were quantified by introducing the cost-sensitive evaluation. Some results may provide a little reference to research on the cost-sensitive modeling of decision support system in the future works.

Acknowledgements. This work is supported by the National Science Foundation of P. R. China under Grant (No. 61309033, No. 61402147), the Scientific Research Foundation of the Higher Education Institutions of Hebei Province of China (QN20131048) and the Open Fund of the State Key Laboratory of Virtual Reality Technology and Systems (No. BUAA-VR-16KF-02).

References

1. Witten, I.H., Frank, E., Hall, M.A.: Data Mining: Practical Machine Learning Tools and Techniques, 3rd edn. Morgan Kaufmann Publishers Inc., San Francisco (2011)
2. Khoshgoftaar, T.M., Allen, E.B., Jones, W.D., et al.: Cost-benefit analysis of software quality models. Software Qual. J. **9**(1), 9–30 (2001)
3. Quinlan, R.: C4.5: Programs for Machine Learning. Morgan Kaufmann Publishers Inc., San Francisco (1993)
4. Cohen, W.W.: Fast effective rule induction. In: Proceedings of the Twelfth International Conference on Machine Learning, Tahoe City, California, pp. 115–123 (1995)

5. Frank, E., Witten, I.H.: Generating accurate rule sets without global optimization. In: Proceedings of the Fifteenth International Conference on Machine Learning, pp. 144–151 (1998)
6. Huang, S.H., et al.: Identifying a small set of marker genes using minimum expected cost of misclassification. Artif. Intell. Med. **55**(1), 51–59 (2012)
7. Hua, Z.S., Zhang, X.M., Xu, X.Y.: Asymmetric support vector machine for the classification problem with asymmetric cost of misclassification. Int. J. Innovative Comput. Inf. Control **6** (12), 5597–5608 (2010)

Quality of Company D Welding Workshop

Yanhua Ma[✉], Lingyu Li, TianRong Bai, and Chao Jin

Institute of Mechanical Science and Engineering, Jilin University,
Changchun 130022, China
yhma@jlu.edu.cn

Abstract. In this paper, the author made statistical analysis to quality problems of welding workshop in company D and then found out the main factors affecting product defects. After analyzing the results, this paper put forward some improvement measures. Initially, the paper presented the background and significance of this research and the status of the quality management of domestic and foreign enterprises. Secondly company D and its quality management system were introduced briefly. Thirdly the quality problems in welding workshop were analyzed to find out the factors that mainly affect the product quality. Finally some corresponding improvement measures were put forward in terms of quality problems in the welding workshop to decrease the rate of defect.

Keywords: Welding workshop · Factors affecting quality · Statistical methods · Analytic Hierarchy Process (AHP) · 6σ · Quality improvement

1 The Use of Statistical Techniques for the Analysis of Factors Affecting the Quality of the Welding Workshop of D Company

The quality problem of D company in 2014 January or February of the welding workshop summary to Table 1:

It can be observe that a number of undesirable for welding workshop occurrence in T80 section, the defective products were pooled about T80 section in 2014 in January or February, as shown in Table 2. We can see the main quality problems in welding workshop for bad solder joint.

In order to further find out the reasons for poor solder joints, from the human, machine, material, method, ring five items to re count, draw the arrangement chart to a more intuitive representation of the proportion of all kinds of factors, as shown in Fig. 1:

It can be seen, the maximum number of defective products produced by the process factors, followed by human factors. Each factor refinement, Summarized in the Table 3, and then can draw cause and effect diagram.

© Springer International Publishing AG 2017
F. Xhafa et al. (eds.), *Recent Developments in Intelligent Systems and Interactive Applications*,
Advances in Intelligent Systems and Computing 541, DOI 10.1007/978-3-319-49568-2_62

Table 1. Welding shop in January and February of quality problems Summary

Serial number	Section	Poor number product	Frequency (percentage)	Cumulative frequency (percentage)
1	T80 Sections	1958	80.34	80.34
2	V70 Section	199	8.17	88.51
3	one Section	180	7.39	95.90
4	two Section	100	4.10	100.00
	Total	2437	100.00	100.00

Table 2. T80 body shop section in January or February of quality problems Summary

Serial numbers	Defective item	Frequency (a)	Frequency (percentage)	Cumulative frequency (percentage)
1	Poor solder joint	1214	62	62
2	Poor assembly	274	14	76
3	Poor in sheet metal	235	12	88
4	hand Repair Poor	157	8	96
5	Other bad	78	4	100
	Total	1958	100	100

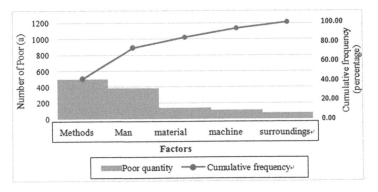

Fig. 1. T80 section in January or February 2014 of defective rate classification chart

Table 3. T80 section summarizes the causes of waste

Reason	Category
Entry time is short, the difference between the technical level	Man
Poor employee responsibility	
Poor physical condition of the operator	
Poor equipment maintenance situation	Machine
Equipment commissioning	
Gun damage	
Material damage during transportation	Material
Raw material quality problems during storage	
Raw material quality problems	
Operation unreasonable	Method
Welding parameter setting unreasonable	
New Test Method	
Low level of implementation of standard operating instructions	
Operating space is small, convenient operation	Surroundings
Poor surroundings	

2   Using the Improved Analytic Hierarchy Process to Analyze the Factors Influencing the Quality of T80 Section

The Improved Analytic Hierarchy Process to determine the cause of the weight, use more in line with people's things compare to quantify the results of the index scale method instead of the 1–9 scale. Reference to the fishbone diagram, to establish the improved AHP analysis model of T80 section. Experts questionnaires to calculate the index weight, total consistency test passed, eventually got the two most vital factors for the operating method of reasonableness and quality awareness of the quality of employees. In addition, methods and human factors together are account for a total of 52.26 %, consistent with the results of statistical methods, indicating the reliability of the results.

3   Quality Improvement Company of D Welding Shop

3.1   Pairs of Results of Statistical Analysis and Analytic Hierarchy Process to Was Subjected to Quality Improvement

(1) Methods of operation unreasonable
Improvement measures :
(1) The formation of QC team to address field problems, combined with the views of frontline workers be improved.

(2) The production of prototypes, and the quality of the process samples to monitor, if stable production process, improve the standard operating instructions. Otherwise, continue to improve.

(2) Poor quality awareness of employees

   Improvement measures :

   (1) Strengthen the quality of propaganda;

   (2) Implementation of quality improvement activities on a regular basis, to allow more employees to participate;

   (3) Reduce defective index to reach the goal constraint. While strengthening Incentive measures, encourage play a role;

   (4) Responsibilities to individuals.

3.2   Application of Six Sigma Team for T80 Main Body for Quality Improvement

(1) The definition phase

   As can be seen from Table 4 occurred in the main body of the team accounted for the largest proportion of poor solder joints. Therefore, to reduce the defect rate as the main body of this team improvement goals.

(2) Measurement phase

   According to data collected in January and February, the main body welding quality Teams and groups in March to record the different types of solder joint issues classified statistical summary, as shown in Table 5.

Table 4. Each team solder joint incidence of adverse Statistics

Serial numbers	Teams and groups	Frequency (a)	Frequency (percentage)	Cumulative frequency (percentage)
1	Main body classes	522	43.00	43.00
2	Side of the confining classes	218	17.96	60.96
3	Door classes	146	12.03	72.98
4	Under the car classes	109	8.98	81.96
5	Former machine classes	97	7.99	89.95
6	Assembly classes	74	6.01	95.96
7	Hand repair classes	49	4.04	100
	Totals	1214	100	100

Table 5. Welding workshop in March 2014 in the form of all kinds of Poor solder joint classification tables

Serial number	Defective item	Frequency (a)	Frequency (percentage)	Cumulative frequency (percentage)
1	Distorting	230	37.52	37.52
2	Deviates	150	24.47	61.99
3	Cold solder joint	120	19.58	81.57
4	Leakage welding	70	11.42	92.99
5	glitch	18	2.94	95.92
6	The breakdown	15	2.45	98.37
7	Pinhole	10	1.63	100.00
	Total	613	100	100.00

(3) The analysis phase

In this paper, the main body of the process of FMEA team to analyze the several factors that affect product defects, the distinction between the main cause and secondary cause.

The joints twisted, deviate, Weld and leakage welding as potential failure modes. By analyzing the severity of occurrence and detectability obtain the failure mode and effects analysis summary Table.

(4) Improve phase

An investigation for four main failure reason and improving advice.

(1) welding work is not in the visible range

The operator when the job is to bend over to find the location of the solder joint, a waste of time, and could easily lead to operator fatigue, but also can not guarantee the accuracy of the solder joint position.

Improvement measures: defining the welding area.

(2) The electrode arm into contact with the workpiece to produce shunt

Electrode arm contact with the workpiece to produce shunt capacitance caused by Weld.

Improvement measures: making insulation baffles, welding electrode arm height restrictions.

(3) The workers before the end of butt welding tongs swing

Squat operator to operate, easy to cause fatigue and cause welding tongs shaking, resulting in twisted joints.

Improvement measures: allowing the operator to stand operation, the welding tongs into portable, and set the switch on the handle, at the same time through a link with the limit switch is connected to the control limit switch.

(4) The operator leakage welding

Improvement measures: making pads like cars, by visual methods of training employees and by self and mutual inspection, conducted leakage welding control.

(5) Control phase

Implementing improvement programs to improve the effect of track record, if achieved remarkable results, improvement activities will continue to be documented and its effectiveness.

4 Conclusion

Improved suggestion has been presented in this article, there are several measures have been accepted and implemented in the workshop, and achieved significant results. This article has described the comments made by certain theoretical and practical value of the company D welding shop. Of course, the research work of this paper, have some limitations, not to improve the glitch, breakdown, pinholes and other negative phenomena, which are worth for further thought and study.

References

1. Lee, C.Y., Zhou, X.: Quality management and manufacturing strategies in China. Int. J. Qual. Reliab. Manage. 17(8), 876–899 (2000)
2. Lee, C.Y.: TQM in small manufacturers: an exploratory study in China. Int. J. Qual. Reliab. Manage. 21(2), 175–197 (2004)
3. Flynn, B.B., Schroeder, R.G., Sakakibara, S.: A framework for quality management research and an associated measurement instrument. J. Oper. Manage. 11(4), 339–366 (1994)
4. Dow, D., Samson, D., Ford, S.: Exploding the myth: do all quality management practices contribute to superior quality performance? Prod. Oper. Manage. 8(1), 1–27 (1999)
5. Tari, J.J., Molina, J.F., Castejon, J.L.: The relationship between quality management practices and their effects on quality outcomes. Eur. J. Oper. Res. 183(2), 483–501 (2007)

Design of ANFIS Based E-Health Care System for Cardio Vascular Disease Detection

Lokanath Sarangi[(✉)], Mihir Narayan Mohanty[(✉)],
and Srikanta Patnaik

ITER, Siksha 'O' Anusandhan University, Bhubaneswar, Odisha, India
lokanathsarangi@yahoo.com, {mihirmohanty,
srikantapatnaik}@soauniversity.ac.in

Abstract. As the society is becoming superior day-by- day, loads of smart devices are used in different application areas. This is the challenge to the technocrats for forming the intelligent and smart social systems. It requires easy access and fast processing, which is the main focus of any application. In this work, an attempt has been taken into consideration to develop an intelligent e-healthcare system. In e-healthcare system the entities are considered as the patient, the physician, the pathological centre and result as diagnosis, treatment and post care. This paper uses an ANFIS structure for e-healthcare system. Further the ANFIS system is used for disease diagnosis and support to the patient as well as for physicians. For the management of multi-agent system has been satisfied by, using rule based fuzzy parameters. The service can be provided through internet to the patient as well as by the physician. The different situation of patient automatically informs to the doctor similarly the prescription from the doctor for diagnosis can inform to the pathology centre and vice versa. The result of detection communicated to both for desired medicine, monitoring and post care purpose. The performance found to be excellent to satisfy this part of intelligent system.

Keywords: Multiagent system · Disease · Diagnosis · E-healthcare · ANFIS

1 Introduction

E-healthcare is an advanced technology to support health care units. Electronic data storage and efficient transmission has brought a noticeable change in traditional service [1, 2]. Patient health records, telemedicine healthcare informatics, in digital form provide medical care at remote places. Acute shortage of physicians in comparison to the growing population in the society, the health care sector indispensably depends on new models. These can support information transformation with ensuring smart health care service to remote area people as well as to the aged people in the society. So to provide advanced health care at affordable cost with an easy of accessibility of information and communication between different facilities is the need of the society. Regular health check up can aware them about specific diseases diagnosed by doctors measuring some of the signs and symptoms like ECG, Blood pressure, etc. But the trend has been changed because of enhancement in medicine and technology. It can

© Springer International Publishing AG 2017
F. Xhafa et al. (eds.), *Recent Developments in Intelligent Systems and Interactive Applications*,
Advances in Intelligent Systems and Computing 541, DOI 10.1007/978-3-319-49568-2_63

also helps in post care and monitoring their health on routine basis at their home comfortably. The systems are cost effective and time saving for both physicians and patients. This information can be assessed by the health care professionals regularly through wirelesscommunication at their own time and schedule. An appointment at the time of need can be fixed by the patients.

The complex healthcare system consists of patient, hospital, physician, diagnostic centre etc. The co-ordination among each entity in right time can save the life of a patient. Multiagent system environment provides the platform to maintain co-ordination among all the agents. Each entity of healthcare system can perform in form of an agent. Each agent is designed and assigned with a particular task. The whole task is not solved by a single agent. The co-ordination and co-operation among all the agents in MAS solve the healthcare tasks efficiently.

People are suffering from chronic diseases such as cardiac disease, cancer, diabetes, asthma etc. These diseases increases the healthcare burden in the society. Constant monitoring is vital for these diseases. Collection of patient symptoms, advice for time to time diagnostic tests, thorough investigation of the reports for perfect prediction of the disease, prescribing the required medicines and post operative care are highly essential. The smart solution of all these requirements can be realized with the help of multiagent based e-healthcare system.

In this paper our objective is to develop a cost effective, time saving and informative monitor which can diagnose irregular signs and can analyse the data measured by implements used by patients and to determine the health status of them with comparison to normal range data. The hybrid system ANFIS consisting of both the ANN and FIS performs like a simple FIS whose inputs are trained adaptively to change using an ANN [3]. The IF-THEN rule of FIS to take decisions just like human brain thinking and reasoning along with the support of ANN to optimize the inputs through learning provides better accuracy than the result obtained if these are used individually.

The paper is structured as follows. In Sect. 2, a survey on related work is discussed. Section 3, explains the proposed work where the processes involved in neuro-fuzzy based classification are discussed. Results are discussed in Sect. 4 followed by conclusions and future work to be carried out in Sect. 5.

2 Literature Survey

An economic disease detection method was established using cost-sensitive k nearest neighbor (KNN) algorithm for the patients under high-risk of having heart disease [4, 5]. Extraction of features from abnormal heart sounds were detected by using various software tools for easy detection of the cardiac diseases [6, 7]. Authors compared the performance of cardiac disease detection using various types of predictive data mining techniques [8]. The implementation of an android app for early detection of heart disease was trieled [9]. Various classifiers like random forest classifier and a support vector machine classifier were used to the ECG based heart beat signal for Cardiomyopathy Identification [10]. Mobile Phone Based e-Health Monitoring Application for display of ECG signal on the mobile screen. Continuous and intelligent monitoring of ECG signal and its' classification alerts the patient, medical personnel which can save the life of a heart patient [11, 12]. In [13] a system was designed for monitoring heart rate, respiration rate and movement behavior of at-home elderly people who are living alone.

The fuzzy Logic Controller receives the data from the wearable sensors and after fuzzification of the data converted into fuzzy variables and made suitable for under-standing for physicians or patients. The fuzzy modeling and fuzzy membership func-tions reduce the human error to a greater extent [14, 15]. Artificial Neural Network (ANN) and Decision Tree Method [16–18]. Fuzzy systems, on the other hand, have shown significant ability to solve many types of real world problems, especially in a system that is complex and difficult to model mathematically, controlled by a human operator or expert, and a system in which ambiguity or vagueness is common. A fuzzy inference system is a system that uses the concept of fuzzy logic in formulating the mapping from a given input to an output. The inputs and outputs of fuzzy systems are expressed by variables that range between [0, 1].

3 Proposed Method

ANFIS is suitable for solution of engineering as well as biomedical problems widely. In this work, we have used the ANFIS for heart disease diagnosis and detection. The system works on internet to provide health service. Various agents use internet to have communication among each other. The proposed system is shown with the help of blocks in Fig. 1.

It consists of various agents represented as blocks and communication paths are shown with the help of lines. The user agent is the patient who having an user id transfer his parameters i.e. symptoms as well as pathological data to the server with the help of a PC or mobile SMS. All the data uploaded to the server. As the patient uploads the data to the server the diagnostic agent collects the data from the server. The same data is stored in the data base for reference. The diagnostic agent starts analysis. The data analysis is done by ANFIS for diagnosis and detection. If the patient found normal, an email is sent to the patient regarding his health condition along with some remedial measures to take. If any abnormality is detected, an email is sent to the

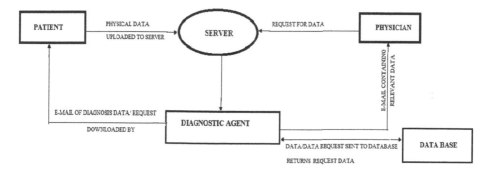

Fig. 1. Proposed intelligent E-healthcare system

physician for consultation. If the Physician desires to check patient's medical history, he can be able to access the database at any time.

In this work two vital parameters i.e. ECG and BP are taken into consideration to detect the cardiovascular diseases. These parameters are fed as the input to the proposed model and output is the detection of the disease.

Adaptive Neuro-Fuzzy Inference System Structure. In this case an adaptive system ANFIS (consists of 6 layers feed forward Artificial Neural Network) is described as fuzzy Sugeno model. This adaptive model is simpler, less dependent on expert knowledge. The advantage of this model is that, it does not require defuzzification. So this can be used for objective non-linear fuzzy modeling. This structure of ANFIS can be presented as first order Sugeno type fuzzy model. Two fuzzy IF-THEN rules with two inputs and a single output are implemented.

Rule 1: IF BP is X_1 AND ECG abnormality is Y_1 THEN $f_1(bp,ea) = r_1 (bp) + s_1 (e\,a) + t_1$

Rule 2: IF BP is X_2 AND ECG abnormality is Y_2 THEN $f_2(bp,ea) = r_2(bp) + s_2(ea) + t_2$

In this case BP and ECG abnormality are taken as two input variables where as X_i and Y_i for i = 1, 2 are the fuzzy sets associated with inputs bp and ea, respectively. $f_i(bp,ea)$ (for i = 1,2) are the outputs (linear combination of input variables) within the fuzzy region specified by rule 1, r_i and s_i are the design parameters associated with rule i which are determined during the learning process. The ANFIS structure for the above said two IF THEN rules consists of six layers where a circle represents a fixed node and a square represents an adaptive node as shown in Fig. 2.

The function of each layer is described as follows:

Layer 1: This layer is free from any type of computations. The Input variable is feed to the input node. Through the input node the variable moves to the first hidden layer.

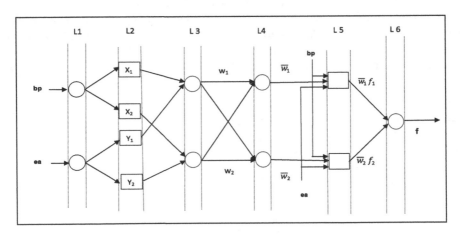

Fig. 2. Proposed ANFIS model

Layer 2: (the fuzzification layer) includes adaptive nodes in which each generates membership grades of an input variable. The outputs of the nodes belonging to this layer (O_i^2) are given by

$$O_i^2 = w_i = \mu_{X_i} \text{ (bp)} \qquad i = 1, 2 \tag{1}$$

$$O_i^2 = w_i = \mu_{Y_{i-2}}(ea) \qquad i = 3, 4 \tag{2}$$

The membership functions can be any continuous, piecewise differentiable functions (e.g. Gaussian, generalized bell shaped and triangular). Assuming a Gaussian membership function, the output of the node $\left(O_i^2\right)$ can be computed as

$$\mu_{X_i}(\text{bp}) = e^{-\frac{1}{2}\left(\frac{bp-ci}{\sigma i}\right)^2}, \; i = 1, 2 \tag{3}$$

$$\mu_{Y_{i-2}}(\text{ea}) = e^{-\frac{1}{2}\left(\frac{ea-ci}{\sigma i}\right)^2}, \; i = 3, 4 \tag{4}$$

Where c_i and σ_i are the parameter (centers and the width respectively) of the Gaussian membership function characterizing the fuzzy sets describing each input variable.

Layer 3: (rule antecedent layer) includes fixed nodes and each node represents the antecedent part of the associate rule. The product t-norm operator used by each node calculates the firing strength of the associated rule. As a result, the output of each node is given by

$$O_i^3 = w_i = \mu_{X_i}(bp), \; \mu_{Y_i}(ea) \quad i = 1, 2 \tag{5}$$

Layer 4: (strength normalization layer) The firing strength obtained from the layer 3 is normalized by the fixed nodes present in this layer. The addition result of all rules firing strength and the firing strength at the concerned node is compared. The output result at each node of this layer can be expressed as

$$O_i^4 = \bar{w}_1 = \frac{w_i}{w_1 + w_2} \quad i = 1, 2 \tag{6}$$

Layer 5: (consequent layer) The adaptive nodes included in this layer are the node functions which can be expressed as the products of the normalized firing strength (i.e. the output of 4$^\text{th}$ layer) resulting in a first order polynomial:

$$O_i^5 = \bar{w}_1 f_i = \bar{w}_1 (r_i(bp) + s_i(ea) + t_i) \quad i = 1, 2 \tag{7}$$

Layer 6: (inference layer) It is a fixed node that calculates the overall output given by

$$O^6 == \sum_{i=1}^{2} \bar{w}_1 f_i = \frac{\sum_{i=1}^{2} w_i f_i}{\sum_{i=1}^{2} wi} \tag{8}$$

The above ANFIS model consists two adaptive layers, i.e. layer 2 & 5. In layer 2, eight modifiable parameters are present{c_i and σ_i: i = 1, 2, 3, 4}, these parameters are communicated through the input membership functions called, "antecedent parameters". In layer 5 there are also six modifiable parameters {r_i, s_i and t_i :i = 1,2}. These parameters are linked with the first order polynomial called, "consequent parameters". All these modifiable parameters in this ANFIS model are considered for optimization which helps a lot for smart detection of cardiovascular disease.

4 Experimental Results

It is accurate to provide the information to the patient and adaption of different information. Multiple agent work collectively to resolve the transportation among them. Also this technique maximizes the accuracy and consistency in a short period of time. The result of detection can be specified from the following table and that can be communicated to the physician as well as the patient (Table 1).

Figure 3 shows the training data used for ANFIS modeling with BP and ECG feature sets of 104 patients. Each input consists of three membership function, low medium and high for BP and poor, good and moderate for ECG. With different rule formation of these membership functions connected to these inputs provide outputs as the cardio vascular disease.

Figure 4 shows the corresponding FIS output. Grid partitioning is used for dividing the data. The data are trained using 50 number of epochs.

Figure 5 shows the formed rules for BP and ECG for detection of cardio vascular disease.

Table 1. .

Sl. No. of Patient	Test		Fitness range				Abnormality	
BP	Result						Suggestion	ECG
BP	ECG	BP	ECG					
1	65/110	Poor	80/130	Good	A	A	Detected	Consult physician
2	62/105	Moderate	80/130	Good	A	A	Detected	Consult physician
3	70/110	Good	80/130	Good	A	N	Undefined	No risk
4	80/130	Poor	80/130	Good	N	A	Undefined	No risk
5	85/125	Moderate	80/130	Good	N	N	Not detected	No risk
6	80/135	Good	80/130	Good	N	N	Not detected	No risk
7	90/140	Poor	80/130	Good	A	A	Detected	High risk
8	95/145	Moderate	80/130	Good	A	A	Detected	High risk
9	100/150	Good	80/130	Good	A	N	Detected	Consult physician

'A' stands for abnormality and 'N' stands for Normal

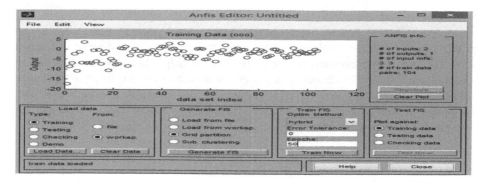

Fig. 3. Plot of training data for BP and ECG using ANFIS

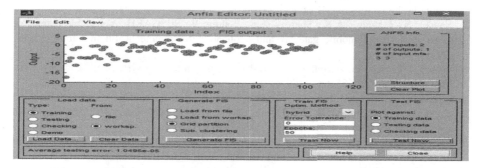

Fig. 4. Plot of FIS output for BP and ECG using ANFIS

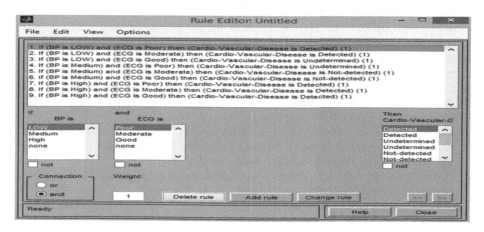

Fig. 5. ANFIS rule formation for BP and ECG for detection of Cardio-Vascular disease

5 Conclusion

In this paper, an idea is implemented to detect the cardiovascular disease in a smart way. The two vital parameters of the above disease help efficiently for detection in e-health care system. This detection has been communicated to physician as well as to the patient for preferred medicine, monitoring and post care purpose. Addition of more symptoms as well as more parameters and implementation of the same model for detection of more diseases can be helpful in the e-health care system.

References

1. Poorani, D., Ganapathy, K., Vaidehi, V.: Sensor based decision making inference system for remote health monitoring. IEEE (2012)
2. Nawka, N., Maguliri, A.K., Sharma, A.K., Saluja, P.: SESGARH: A scalable extensible smart-phone based mobile gateway and application for remote health monitoring. IEEE (2011)
3. Jang, J.-S.R.: ANFIS: Adaptive-network-based fuzzy inference system. IEEE Trans. Syst. Man Cybern. **23**(3), 665–685 (1993)
4. Uguroglu, S., Carbonell, J., Doyle, M., Biederman, R.: Cost-sensitive risk stratification in the diagnosis of heart disease. In: Proceedings of the Twenty-Fourth Innovative Applications of Artificial Intelligence Conference, pp. 2335–2340 (2012)
5. Shouman, M., Turner, T., Stocker, R.: Applying k-nearest neighbour in diagnosing heart disease patients. Int. J. Inf. Educ. Technol. **2**(3), 220–223 (2012)
6. Mandal, D., Chattopadhyay, I.M., Mishra S.: A low cost non-invasive digital signal processor based (TMS320C6713) heart diagnosis system. In: 1st International Conference on Recent Advances in Information Technology | RAIT (2012)
7. Perera, I.S., Muthalif, F.A., Selvarathnam, M.: Automated diagnosis of cardiac abnormalities using heart sounds. In: 2013 IEEE Point-of-Care Healthcare Technologies (PHT), Bangalore, India, pp. 252–255, 16–18 January 2013
8. Venkatalakshmi, B., Shivsankar, M.V.: Heart disease diagnosis using predictive data mining. Int. J. Innovative Res. Sci. Eng. Technol. **3**(3), 1873–1877 (2014)
9. Zennifa, F., Fitrilina, Kamil, H., Iramina, K.: Prototype Early Warning System for Heart Disease Detection Using Android Application 978-1-4244-7929-0/14/$26.00 ©2014, pp. 3468–3471. IEEE (2014)
10. Rahman, Q.A., Tereshchenko, L.G., Kongkatong, M., Abraham, T., Abraham, M.R., Shatkay, H.: Utilizing ECG-based heartbeat classification for hypertrophic cardiomyopathy identification. IEEE Trans. Nanobiosci. **14**(5), 505–512 (2015)
11. Forkan, A., Khalil, I., Tari, Z.: Context-aware cardiac monitoring for early detection of heart diseases. Comput. Cardiol. **40**, 277–280 (2013)
12. Sani, A.S., Islam, A.K.M.M., Mahrin, M.N., Mamoon, I.A., Baharun, S., Komaki, S., Imai, M.: A framework for remote monitoring of early heart attack diagnosis system for ambulatory patient. In: 2014 IEEE Conference on Biomedical Engineering and Sciences, Miri, Sarawak, Malaysia, pp. 159–164, 8–10 December 2014
13. Mukai, K., Yonezawa, Y., Ogawa, H., Maki, H., Morton Caldwell, W.: A remote monitor of bed patient cardiac vibration, respiration and movement. In: 31st Annual International Conference of the IEEE EMBS Minneapolis, Minnesota, USA, 2–6 September 2009

14. Al-Sakran, H.O.: Framework architecture for improving healthcare information systems using agent technology. Int. J. Managing Inf. Technol. (IJMIT) 7(1), 17–31 (2015)
15. Devi, C.S., Ramani, G.G., Pandian, J.A.: Intelligent E-healthcare management system in medicinal science. Int. J. PharmTech Res. 6(6), 1838–1845 (2014)
16. Wang, Y.L., Li, G.Z., Xu, S.W., Liu, G.P., Wang, Y.Q.: Symptom selection of inquiry diagnosis data for coronary heart disease in traditional Chinese medicine by using social network techniques. IEEE International Conference on Bioinformatics and Biomedicine Workshops, pp. 785–789 (2010)
17. Hannan, S.A., Mane, A.V., Manza, R.R., Ramteke, R.J.: Prediction of heart disease medical prescription using radial basis function. 978-1-4244-5967-4/10/$26.00 ©2010. IEEE (2010)
18. Shouman, M., Turner, T., Stocker, R.: Using decision tree for diagnosing heart disease patients. In: Conferences in Research and Practice in Information Technology (CRPIT), vol. 121, pp. 23–29 (2011)

CGSA-CFAR Detector in Urban Traffic Environments

Guiru Llu[1(✉)], Lulin Wang[2], Jun Wang[1], and Jun Qiang[1]

[1] School of Computer and Information, Anhui Polytechnic University,
Wuhu 241000, China
liuguiru_yunnan@163.com
[2] Prospective Technology Research Institute, Chery Automobile Co., Ltd.,
Wuhu 241006, China

Abstract. In order to improve the detection performance of the current detector in urban traffic environments, a cell with greatest, smallest and averaging constant false alarm rate (CGSA-CFAR) detector was proposed. By adjusting threshold in time, based on the noise intensity, which was estimated according to the mean and standard deviation. Digital filter banks were used to restrain noise effectively by reducing the digital signal sidelobe powers. According to simulation and analysis results with other detectors, the proposed detector had the best detection performance, its detection rate was up to 97.70 %. The detector was applied to a vehicle blind spot detection and warning system (BSDWS), which was calibrated on the Chery Arrizo7 car and tested under daytime and nighttime conditions, the average early warning rate was up to 97.84 % and 98.34 %, false alarm rate was reduced to 2.68 % and 2.40 %. The result shows that the detector has a good detection performance in urban traffic environments.

Keywords: Signal processing · CFAR · Target detection

1 Introduction

The aim of the radar system is to determine whether a target is present or not in the noise background environment, including thermal noise, clutter edge signals, a fixed threshold detection scheme cannot be applied, since they might have excessive false alarms or lower detection performance [1]. Thus, the Constant false alarm rate (CFAR) detectors was proposed, which was used to regulate the background noise power level based on the probability of false alarm in varying background environments. The background power level is estimated by averaging the amplitude of the nearby cells. The detection threshold is obtained by scaling the noise background power level estimate value with a constant scale based on the desired probability of false alarm. At present, CFAR detectors have been presented in the literature, which was designed to work satisfactorily in special environments, either the clutter or multiple interfering target environments, and their performance degraded significantly in the complex environments [1, 2].

© Springer International Publishing AG 2017
F. Xhafa et al. (eds.), *Recent Developments in Intelligent Systems and Interactive Applications*,
Advances in Intelligent Systems and Computing 541, DOI 10.1007/978-3-319-49568-2_64

In this paper, the authors have proposed a cell having greatest, smallest and averaging constant false alarm rate (CGSA-CFAR) detector to overcome the problems of the current detectors [3].

2 CGSA-CFAR Detector

The signal detection algorithm has a significant impact on the performance and functionality of the radar sensor system. Algorithm selection must meet the application scene and product requirements. The first step is to define the initial conditions, noise analysis. Then, Signal detection algorithm is designed to confirm the performance improvement of the radar sensor system in comparison with modern radar sensor systems for BSD applications. We will analyze the advantages and disadvantages of each signal detection and processing algorithm [4].

In radar systems, the received signal was sampled by the range resolution cells. The noise background power level in the cell under test is estimated by averaging the amplitude of the nearby resolution cells. The detection threshold is obtained by scaling the noise power level estimate value with a constant scale which was computed based on a desired probability of false alarm [5]. Some improved detection algorithms have been proposed. The smallest of constant false alarm rate (SO-CFAR) detector can obtain high detection performance [6]. The greatest of constant false alarm rate (GO-CFAR) can get lower false alarm rate in the present of clutter edge [7]. In these detectors, the threshold is obtained by the mean power of the noise background power. Hence, the knowledge of the clutter edge signal statistics distribution is essential for detector. In real environments, the clutter distribution presents non-Gaussian distribution. In many literature, the clutter presented Weibull distribution [8]. Other literatures showed that the clutter show compound Gaussian distribution [9]. When high resolution radar system which operates at low grazing angles, the clutter presented the alpha-stable distribution [10]. In the case of multi-target interfering environment, the clutter shows Rayleigh distribution. CA-CFAR, SO-CFAR and GO-CFAR detectors were derived based on the Rayleigh clutter distribution.

In this study, based on clutter distribution model and the most detectors, a cell greatest, smallest and averaging constant false alarm rate (CGSA-CFAR) detector was proposed. The proposed detector was shown in Fig. 1. At first the signals were filtered by Digital filter banks that suppressed noise effectively by lowering the digital signal side lobe power. The filtered digital signals were sent to the CGSA-CFAR detector. The greatest and smallest cells for a & b in the sorted reference window were selected. The estimate power threshold T was obtained from the average of $T(m)$. In order to get the proper threshold, using $\beta \cdot T$ replace T. The estimated threshold was compared with the amplitude of the cell. The parameter β varying from 0.7 to 1.8 and size windows is $N = 16$. The scale β was set 1.2. The optimization was carried out for a SNR value equal to 20 dB.

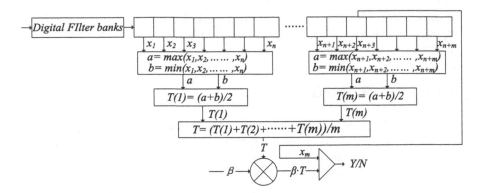

Fig. 1. CGSA-CFAR algorithm

3 Simulation and Analysis

The detection probability was increased significantly when a huge number of cells were included. The detector performance of the CGSA-CFAR detector is compared to that of the conventional detectors in homogenous environments, the results were shown in Table 1.

Table 1. Performance simulation comparison of various detectors

Detector	Time complexity/ $T(n)$	Space complexity/ $S(n)$	False detection rate	Detection rate
CA-CFAR	$O(n*1.0)$	$O(n)$	4.80 %	91.70 %
GO-CFAR	$O(n*1.2)$	$O(n)$	1.10 %	73.40 %
SO-CFAR	$O(n*1.2)$	$O(n)$	17.20 %	99.50 %
CGSA-CFAR	$O(n*1.4)$	$O(n*1.3)$	2.60 %	97.70 %

It could be observed from Table 1, the CGSA-CFAR detector performs better than other detectors. For the same clutter edge samples and distribution model, different average power level estimation method have same time and space complexity for SO-CFAR and GO-CFAR, but GO-CFAR detection performance descended heavily in order to keep lower false alarm rate, its detection rate was lower to 73.40 %, but false detection rate is lower to 1.10 %. SO-CFAR has higher false alarm rate in order to keep better detection performance, its false detection rate was higher to 17.20 %, but detection rate is higher to 99.50 %. CA-CFAR detection performance is stable, its false detection rate was 4.80 % and detection rate was 91.70 %. CA-CFAR presented better detection performance in the homogeneous Gaussian background, but the detection performance descends heavily in the presence of a clutter or interfering targets environments. CGSA-CFAR has the advantage of the above detectors and presents better detection performance in the homogeneous and non-homogeneous clutter background, its false detection rate is lower to 2.60 % and detection rate is up to 97.70 % which

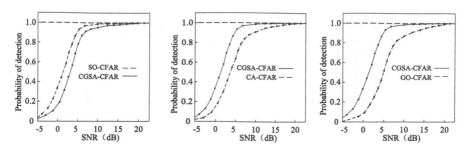

Fig. 2. Detection performance comparison between the detectors

could meet the most application or the product requirements. Detection performance comparison between the detectors and CGSA-CFAR in the same background noise was shown in Fig. 2.

The detectors is evaluated in the presence of multiple interfering targets and clutter edge environments. It can be observed in Fig. 2 that the detectors detection probability have little difference for SNR > 20 dB or SNR < −5 dB, but have obvious difference for SNR verify from 5 dB to 30 dB. CGSA-CFAR detector detection performance improved 5 dB compared with other detectors, its detection probability is still up to 97.70 % and false detection probability is lower to 2.60 %.

4 Experimental Results and Discussion

The implementation of the proposed detector was described and applied in blind spot detection and warning system (BSDWS).

4.1 Experimental Environments

The proposed detector run on TI TMS320F28335 embedded platform and was tested in real scenes under daytime and nighttime in Wuhu. Hardware platform has been shown in Fig. 3(a). The two radars were installed in rear left and right and used to detect the targets into rear blind spot zone and provide the warning information for the driver and passenger, Radars installation has been shown in Fig. 3(b). The warning equipment will alarm the driver and passenger when the targets coming into the blind spot zone.

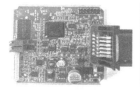   

(a) Radar system hardware platform   (b) The radars installation

Fig. 3. Radar system hardware platform and installation

(a) 1st level warning under daytime   (b) 2nd level warning under daytime   (c) 1st level warning under nighttime

Fig. 4. Test results of the system under daytime and nighttime conditions

Table 2. Test results under day and night environments

Conditions	Actual scenes	Warning times	False warning times	Warning probability/%	False warning probability/%
daytime	186	182	5	97.84	2.68
nighttime	121	119	3	98.34	2.47

4.2   Scenario Testing and Performance Comparisons

The real scenes conditions are shown in Fig. 4. Figure 4a shows that a tricycle is detected in the left rear blind spot area, system gives first level warning: led on. Figure 4b shows that a truck is detected in the left blind spot zone when the driver turn on left turn light on, system gives second level warning: led and buzzer on. Figure 4c shows that a car is detected in the left rear blind spot zone under rainy day, system gives first level warning to the driver: LED is bright. On-road experimental results are shown in Table 2. Detection rates were up to 97.84 % and 98.34 %, false alarm rates were down to 2.68 % and 2.47 %. Compared with other detector, the proposed detector achieves high detection rate. The proposed detector has a good detection performance under daytime and nighttime conditions in urban traffic environments, shown in Table 2.

5   Conclusions

In this paper, we have proposed a CGSA-CFAR detector and compared to GO-CFAR and SO-CFAR detector in homogeneous and heterogeneous environment. The detection performance of the proposed detector is compared with the other CFAR detectors such as CA-CFAR, SO-CFAR & GO-CFAR. It is observed that the CGSA-CFAR detector not only performs like the CA-CFAR detectors in the homogeneous environments but also performs robustly in non-homogeneous environments.

The complexity of the proposed detector is higher than the other CFAR detectors, but its detection performance is better. The detector was applied on BSDWS which was calibrated and tested on the Chery Arrizo7. Under daytime and nighttime conditions, detection rates were up to 98.34 %, false alarm rates were down to 2.47 %. Compared with other detector, the proposed detector achieves high detection rate in vehicle detection performance. The experimental results show that the proposed detector can present high vehicle detection performance in urban traffic environments.

Acknowledgements. This work was supported by Anhui Provincial Natural Science Foundation (KZ00215072), Six Talent Peaks Project in Jiangsu Province (2014-DZXX-040), Key Laboratory of Computer Application Technology, Computer and Information Science, Anhui Polytechnic University (JSJKF201-514).

References

1. Shtarkalev, B., Mulgrew, B.: Multistatic moving target detection in unknown coloured Gaussian interference. Sig. Process. **115**, 130–143 (2015)
2. Kennedy, H.L.: Multidimensional digital smoothing filters for target detection. Signal Process. **114**, 251–264 (2015)
3. Bahrampour, S., Ray, A., Sarkar, S., Damarla, T., Nasrabadi, N.M.: Performance comparison of feature extraction algorithms for target detection and classification. Pattern Recogn. Lett. **34**, 2126–2134 (2013)
4. Chengpeng, H., Orlando, D., Foglia, G., Ma, X.C., Yan, S.F., Hou, C.H.: Persymmetric adaptive detection of distributed targets in partially-homogeneous environment. Digit. Signal Process. **24**, 42–51 (2014)
5. Zaimbashi, A.: An adaptive cell averaging-based CFAR detector for interfering targets and clutter-edge situations. Digit. Signal Proc. **31**, 59–68 (2014)
6. Weinberg, G.V., Kyprianou, R.: Optimised binary integration with Order Statistic CFAR in Pareto distributed clutter. Digit. Signal Proc. **42**, 50–60 (2015)
7. Zhang, R.L., Sheng, W.X., Ma, X.F., Han, Y.B.: Constant false alarm rate detector based on the maximal reference cell. Digit. Signal Proc. **23**, 1974–1988 (2013)
8. Gurakan, B., Candan, C., Ciloglu, T.: CFAR processing with switching exponential smoothers for nonhomogeneous environments. Digit. Signal Proc. **22**, 407–416 (2012)
9. David, M.M., Nerea, D.R.M., Victor, M.P.S., Jarabo, M.P., Jaime, M.D.N.: MLP-CFAR for improving coherent radar detectors robustness in variable scenarios. Expert Syst. Appl. **42**, 4878–4891 (2015)
10. Weinberg, G.V.: Management of interference in Pareto CFAR processes using adaptive test cell analysis. Signal Process. **104**, 264–273 (2014)

Information Navigation System of Pulse Radar Employing Augmented Reality

Chen Kai[(⊠)], Zhou Lujun, and Yuan Chenghong

China Satellite Maritime Tracking and Controlling Department,
Jiangyin 214431, Jiangsu, China
ckzgwxhsckb@163.com

Abstract. A method presenting equipment information is proposed, based on augmented reality. Firstly, equipment drawing is automatic generated using Microsoft Visio Data Graphics. Then drawing information is embedded using XML file in android platform. Finally, drawing information is superimposed on the real-time image employing Vuforia SDK. Experimental results indicate that the augmented reality technology employed in information navigation system of equipment is reliable and feasible.

Keywords: Augmented reality · Operating instructions · Feature matching

1 Introduction

In the practice of information management and application, interactive electronic technical manual, (IETM) technology has done so much, that a suit of mature commercial standards is developed [1–6]. However, according to current IETM standards, management and maintenance information of equipment are stored in the form of text, graphics, and animations in the software. So the information and physical equipment is isolated, then people who want to access information, have to search the IETM file. However, traditional drawing is designed around cable. Such form of drawing is conducive to assemble, and to calculate the cost, but this is not convenient for the equipment user. For example, to query a flow of signal is a hard work to search.

Augmented reality technology is expected to improve the progress. It tracks camera image, shows superimposed frame layers integrated the computer graphics and real image, brings with immersive experience. So a method of information organization is suggested by applying augmented reality technology: equipment navigation information and equipment is displayed simultaneously on tablet PC screen [7–12]. By this means, the drawing is ready in touch for equipment users.

2 Transformations Between Real-World Coordinates and Virtual Screen Coordinate

In order to superpose the equipment information with camera image, the current relative position of real-world has to be calculated in the screen coordinate, this relative position usually means a group of three-dimensional coordinate translation and

© Springer International Publishing AG 2017
F. Xhafa et al. (eds.), *Recent Developments in Intelligent Systems and Interactive Applications*,
Advances in Intelligent Systems and Computing 541, DOI 10.1007/978-3-319-49568-2_65

rotation. In addition, if a staff interaction click is happened, then the coordinate of click object requires to be calculate. The following detailed description of a coordinate transformation process.

During the rotation axis it can be understood as: rotate around x, y, z-axis 3 times. After rotation, coordinate of $O - X_0Y_0Z_0$ coincide with $O - X_nY_nZ_n$. Therefore, the new coordinates can be expressed as:

$$
\begin{bmatrix} x_n \\ y_n \\ z_n \end{bmatrix} = \begin{bmatrix} 1 & 0 & 0 \\ 0 & \cos\gamma & \sin\gamma \\ 0 & -\sin\gamma & \cos\gamma \end{bmatrix} \begin{bmatrix} \cos\beta & 0 & -\sin\beta \\ 0 & 1 & 0 \\ \sin\beta & 0 & \cos\beta \end{bmatrix} \begin{bmatrix} \cos\alpha & \sin\alpha & 0 \\ -\sin\alpha & \cos\alpha & 0 \\ 0 & 0 & 1 \end{bmatrix} \begin{pmatrix} x_0 \\ y_0 \\ z_0 \end{pmatrix}
$$
$$
= R(\alpha, \beta, \gamma) \begin{pmatrix} x_0 \\ y_0 \\ z_0 \end{pmatrix} \tag{1}
$$

Besides the coordinate rotation, there exist coordinate translation usually: suggest the t_x, t_y, and t_z denote the offset along x, y, z axes respectively, and (x_1, y_1, z_1) denotes the new coordinates, then the formula is like this:

$$
\begin{bmatrix} x_1 \\ y_1 \\ z_1 \end{bmatrix} = \begin{bmatrix} & t_x \\ \mathbf{I} & t_y \\ & t_z \end{bmatrix} \begin{bmatrix} x \\ y \\ z \\ 1 \end{bmatrix} \tag{2}
$$

In the Formula (2), I denotes a third-order unit matrix. substitute Formula (2) into Eq. (1), it can be obtained:

$$
\begin{bmatrix} x_1 \\ y_1 \\ z_1 \end{bmatrix} = \begin{bmatrix} & & t_x \\ R(\alpha, \beta, \gamma) & t_y \\ & & t_z \end{bmatrix} \begin{bmatrix} x_0 \\ y_0 \\ z_0 \\ 1 \end{bmatrix} \tag{3}
$$

Of Formula (3), (x_1, y_1, z_1) is the coordinate of (x_0, y_0, z_0) after coordinate transformation.

For a point (x_s, y_s) on the two-dimensional screen, its coordinate is correlated to camera coordinate system by camera internal parameter matrix Q [9]:

$$
h \begin{bmatrix} x_s \\ y_s \\ 1 \end{bmatrix} = \begin{bmatrix} k_x f & k & x_0 \\ 0 & k_y f & y_0 \\ 0 & 0 & 1 \end{bmatrix} \begin{bmatrix} X \\ Y \\ Z \end{bmatrix} = Q \begin{bmatrix} X \\ Y \\ Z \end{bmatrix} \tag{4}
$$

Formula (4), h denotes affine matrix of scale factors, f denotes the focal length of the camera, x_0 and y_0 are the center coordinates of camera, k_x and k_y denote the ratio of the focal length factor along coordinate axes respectively, k denotes the distortion factor. Correlate Formulas (3) and (4), the relationship matrix between two-dimensional screen coordinate and three-dimensional coordinates of real-world is obtained:

$$h \begin{bmatrix} x_s \\ y_s \\ 1 \end{bmatrix} = Q \begin{bmatrix} & & t_x \\ R(\alpha, \beta, \gamma) & & t_y \\ & & t_z \end{bmatrix} \begin{bmatrix} X \\ Y \\ Z \\ 1 \end{bmatrix} \tag{5}$$

3 New Form of Data Presentation

Table 1 is a classical drawing. It's obvious that this kind of drawing is cable-centric. And it is difficult to search. Then a new drawing style is proposed. Drawing is generated based on the composite of Microsoft's Visio and Excel. And it can be automatically updated with Excel data. Then by visual registration technology, real equipment image and augmented reality information are superimposed on the flat panel.

Table 1. Classical list drawing of extracted from IETM

Cable no.	Electrical characteristic	Term1		Term2		Cable spec.	Length (cm)
		Rack	Pin	Rack	pin		
		204		204			
1	220 V-L	XT1-XS1	:3	XT1-K1	:1	Wire BVR 2.5(Red)	20
2	220 V-L	XT1-K1	:2	T1	:1	Wire BVR 2.5(Red)	100
3	220 V-N	XT1-XS1	:2	T1	:2	Wire BVR 2.5(Blue)	100

All wire Jacket in brown silk sleeve

3.1 Generation of Data Graphics

In Visio, it use the "Data Graphics' technology, which can facilitate the realization of automated data presentation.

(1) Storage of connection point in an Excel spreadsheet. The key elements of this step is design of column. In a cable list drawing, the interest of user includes an electrical connector signal definition, rating, signal source, destination, and the mounting position. Based on this analysis, Excel spreadsheet column is designed as shown Table 2.

(2) Visio Data Graphics
   (a) To import excel data into visio drawing.
   (b) To draw the schematic diagram which marks relative position of real electrical connector.

Table 2. Column description of excel spreadsheet

Table column names	Column meaning	Table column names	Column meaning
Outer panel	Location automatically generated signal leads outside the enclosure	Assignment	Signal Description
Pin position	Local location pins automatically generated	Input Output	
Inner panel	Location automatically generated signal leads inside the cabinet	Rated conditional function	Signal Ratings

(c) To add a new "data graphics" layer in Visio, then create "data graphic" model which defines the relative position between information and the schematic diagram, and designs display style.

(d) To link the information data in Excel datasheet with the data electrical connector drawn in (b) step.

Compared with Fig. 1, the drawing is organized like the real rack interface. This presentation makes the representation of the drawings is consistent with the actual rack style, can significantly improve the efficiency of the user to view. In addition, in order to facilitate the application of augmented reality technology will combine drawing and reality, we need to export the drawing Fig. 1 to a transparent background image.

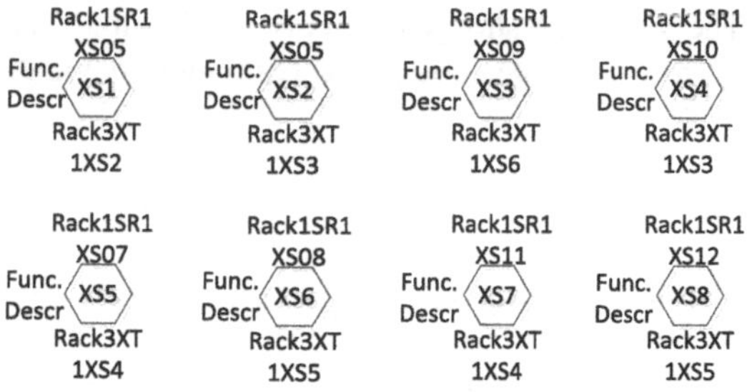

Fig. 1. Adds equipment drawing information data pattern

3.2 Vision Register

Employing Visual registration technology, the camera captured images of reality and drawing information is superimposed together. In the android platform, these situation is realized as this: to state resource permission, to use the XML file store equipment information, and to use OpenGL technology to implement image tracking and superimposed layers rendering.

(1) The statement of permission

The permission statement is required in Android operating system. And user would authorize these permissions. Permissions needed by the project include three, respectively, using the camera, read and write external memory. These permission applications is realized by editing AndroidManifest.xml: "android. permission.CAMERA", "android.permission. MOUNT_UNMOUNT_FILESYS-TEMS" and "android.permission. WRITE_EXTERNAL_STORAGE".

(2) Equipment information storage in XML file

Android itself provides class "SharedPreferences" to read xml file. This class provides a method getStrings ("name", "default value") to read xml file, and subclasses "Editor()" of "putString()" method to update/modify xml file.

(3) Visual Registration Process

(a) In the class constructor function, modify the initialization code: Initialize renderer and tracker, then and load data.

(b) To create and set OpenGL view, add to a Activity android application program.

(c) To start the camera and tracker.

4 User Interface Design

Software interface is divided into 3 Activity: (1) entry interface, the name and version of the application tips. Application entry page (similar to C in the main program interface) is configured in "AndroidManifest.xml". Activity specify a property for the "android. intent.action.MAIN" and "android.intent.category.LAUNCHER". (2) provide the rack number selection button and prompt basic information about the selected rack. (3) To track and display image, then superimposed the equipment information. Figure 2 is the experiment result. The Fig. 2(a) shows the real original image of equipment rack. Comparatively, Fig. 2(b) is a image superposed equipment navigation information.

5 Summary

An equipment information navigation system is realized by utilizing augmented reality technology. The equipment is stored in a portable tablet, by amalgamation of Visio, Excel and Vuforia SDK, equipment navigation information is superposed on the real image.

(1) The system changes traditional IETM presentation.

(2) The system changes the factory provided cable-centric presentation drawings, to the user visible panel-centric approach, in favor of the query data. By Excel and Visio data graphic, one can automatically update the drawing information easily by modifying the Excel spreadsheet.

 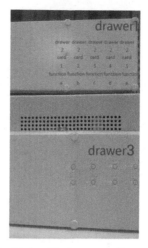

(a) Real original image of equipment rack (b) Applications actually run interface

Fig. 2. Experiment result

References

1. Yuxin, Yu.: Research on interactive electronic technical manual based on handheld devices. Ship Electron. Eng. **32**(9), 15–17 (2012)
2. Wu, Z., Fan, C.T., Hong, X.: Fifth grade IETM library based on ontology. East China Shipbuilding Inst. (Nat. Sci.) **18**(6), 90–96 (2004)
3. Fei, T.T.: Double meter mountain, Liu Peng far, and so on. IETM production process research and application based on S1000D standards. Comput. Measur. Control **19**(6), 1426–1428 (2011)
4. Zhen, Z., Jingjing, Z., Huizhen, L.: Design and implementation of a portable radar maintenance of auxiliary equipment. Mod. Radar **35**(7), 63–66 (2013)
5. Chun-hui, Yu., Yong, W.: Application of XML in IETM system. Comput. Technol. Dev. **23**(9), 199–202 (2013)
6. Zhu, X., Kui, H., Wang.: S1000D of IETM reader based on. Comput. Eng. **36**(13), 288–230 (2010)
7. Bimber, O., Raskar, R., Inami, M.: Spatial Augmented Reality. AK Peters, USA (2005)
8. Bin, X., Yi, J., Fan, L.: Realistic 3D registration algorithm and enhanced mobile ORB KLT based. Comput. Modernization **223**, 57–61 (2014)
9. Tian, X.: Guidelines for the operation of augmented reality system. Ph.D. thesis, Zhejiang University, Hangzhou (2013)
10. Wen, G., Yong-tian, W., Yue, L., et al.: Application of two-dimensional code in mobile augmented reality. Comput. Aided Des. Comput. Graph. **26**(1), 34–39 (2014)
11. Zhixiang, C., Liming, W.: High WPI. Mobile FAST-SURF algorithm based on augmented reality tracking technology. Comput. Modernization **09**, 105–108 (2013)
12. Fang, L.: Tablet PC based augmented reality system. Shanghai Univ. Electr. Power **30**(2), 165–169 (2014)

Author Index

© Springer International Publishing AG 2017
F. Xhafa et al. (eds.), *Recent Developments in Intelligent Systems and Interactive Applications*, Advances in Intelligent Systems and Computing 541, DOI 10.1007/978-3-319-49568-2

467

Printed in the United States
By Bookmasters